JAMES GLEICK

THE INFORMATION

James Gleick is the author of *Chaos* and *Genius*, both nominated for the National Book Award, *Faster*, *What Just Happened*, and *Isaac Newton*, which was shortlisted for the Pulitzer Prize.

www.around.com

THE INFORMATION

THE
INFORMATION

A History
A Theory
A Flood

JAMES GLEICK

VINTAGE BOOKS
A DIVISION OF RANDOM HOUSE, INC.
NEW YORK

FIRST VINTAGE BOOKS EDITION, MARCH 2012

Copyright © 2011 by James Gleick

All rights reserved. Published in the United States by Vintage Books, a division of
Random House, Inc., New York, and in Canada by Random House of Canada Limited,
Toronto. Originally published in hardcover in the United States by Pantheon Books,
a division of Random House, Inc., New York, in 2011.

Vintage and colophon are registered trademarks of Random House, Inc.

The Library of Congress has cataloged the Pantheon edition as follows:
Gleick, James.
The information : a history, a theory, a flood / James Gleick.
p.   cm.
Includes bibliographical references and index.
1. Information science—History.   2. Information society.   I. Title.
Z665.G547 2011   020.9—dc22   2010023221

Vintage ISBN: 978-1-4000-9623-7

Author photograph © Phyllis Rose
Book design by M. Kristen Bearse

www.vintagebooks.com

Manufactured in the United States of America
10   9   8   7

FOR CYNTHIA

Anyway, those tickets, the old ones, they didn't tell you where you were going, much less where you came from. He couldn't remember seeing any dates on them, either, and there was certainly no mention of time. It was all different now, of course. All this information. Archie wondered why that was.

—Zadie Smith

What we call the past is built on bits.

—John Archibald Wheeler

Contents

THE INFORMATION

PROLOGUE

The fundamental problem of communication is that of reproduc-
ing at one point either exactly or approximately a message selected
at another point. Frequently the messages have meaning.
                                                    —Claude Shannon (1948)

AFTER 1948, which was the crucial year, people thought they could see
the clear purpose that inspired Claude Shannon's work, but that was
hindsight. He saw it differently: *My mind wanders around, and I conceive*
of different things day and night. Like a science-fiction writer, I'm thinking,
"What if it were like this?"

As it happened, 1948 was when the Bell Telephone Laboratories
announced the invention of a tiny electronic semiconductor, "an amazingly simple device" that could do anything a vacuum tube could do
and more efficiently. It was a crystalline sliver, so small that a hundred
would fit in the palm of a hand. In May, scientists formed a committee
to come up with a name, and the committee passed out paper ballots
to senior engineers in Murray Hill, New Jersey, listing some choices:
semiconductor triode . . . iotatron . . . transistor (a hybrid of *varistor* and
transconductance). *Transistor* won out. "It may have far-reaching significance in electronics and electrical communication," Bell Labs declared in
a press release, and for once the reality surpassed the hype. The transistor
sparked the revolution in electronics, setting the technology on its path

of miniaturization and ubiquity, and soon won the Nobel Prize for its three chief inventors. For the laboratory it was the jewel in the crown. But it was only the second most significant development of that year. The transistor was only hardware.

An invention even more profound and more fundamental came in a monograph spread across seventy-nine pages of *The Bell System Technical Journal* in July and October. No one bothered with a press release. It carried a title both simple and grand—"A Mathematical Theory of Communication"—and the message was hard to summarize. But it was a fulcrum around which the world began to turn. Like the transistor, this development also involved a neologism: the word *bit*, chosen in this case not by committee but by the lone author, a thirty-two-year-old named Claude Shannon. The bit now joined the inch, the pound, the quart, and the minute as a determinate quantity—a fundamental unit of measure.

But measuring what? "A unit for measuring information," Shannon wrote, as though there were such a thing, measurable and quantifiable, as information.

Shannon supposedly belonged to the Bell Labs mathematical research group, but he mostly kept to himself. When the group left the New York headquarters for shiny new space in the New Jersey suburbs, he stayed behind, haunting a cubbyhole in the old building, a twelve-story sandy brick hulk on West Street, its industrial back to the Hudson River, its front facing the edge of Greenwich Village. He disliked commuting, and he liked the downtown neighborhood, where he could hear jazz clarinetists in late-night clubs. He was flirting shyly with a young woman who worked in Bell Labs' microwave research group in the two-story former Nabisco factory across the street. People considered him a smart young man. Fresh from MIT he had plunged into the laboratory's war work, first developing an automatic fire-control director for antiaircraft guns, then focusing on the theoretical underpinnings of secret communication—cryptography— and working out a mathematical proof of the security of the so-called X System, the telephone hotline between Winston Churchill and President

Roosevelt. So now his managers were willing to leave him alone, even though they did not understand exactly what he was working on.

AT&T at midcentury did not demand instant gratification from its research division. It allowed detours into mathematics or astrophysics with no apparent commercial purpose. Anyway so much of modern science bore directly or indirectly on the company's mission, which was vast, monopolistic, and almost all-encompassing. Still, broad as it was, the telephone company's core subject matter remained just out of focus. By 1948 more than 125 million conversations passed daily through the Bell System's 138 million miles of cable and 31 million telephone sets. The Bureau of the Census reported these facts under the rubric of "Communications in the United States," but they were crude measures of communication. The census also counted several thousand broadcasting stations for radio and a few dozen for television, along with newspapers, books, pamphlets, and the mail. The post office counted its letters and parcels, but what, exactly, did the Bell System carry, counted in what units? Not *conversations*, surely; nor *words*, nor certainly *characters*. Perhaps it was just electricity. The company's engineers were electrical engineers. Everyone understood that electricity served as a surrogate for sound, the sound of the human voice, waves in the air entering the telephone mouthpiece and converted into electrical waveforms. This conversion was the essence of the telephone's advance over the telegraph—the predecessor technology, already seeming so quaint. Telegraphy relied on a different sort of conversion: a code of dots and dashes, not based on sounds at all but on the written alphabet, which was, after all, a code in its turn. Indeed, considering the matter closely, one could see a chain of abstraction and conversion: the dots and dashes representing letters of the alphabet; the letters representing sounds, and in combination forming words; the words representing some ultimate substrate of meaning, perhaps best left to philosophers.

The Bell System had none of those, but the company had hired its first mathematician in 1897: George Campbell, a Minnesotan who had studied in Göttingen and Vienna. He immediately confronted a crippling problem of early telephone transmission. Signals were distorted as they passed across the circuits; the greater the distance, the worse the distortion. Campbell's solution was partly mathematics and partly electrical engineering. His employers learned not to worry much about the distinction. Shannon himself, as a student, had never been quite able to decide whether to become an engineer or a mathematician. For Bell Labs he was both, willy-nilly, practical about circuits and relays but happiest in a realm of symbolic abstraction. Most communications engineers focused their expertise on physical problems, amplification and modulation, phase distortion and signal-to-noise degradation. Shannon liked games and puzzles. Secret codes entranced him, beginning when he was a boy reading Edgar Allan Poe. He gathered threads like a magpie. As a first-year research assistant at MIT, he worked on a hundred-ton proto-computer, Vannevar Bush's Differential Analyzer, which could solve equations with great rotating gears, shafts, and wheels. At twenty-two he wrote a dissertation that applied a nineteenth-century idea, George Boole's algebra of logic, to the design of electrical circuits. (Logic and electricity—a peculiar combination.) Later he worked with the mathematician and logician Hermann Weyl, who taught him what a theory was: "Theories permit consciousness to 'jump over its own shadow,' to leave behind the given, to represent the transcendent, yet, as is self-evident, only in symbols."

In 1943 the English mathematician and code breaker Alan Turing visited Bell Labs on a cryptographic mission and met Shannon sometimes over lunch, where they traded speculation on the future of artificial thinking machines. ("Shannon wants to feed not just *data* to a Brain, but cultural things!" Turing exclaimed. "He wants to play music to it!") Shannon also crossed paths with Norbert Wiener, who had taught him at MIT and by 1948 was proposing a new discipline to be called "cybernetics," the study of communication and control. Meanwhile Shannon began paying special attention to television signals, from a peculiar point of view: wondering whether their content could be somehow compacted or compressed to allow for faster transmission. Logic and circuits cross-bred to make a new, hybrid thing; so did codes and genes. In his solitary way, seeking a framework to connect his many threads, Shannon began assembling a theory for information.

The raw material lay all around, glistening and buzzing in the landscape of the early twentieth century, letters and messages, sounds and images, news and instructions, figures and facts, signals and signs: a hodgepodge of related species. They were on the move, by post or wire or electromagnetic wave. But no one word denoted all that stuff. "Off and on," Shannon wrote to Vannevar Bush at MIT in 1939, "I have been working on an analysis of some of the fundamental properties of general systems for the transmission of intelligence." *Intelligence:* that was a flexible term, very old. "Nowe used for an elegant worde," Sir Thomas Elyot wrote in the sixteenth century, "where there is mutuall treaties or appoyntementes, eyther by letters or message." It had taken on other meanings, though. A few engineers, especially in the telephone labs, began speaking of *information*. They used the word in a way suggesting something technical: quantity of information, or measure of information. Shannon adopted this usage.

For the purposes of science, *information* had to mean something special. Three centuries earlier, the new discipline of physics could not proceed until Isaac Newton appropriated words that were ancient and vague—*force, mass, motion*, and even *time*—and gave them new meanings. Newton made these terms into quantities, suitable for use in mathematical formulas. Until then, *motion* (for example) had been just as soft and inclusive a term as *information*. For Aristotelians, motion covered a far-flung family of phenomena: a peach ripening, a stone falling, a child growing, a body decaying. That was too rich. Most varieties of motion had to be tossed out before Newton's laws could apply and the Scientific Revolution could succeed. In the nineteenth century, *energy* began to undergo a similar transformation: natural philosophers adapted a word meaning vigor or intensity. They mathematicized it, giving energy its fundamental place in the physicists' view of nature.

It was the same with information. A rite of purification became necessary.

And then, when it was made simple, distilled, counted in bits, information was found to be everywhere. Shannon's theory made a bridge between information and uncertainty; between information and entropy; and between information and chaos. It led to compact discs and fax machines, computers and cyberspace, Moore's law and all the world's Silicon Alleys. Information processing was born, along with information storage and information retrieval. People began to name a successor to the Iron Age and the Steam Age. "Man the food-gatherer reappears incongruously as information-gatherer," remarked Marshall McLuhan in 1967.* He wrote this an instant too soon, in the first dawn of computation and cyberspace.

We can see now that information is what our world runs on: the blood and the fuel, the vital principle. It pervades the sciences from top to bottom, transforming every branch of knowledge. Information theory began as a bridge from mathematics to electrical engineering and from there to computing. What English speakers call "computer science" Europeans have known as *informatique*, *informatica*, and *Informatik*. Now even biology has become an information science, a subject of messages, instructions, and code. Genes encapsulate information and enable procedures for reading it in and writing it out. Life spreads by networking. The body itself is an information processor. Memory resides not just in brains but in every cell. No wonder genetics bloomed along with information theory. DNA is the quintessential information molecule, the most advanced message processor at the cellular level—an alphabet and a code, 6 billion bits to form a human being. "What lies at the heart of every living thing is not a fire, not warm breath, not a 'spark of life,'" declares the evolutionary theorist Richard Dawkins. "It is information, words, instructions. . . . If you want to understand life, don't think about

* And added drily: "In this role, electronic man is no less a nomad than his Paleolithic ancestors."

vibrant, throbbing gels and oozes, think about information technology." The cells of an organism are nodes in a richly interwoven communications network, transmitting and receiving, coding and decoding. Evolution itself embodies an ongoing exchange of information between organism and environment.

"The information circle becomes the unit of life," says Werner Loewenstein after thirty years spent studying intercellular communication. He reminds us that *information* means something deeper now: "It connotes a cosmic principle of organization and order, and it provides an exact measure of that." The gene has its cultural analog, too: the meme. In cultural evolution, a meme is a replicator and propagator—an idea, a fashion, a chain letter, or a conspiracy theory. On a bad day, a meme is a virus.

Economics is recognizing itself as an information science, now that money itself is completing a developmental arc from matter to bits, stored in computer memory and magnetic strips, world finance coursing through the global nervous system. Even when money seemed to be material treasure, heavy in pockets and ships' holds and bank vaults, it always was information. Coins and notes, shekels and cowries were all just short-lived technologies for tokenizing information about who owns what.

And atoms? Matter has its own coinage, and the hardest science of all, physics, seemed to have reached maturity. But physics, too, finds itself sideswiped by a new intellectual model. In the years after World War II, the heyday of the physicists, the great news of science appeared to be the splitting of the atom and the control of nuclear energy. Theorists focused their prestige and resources on the search for fundamental particles and the laws governing their interaction, the construction of giant accelerators and the discovery of quarks and gluons. From this exalted enterprise, the business of communications research could not have appeared further removed. At Bell Labs, Claude Shannon was not thinking about physics. Particle physicists did not need bits.

And then, all at once, they did. Increasingly, the physicists and the information theorists are one and the same. The bit is a fundamental particle of a different sort: not just tiny but abstract—a binary digit, a flip-flop, a yes-or-no. It is insubstantial, yet as scientists finally come to understand information, they wonder whether it may be primary: more fundamental than matter itself. They suggest that the bit is the irreducible kernel and that information forms the very core of existence. Bridging the physics of the twentieth and twenty-first centuries, John Archibald Wheeler, the last surviving collaborator of both Einstein and Bohr, put this manifesto in oracular monosyllables: "It from Bit." Information gives rise to "every it—every particle, every field of force, even the spacetime continuum itself." This is another way of fathoming the paradox of the observer: that the outcome of an experiment is affected, or even determined, when it is observed. Not only is the observer observing, she is asking questions and making statements that must ultimately be expressed in discrete bits. "What we call reality," Wheeler wrote coyly, "arises in the last analysis from the posing of yes-no questions." He added: "All things physical are information-theoretic in origin, and this is a participatory universe." The whole universe is thus seen as a computer—a cosmic information-processing machine.

A key to the enigma is a type of relationship that had no place in classical physics: the phenomenon known as entanglement. When particles or quantum systems are entangled, their properties remain correlated across vast distances and vast times. Light-years apart, they share something that is physical, yet not only physical. Spooky paradoxes arise, unresolvable until one understands how entanglement encodes information, measured in bits or their drolly named quantum counterpart, qubits. When photons and electrons and other particles interact, what are they really doing? Exchanging bits, transmitting quantum states, processing information. The laws of physics are the algorithms. Every burning star, every silent nebula, every particle leaving its ghostly trace in a cloud chamber is an information processor. The universe computes its own destiny.

How much does it compute? How fast? How big is its total information capacity, its memory space? What is the link between energy and information; what is the energy cost of flipping a bit? These are hard questions, but they are not as mystical or metaphorical as they sound. Physicists and quantum information theorists, a new breed, struggle with them together. They do the math and produce tentative answers. ("The bit count of the cosmos, however it is figured, is ten raised to a very large power," according to Wheeler. According to Seth Lloyd: "No more than 10^{120} ops on 10^{90} bits.") They look anew at the mysteries of thermodynamic entropy and at those notorious information swallowers, black holes. "Tomorrow," Wheeler declares, "we will have learned to understand and express *all* of physics in the language of information."

As the role of information grows beyond anyone's reckoning, it grows to be too much. "TMI," people now say. We have information fatigue, anxiety, and glut. We have met the Devil of Information Overload and his impish underlings, the computer virus, the busy signal, the dead link, and the PowerPoint presentation. All this, too, is due in its roundabout way to Shannon. Everything changed so quickly. John Robinson Pierce (the Bell Labs engineer who had come up with the word *transistor*) mused afterward: "It is hard to picture the world before Shannon as it seemed to those who lived in it. It is difficult to recover innocence, ignorance, and lack of understanding."

Yet the past does come back into focus. *In the beginning was the word*, according to John. We are the species that named itself *Homo sapiens*, the one who knows—and then, after reflection, amended that to *Homo sapiens sapiens*. The greatest gift of Prometheus to humanity was not fire after all: "Numbers, too, chiefest of sciences, I invented for them, and the combining of letters, creative mother of the Muses' arts, with which to hold all things in memory." The alphabet was a founding technology of information. The telephone, the fax machine, the calculator, and, ultimately, the computer are only the latest innovations devised for saving, manipulating, and communicating knowledge. Our culture has absorbed a working vocabulary for these useful inventions. We speak of compressing data, aware that this is quite different from compressing a gas. We know about streaming information, parsing it, sorting it, matching it, and filtering it. Our furniture includes iPods and plasma displays, our skills include texting and Googling, we are endowed, we are expert, so we see information in the foreground. But it has always been there. It pervaded our ancestors' world, too, taking forms from solid to ethereal, granite gravestones and the whispers of courtiers. The punched card, the cash register, the nineteenth-century Difference Engine, the wires of telegraphy all played their parts in weaving the spiderweb of information to which we cling. Each new information technology, in its own time, set off blooms in storage and transmission. From the printing press came new species of information organizers: dictionaries, cyclopaedias, almanacs—compendiums of words, classifiers of facts, trees of knowledge. Hardly any information technology goes obsolete. Each new one throws its predecessors into relief. Thus Thomas Hobbes, in the seventeenth century, resisted his era's new-media hype: "The invention of printing, though ingenious, compared with the invention of letters is no great matter." Up to a point, he was right. Every new medium transforms the nature of human thought. In the long run, history is the story of information becoming aware of itself.

Some information technologies were appreciated in their own time, but others were not. One that was sorely misunderstood was the African talking drum.

1 | DRUMS THAT TALK

(When a Code Is Not a Code)

Across the Dark Continent sound the never-silent drums:
the base of all the music, the focus of every dance;
the talking drums, the wireless of the unmapped jungle.
— Irma Wassall (1943)

NO ONE SPOKE SIMPLY ON THE DRUMS. Drummers would not say, "Come back home," but rather,

Make your feet come back the way they went,
make your legs come back the way they went,
plant your feet and your legs below,
in the village which belongs to us.

They could not just say "corpse" but would elaborate: "which lies on its back on clods of earth." Instead of "don't be afraid," they would say, "Bring your heart back down out of your mouth, your heart out of your mouth, get it back down from there." The drums generated fountains of oratory. This seemed inefficient. Was it grandiloquence or bombast? Or something else?

For a long time Europeans in sub-Saharan Africa had no idea. In fact they had no idea that the drums conveyed information at all. In their own cultures, in special cases a drum could be an instrument of signaling, along with the bugle and the bell, used to transmit a small set of messages: *attack;*

retreat; come to church. But they could not conceive of talking drums. In 1730 Francis Moore sailed eastward up the Gambia River, finding it navigable for six hundred miles, all the way admiring the beauty of the country and such curious wonders as "oysters that grew upon trees" (mangroves). He was not much of a naturalist. He was reconnoitering as an agent for English slavers in kingdoms inhabited, as he saw it, by different races of people of black or tawny colors, "as Mundingoes, Jolloiffs, Pholeys, Floops, and Portuguese." When he came upon men and women carrying drums, carved wood as much as a yard long, tapered from top to bottom, he noted that women danced briskly to their music, and sometimes that the drums were "beat on the approach of an enemy," and finally, "on some very extraordinary occasions," that the drums summoned help from neighboring towns. But that was all he noticed.

A century later, Captain William Allen, on an expedition to the Niger River,* made a further discovery, by virtue of paying attention to his Cameroon pilot, whom he called Glasgow. They were in the cabin of the iron paddle ship when, as Allen recalled:

> Suddenly he became totally abstracted, and remained for a while in the attitude of listening. On being taxed with inattention, he said, "You no hear my son speak?"As we had heard no voice, he was asked how he knew it. He said, "Drum speak me, tell me come up deck." This seemed to be very singular.

The captain's skepticism gave way to amazement, as Glasgow convinced him that every village had this "facility of musical correspondence." Hard though it was to believe, the captain finally accepted that detailed messages of many sentences could be conveyed across miles. "We are often surprised," he wrote, "to find the sound of the trumpet so well understood in our military evolutions; but how far short that falls of the result arrived at by those untutored savages." That result was a technology much sought

* The trip was sponsored by the Society for the Extinction of the Slave Trade and the Civilization of Africa for the purpose of interfering with slavers.

in Europe: long-distance communication faster than any traveler on foot or horseback. Through the still night air over a river, the thump of the drum could carry six or seven miles. Relayed from village to village, messages could rumble a hundred miles or more in a matter of an hour.

A birth announcement in Bolenge, a village of the Belgian Congo, went like this:

> *Batoko fala fala, tokema bolo bolo, boseka woliana imaki tonkilingonda, ale nda bobila wa fole fole, asokoka l'isika koke koke.*

> The mats are rolled up, we feel strong, a woman came from the forest, she is in the open village, that is enough for this time.

A missionary, Roger T. Clarke, transcribed this call to a fisherman's funeral:

> *La nkesa laa mpombolo, tofolange benteke biesala, tolanga bonteke bolokolo bole nda elinga l'enjale baenga, basaki l'okala bopele pele. Bojende bosalaki lifeta Bolenge wa kala kala, tekendake tonkilingonda, tekendake beningo la nkaka elinga l'enjale. Tolanga bonteke bolokolo bole nda elinga l'enjale, la nkesa la mpombolo.*

> In the morning at dawn, we do not want gatherings for work, we want a meeting of play on the river. Men who live in Bolenge, do not go to the forest, do not go fishing. We want a meeting of play on the river, in the morning at dawn.

Clarke noted several facts. While only some people learned to communicate by drum, almost anyone could understand the messages in the drumbeats. Some people drummed rapidly and some slowly. Set phrases would recur again and again, virtually unchanged, yet different drummers would send the same message with different wording. Clarke decided that the drum language was at once formulaic and fluid. "The signals represent the tones of the syllables of conventional phrases of a traditional and highly poetic character," he concluded, and this was correct, but he could not take the last step toward understanding why.

These Europeans spoke of "the native mind" and described Africans as "primitive" and "animistic" and nonetheless came to see that they had achieved an ancient dream of every human culture. Here was a messaging system that outpaced the best couriers, the fastest horses on good roads with way stations and relays. Earth-bound, foot-based messaging systems always disappointed. Their armies outran them. Julius Caesar, for example, was "very often arriving before the messengers sent to announce his coming," as Suetonius reported in the first century. The ancients were not without resources, however. The Greeks used fire beacons at the time of the Trojan War, in the twelfth century BCE, by all accounts—that is, those of Homer, Virgil, and Aeschylus. A bonfire on a mountaintop could be seen from watchtowers twenty miles distant, or in special cases even farther. In the Aeschylus version, Clytemnestra gets the news of the fall of Troy that very night, four hundred miles away in Mycenae. "Yet who so swift could speed the message here?" the skeptical Chorus asks.

She credits Hephaestus, god of fire: "Sent forth his sign; and on, and ever on, beacon to beacon sped the courier-flame." This is no small accomplishment, and the listener needs convincing, so Aeschylus has Clytemnestra continue for several minutes with every detail of the route: the blazing signal rose from Mount Ida, carried across the northern Aegean Sea to the island of Lemnos; from there to Mount Athos in Macedonia; then southward across plains and lakes to Macistus; Messapius, where the watcher "saw the far flame gleam on Euripus' tide, and from the high-piled heap of withered furze lit the new sign and bade the message on"; Cithaeron; Aegiplanetus; and her own town's mountain watch, Arachne. "So sped from stage to stage, fulfilled in turn, flame after flame," she boasts, "along the course ordained." A German historian, Richard Hennig, traced and measured the route in 1908 and confirmed the feasibility of this chain of bonfires. The meaning of the message had, of course, to be prearranged, effectively condensed into a single bit. A binary choice, *something* or *nothing:* the fire signal meant *something*, which, just this once, meant "Troy has fallen." To transmit this one bit required immense planning, labor, watchfulness, and firewood. Many years later, lanterns in Old North Church likewise sent Paul Revere a single precious bit, which he carried onward, one binary choice: by land or by sea.

More capacity was required, for less extraordinary occasions. People tried flags, horns, intermitting smoke, and flashing mirrors. They conjured spirits and angels for purposes of communication—angels being divine messengers, by definition. The discovery of magnetism held particular promise. In a world already suffused with magic, magnets embodied occult powers. The lodestone attracts iron. This power of attraction extends invisibly through the air. Nor is it interrupted by water or even solid bodies. A lodestone held on one side of a wall can move a piece of iron on the other side. Most intriguing, the magnetic power appears able to coordinate objects vast distances apart, across the whole earth: namely, compass needles. What if one needle could control another? This idea spread—a "conceit," Thomas Browne wrote in the 1640s,

> whispered thorow the world with some attention, credulous and vulgar auditors readily believing it, and more judicious and distinctive heads, not altogether rejecting it. The conceit is excellent, and if the effect would follow, somewhat divine; whereby we might communicate like spirits, and confer on earth with Menippus in the Moon.

The idea of "sympathetic" needles appeared wherever there were natural philosophers and confidence artists. In Italy a man tried to sell Galileo "a secret method of communicating with a person two or three thousand miles away, by means of a certain sympathy of magnetic needles."

> I told him that I would gladly buy, but wanted to see by experiment and that it would be enough for me if he would stand in one room and I in another. He replied that its operation could not be detected at such a short distance. I sent him on his way, with the remark that I was not in the mood at that time to go to Cairo or Moscow for the experiment, but that if he wanted to go I would stay in Venice and take care of the other end.

The idea was that if a pair of needles were magnetized together—"touched with the same Loadstone," as Browne put it—they would remain in sympathy from then on, even when separated by distance. One might call this "entanglement." A sender and a recipient would take the needles and agree on a time to communicate. They would place their needle in disks with the letters of the alphabet spaced around the rim. The sender would spell out a message by turning the needle. "For then, saith tradition," Browne explained, "at what distance of place soever, when one needle shall be removed unto any letter, the other by a wonderfull sympathy will move unto the same." Unlike most people who considered the idea of sympathetic needles, however, Browne actually tried the experiment. It did not work. When he turned one needle, the other stood still.

Browne did not go so far as to rule out the possibility that this mysterious force could someday be used for communication, but he added one more caveat. Even if magnetic communication at a distance was possible, he suggested, a problem might arise when sender and receiver tried to synchronize their actions. How would they know the time,

> it being no ordinary or Almanack business, but a probleme Mathematical, to finde out the difference of hours in different places; nor do the wisest exactly satisfy themselves in all. For the hours of several places anticipate each other, according to their Longitudes; which are not exactly discovered of every place.

This was a prescient thought, and entirely theoretical, a product of new seventeenth-century knowledge of astronomy and geography. It was the first crack in the hitherto solid assumption of simultaneity. Anyway, as Browne noted, experts differed. Two more centuries would pass before anyone could actually travel fast enough, or communicate fast enough, to experience local time differences. For now, in fact, no one in the world could communicate as much, as fast, as far as unlettered Africans with their drums.

———

By the time Captain Allen discovered the talking drums in 1841, Samuel F. B. Morse was struggling with his own percussive code, the electromagnetic drumbeat designed to pulse along the telegraph wire. Inventing a code was a complex and delicate problem. He did not even think in terms of a code, at first, but "a system of signs for letters, to be indicated and marked by a quick succession of strokes or shocks of the galvanic current." The annals of invention offered scarcely any precedent. How to convert information from one form, the everyday language, into another form suitable for transmission by wire taxed his ingenuity more than any mechanical problem of the telegraph. It is fitting that history attached Morse's name to his code, more than to his device.

He had at hand a technology that seemed to allow only crude pulses, bursts of current on and off, an electrical circuit closing and opening. How could he convey language through the clicking of an electromagnet? His first idea was to send numbers, a digit at a time, with dots and pauses. The sequence ••• •• ••••• would mean 325. Every English word would be assigned a number, and the telegraphists at each end of the line would look them up in a special dictionary. Morse set about creating this dictionary himself, wasting many hours inscribing it on large folios.* He claimed the idea in his first telegraph patent, in 1840:

> The dictionary or vocabulary consists of words alphabetically arranged and regularly numbered, beginning with the letters of the alphabet, so that each word in the language has its telegraphic number, and is designated at pleasure, through the signs of numerals.

Seeking efficiency, he weighed the costs and possibilities across several intersecting planes. There was the cost of transmission itself: the wires would be expensive and would convey only so many pulses per minute.

---

* "A very short experience, however, showed the superiority of the alphabetic mode," he wrote later, "and the big leaves of the numbered dictionary, which cost me a world of labor, . . . were discarded and the alphabetic installed in its stead."

Numbers would be relatively easy to transmit. But then there was the extra cost in time and difficulty for the telegraphists. The idea of code books—lookup tables—still had possibilities, and it echoed into the future, arising again in other technologies. Eventually it worked for Chinese telegraphy. But Morse realized that it would be hopelessly cumbersome for operators to page through a dictionary for every word.

His protégé Alfred Vail, meanwhile, was developing a simple lever key by which an operator could rapidly close and open the electric circuit. Vail and Morse turned to the idea of a coded alphabet, using signs as surrogates for the letters and thus spelling out every word. Somehow the bare signs would have to stand in for all the words of the spoken or written language. They had to map the entire language onto a single dimension of pulses. At first they conceived of a system built on two elements: the clicks (now called dots) and the spaces in between. Then, as they fiddled with the prototype keypad, they came up with a third sign: the line or dash, "when the circuit was closed a longer time than was necessary to make a dot." (The code became known as the dot-and-dash alphabet, but the unmentioned space remained just as important; Morse code was not a binary language.*) That humans could learn this new language was, at first, wondrous. They would have to master the coding system and then perform a continuous act of double translation: language to signs; mind to fingers. One witness was amazed at how the telegraphists internalized these skills:

> The clerks who attend at the recording instrument become so expert in their curious hieroglyphics, that they do not need to look at the printed record to know what the message under reception is; the recording instrument has for them an intelligible articulate language. They understand *its speech*. They can close their eyes and listen to the strange clicking that is going on close to their ear whilst the printing is in progress, and at once say what it all means.

* Operators soon distinguished spaces of different lengths—intercharacter and interword—so Morse code actually employed four signs.

In the name of speed, Morse and Vail had realized that they could save strokes by reserving the shorter sequences of dots and dashes for the most common letters. But which letters would be used most often? Little was known about the alphabet's statistics. In search of data on the letters' relative frequencies, Vail was inspired to visit the local newspaper office in Morristown, New Jersey, and look over the type cases. He found a stock of twelve thousand E's, nine thousand T's, and only two hundred Z's. He and Morse rearranged the alphabet accordingly. They had originally used dash-dash-dot to represent T, the second most common letter; now they promoted T to a single dash, thus saving telegraph operators uncountable billions of key taps in the world to come. Long afterward, information theorists calculated that they had come within 15 percent of an optimal arrangement for telegraphing English text.

No such science, no such pragmatism informed the language of the drums. Yet there had been a problem to solve, just as there was in the design of a code for telegraphers: how to map an entire language onto a one-dimensional stream of the barest sounds. This design problem was solved collectively by generations of drummers in a centuries-long process of social evolution. By the early twentieth century the analogy to the telegraph was apparent to Europeans studying Africa. "Only a few days ago I read in the *Times*," Captain Robert Sutherland Rattray reported to the Royal African Society in London, "how a resident in one part of Africa heard of the death—in another and far remote part of the continent—of a European baby, and how this news was carried by means of drums, which were used, it was stated, 'on the Morse principle'—it is always 'the Morse principle.'"

But the obvious analogy led people astray. They failed to decipher the code of the drums because, in effect, there was no code. Morse had bootstrapped his system from a middle symbolic layer, the written alphabet, intermediate between speech and his final code. His dots and dashes had no direct connection to sound; they represented letters, which formed written words, which represented the spoken words in turn. The drummers could not build on an intermediate code—they could not abstract through a layer of symbols—because the African languages, like all but a few dozen of the six thousand languages spoken in the modern world, lacked an alphabet. The drums metamorphosed speech.

It fell to John F. Carrington to explain. An English missionary, born in 1914 in Northamptonshire, Carrington left for Africa at the age of twenty-four and Africa became his lifetime home. The drums caught his attention early, as he traveled from the Baptist Missionary Society station in Yakusu, on the Upper Congo River, through the villages of the Bambole forest. One day he made an impromptu trip to the small town of Yaongama and was surprised to find a teacher, medical assistant, and church members already assembled for his arrival. They had heard the drums, they explained. Eventually he realized that the drums conveyed not just announcements and warnings but prayers, poetry, and even jokes. The drummers were not signaling but talking: they spoke a special, adapted language.

Eventually Carrington himself learned to drum. He drummed mainly in Kele, a language of the Bantu family in what is now eastern Zaire. "He is not really a European, despite the color of his skin," a Lokele villager said of Carrington. "He used to be from our village, one of us. After he died, the spirits made a mistake and sent him off far away to a village of whites to enter into the body of a little baby who was born of a white woman instead of one of ours. But because he belongs to us, he could not forget where he came from and so he came back." The villager added generously, "If he is a bit awkward on the drums, this is because of the poor education that the whites gave him." Carrington's life in Africa spanned four decades. He became an accomplished botanist, anthropologist, and above all linguist, authoritative on the structure of African language families: thousands of dialects and several hundred distinct languages. He noticed how loquacious a good drummer had to be. He finally published his discoveries about drums in 1949, in a slim volume titled *The Talking Drums of Africa*.

In solving the enigma of the drums, Carrington found the key in a central fact about the relevant African languages. They are tonal languages, in which meaning is determined as much by rising or falling pitch contours as by distinctions between consonants or vowels. This feature is missing from most Indo-European languages, including English, which uses tone only in limited, syntactical ways: for example, to distinguish questions ("you are happy ↗") from declarations ("you are happy ↘"). But for other languages, including, most famously, Mandarin and Cantonese, tone has primary significance in distinguishing words. So it does in most African languages. Even when Europeans learned to communicate in these languages, they generally failed to grasp the importance of tonality, because they had no experience with it. When they transliterated the words they heard into the Latin alphabet, they disregarded pitch altogether. In effect, they were color-blind.

Three different Kele words are transliterated by Europeans as *lisaka*. The words are distinguished only by their speech-tones. Thus *lisaka* with three low syllables is a puddle; *lisaka*, the last syllable rising (not necessarily stressed) is a promise; and *lisaka* is a poison. *Liala* means fiancée and *liala*, rubbish pit. In transliteration they appear to be homonyms, but they are not. Carrington, after the light dawned, recalled, "I must have been guilty many a time of asking a boy to 'paddle for a book' or to 'fish that his friend is coming.'" Europeans just lacked the ear for the distinctions. Carrington saw how comical the confusion could become:

alambaka boili [‒ _ ‒ ‒ _ _ _] = he watched the riverbank

alambaka boili [‒ ‒ ‒ ‒ _ ‒ _] = he boiled his mother-in-law

Since the late nineteenth century, linguists have identified the phoneme as the smallest acoustic unit that makes a difference in meaning. The English word *chuck* comprises three phonemes: different meanings can be created by changing *ch* to *d*, or *u* to *e*, or *ck* to *m*. It is a useful concept but an imperfect one: linguists have found it surprisingly difficult to agree on an exact inventory of phonemes for English or any other language (most estimates for English are in the vicinity of forty-five). The problem is that a stream of speech is a continuum; a linguist may abstractly, and arbitrarily, break it into discrete units, but the meaningfulness of these units varies from speaker to speaker and depends on the context. Most speakers' instincts about phonemes are biased, too, by their knowledge of the written alphabet, which codifies language in its own sometimes arbitrary ways. In any case, tonal languages, with their extra variable, contain many more phonemes than were first apparent to inexperienced linguists.

As the spoken languages of Africa elevated tonality to a crucial role, the drum language went a difficult step further. It employed tone and only tone. It was a language of a single pair of phonemes, a language composed entirely of pitch contours. The drums varied in materials and craft. Some were slit gongs, tubes of padauk wood, hollow, cut with a long and narrow mouth to make a high-sounding lip and a low-sounding lip; others had skin tops, and these were used in pairs. All that mattered was for the drums to sound two distinct notes, at an interval of about a major third.

So in mapping the spoken language to the drum language, information was lost. The drum talk was speech with a deficit. For every village and every tribe, the drum language began with the spoken word and shed the consonants and vowels. That was a lot to lose. The remaining information stream would be riddled with ambiguity. A double stroke on the high-tone lip of the drum [⎺ ⎺] matched the tonal pattern of the Kele word for father, *sango*, but naturally it could just as well be *songe*, the moon; *koko*, fowl; *fele*, a species of fish; or any other word of two high tones. Even the limited dictionary of the missionaries at Yakusu contained 130 such words. Having reduced spoken words, in all their sonic richness, to such a minimal code, how could the drums distinguish them? The answer lay partly in stress and timing, but these could not compensate for the lack of consonants and vowels. Thus, Carrington discovered, a drummer would invariably add "a little phrase" to each short word. *Songe*, the moon, is rendered as *songe li tange la manga*—"the moon looks down at the earth." *Koko*, the fowl, is rendered *koko olongo la bokiokio*—"the fowl, the little one that says kiokio." The extra drumbeats, far from being extraneous, provide context. Every ambiguous word begins in a cloud of possible alternative interpretations; then the unwanted possibilities evaporate. This takes place below the level of consciousness. Listeners are hearing only staccato drum tones, low and high, but in effect they "hear" the missing consonants and vowels, too. For that matter, they hear whole phrases, not individual words. "Among peoples who know nothing of writing or grammar, a word *per se*, cut out of its sound group, seems almost to cease to be an intelligible articulation," Captain Rattray reported.

The stereotyped long tails flap along, their redundancy overcoming ambiguity. The drum language is creative, freely generating neologisms for innovations from the north: steamboats, cigarettes, and the Christian god being three that Carrington particularly noted. But drummers begin by learning the traditional fixed formulas. Indeed, the formulas of the African drummers sometimes preserve archaic words that have been forgotten in the everyday language. For the Yaunde, the elephant is always "the great awkward one." The resemblance to Homeric formulas—not merely Zeus, but Zeus the cloud-gatherer; not just the sea, but the wine-dark sea—is no accident. In an oral culture, inspiration has to serve clarity and memory first. The Muses are the daughters of Mnemosyne.

Neither Kele nor English yet had words to say, *allocate extra bits for disambiguation and error correction*. Yet this is what the drum language did. Redundancy—inefficient by definition—serves as the antidote to confusion. It provides second chances. Every natural language has redundancy built in; this is why people can understand text riddled with errors and why they can understand conversation in a noisy room. The natural redundancy of English motivates the famous New York City subway poster of the 1970s (and the poem by James Merrill),

if u cn rd ths
u cn gt a gd jb w hi pa!

("This counterspell may save your soul," Merrill adds.) Most of the time, redundancy in language is just part of the background. For a telegraphist it is an expensive waste. For an African drummer it is essential. Another specialized language provides a perfect analog: the language of aviation radio. Numbers and letters make up much of the information passed between pilots and air traffic controllers: altitudes, vectors, aircraft tail numbers, runway and taxiway identifiers, radio frequencies. This is critical communication over a notoriously noisy channel, so a specialized alphabet is employed to minimize ambiguity. The spoken letters *B* and *V* are easy to confuse; *bravo* and *victor* are safer. *M* and *N* become *mike* and *november*. In the case of numbers, *five* and *nine*, particularly prone to confusion, are spoken as *fife* and *niner*. The extra syllables perform the same function as the extra verbosity of the talking drums.

After publishing his book, John Carrington came across a mathematical way to understand this point. A paper by a Bell Labs telephone engineer, Ralph Hartley, even had a relevant-looking formula: $H = n \log s$, where H is the amount of information, n is the number of symbols in the message, and s is the number of symbols available in the language. Hartley's younger colleague Claude Shannon later pursued this lead, and one of his touchstone projects became a precise measurement of the redundancy in English. Symbols could be words, phonemes, or dots and dashes. The degree of choice within a symbol set varied—a thousand words or forty-five phonemes or twenty-six letters or three types of interruption in an electrical circuit. The formula quantified a simple enough phenomenon (simple, anyway, once it was noticed): the fewer symbols available, the more of them must be transmitted to get across a given amount of information. For the African drummers, messages need to be about eight times as long as their spoken equivalents.

Hartley took some pains to justify his use of the word *information*. "As commonly used, information is a very elastic term," he wrote, "and it will first be necessary to set up for it a more specific meaning." He proposed to think of information "physically"—his word—rather than psychologically. He found the complications multiplying. Somewhat paradoxically, the complexity arose from the intermediate layers of symbols: letters of the alphabet, or dots and dashes, which were discrete and therefore easily countable in themselves. Harder to measure were the connections between these stand-ins and the bottom layer: the human voice itself. It was this stream of meaningful sound that still seemed, to a telephone engineer as much as an African drummer, the real stuff of communication, even if the sound, in turn, served as a code for the knowledge or meaning below. In any case Hartley thought an engineer should be able to generalize over all cases of communication: writing and telegraph codes as well as the physical transmission of sound by means of electromagnetic waves along telephone wires or through the ether.

He knew nothing of the drums, of course. And no sooner did John Carrington come to understand them than they began to fade from the African scene. He saw Lokele youth practicing the drums less and less, schoolboys who did not even learn their own drum names. He regretted it. He had made the talking drums a part of his own life. In 1954 a visitor from the United States found him running a mission school in the Congolese outpost of Yalemba. Carrington still walked daily in the jungle, and when it was time for lunch his wife would summon him with a fast tattoo. She drummed: "White man spirit in forest come come to house of shingles high up above of white man spirit in forest. Woman with yams awaits. Come come."

Before long, there were people for whom the path of communications technology had leapt directly from the talking drum to the mobile phone, skipping over the intermediate stages.

2 | THE PERSISTENCE OF THE WORD

(There Is No Dictionary in the Mind)

Odysseus wept when he heard the poet sing of his great deeds abroad because, once sung, they were no longer his alone. They belonged to anyone who heard the song.

—Ward Just (2004)

"TRY TO IMAGINE," proposed Walter J. Ong, Jesuit priest, philosopher, and cultural historian, "a culture where no one has ever 'looked up' anything." To subtract the technologies of information internalized over two millennia requires a leap of imagination backward into a forgotten past. The hardest technology to erase from our minds is the first of all: writing. This arises at the very dawn of history, as it must, because the history begins with the writing. The pastness of the past depends on it.

It takes a few thousand years for this mapping of language onto a system of signs to become second nature, and then there is no return to naïveté. Forgotten is the time when our very awareness of words came from *seeing* them. "In a primary oral culture," as Ong noted,

> the expression "to look up something" is an empty phrase: it would have no conceivable meaning. Without writing, words as such have no visual presence, even when the objects they represent are visual. They are sounds. You might "call" them back—"recall" them. But there is nowhere to "look" for them. They have no focus and no trace.

In the 1960s and '70s, Ong declared the electronic age to be a new age of orality—but of "secondary orality," the spoken word amplified and extended as never before, but always in the context of literacy: voices heard against a background of ubiquitous print. The first age of orality had lasted quite a bit longer. It covered almost the entire lifetime of the species, writing being a late development, general literacy being almost an afterthought. Like Marshall McLuhan, with whom he was often compared ("the other eminent Catholic-electronic prophet," said a scornful Frank Kermode), Ong had the misfortune to make his visionary assessments of a new age just before it actually arrived. The new media seemed to be radio, telephone, and television. But these were just the faint glimmerings in the night sky, signaling the light that still lay just beyond the horizon. Whether Ong would have seen cyberspace as fundamentally oral or literary, he would surely have recognized it as transformative: not just a revitalization of older forms, not just an amplification, but something wholly new. He might have sensed a coming discontinuity akin to the emergence of literacy itself. Few understood better than Ong just how profound a discontinuity that had been.

When he began his studies, "oral literature" was a common phrase. It is an oxymoron laced with anachronism; the words imply an all-too-unconscious approach to the past by way of the present. Oral literature was generally treated as a variant of writing; this, Ong said, was "rather like thinking of horses as automobiles without wheels."

> You can, of course, undertake to do this. Imagine writing a treatise on horses (for people who have never seen a horse) which starts with the concept not of "horse" but of "automobile," built on the readers' direct experience of automobiles. It proceeds to discourse on horses by always referring to them as "wheelless automobiles," explaining to highly automobilized readers all the points of difference. . . . Instead of wheels, the wheelless automobiles have enlarged toenails called hooves; instead of headlights, eyes; instead of a coat of lacquer, something called hair; instead of gasoline for fuel, hay, and so on. In the end, horses are only what they are not.

When it comes to understanding the preliterate past, we modern folk are hopelessly automobilized. The written word is the mechanism by which we know what we know. It organizes our thought. We may wish to understand the rise of literacy both historically and logically, but history and logic are themselves the products of literate thought.

Writing, as a technology, requires premeditation and special art. Language is not a technology, no matter how well developed and efficacious. It is not best seen as something separate from the mind; it is what the mind does. "Language in fact bears the same relationship to the concept of mind that legislation bears to the concept of parliament," says Jonathan Miller: "it is a competence forever bodying itself in a series of concrete performances." Much the same might be said of writing—it is concrete performance—but when the word is instantiated in paper or stone, it takes on a separate existence as artifice. It is a product of tools, and it is a tool. And like many technologies that followed, it thereby inspired immediate detractors.

One unlikely Luddite was also one of the first long-term beneficiaries. Plato (channeling the nonwriter Socrates) warned that this technology meant impoverishment:

> For this invention will produce forgetfulness in the minds of those who learn to use it, because they will not practice their memory. Their trust in writing, produced by external characters which are no part of themselves, will discourage the use of their own memory within them. You have invented an elixir not of memory, but of reminding; and you offer your pupils the appearance of wisdom, not true wisdom.

External characters which are no part of themselves—this was the trouble. The written word seemed insincere. Ersatz scratchings on papyrus or clay were far abstracted from the real, the free-flowing sound of language, intimately bound up with thought so as to seem coterminous with it. Writing appeared to draw knowledge away from the person, to place their memories in storage. It also separated the speaker from the listener, by so many miles or years. The deepest consequences of writing, for the individual and for the culture, could hardly have been foreseen, but even Plato could see some of the power of this disconnection. The one speaks to the multitude. The dead speak to the living, the living to the unborn. As McLuhan said, "Two thousand years of manuscript culture lay ahead of the Western world when Plato made this observation." The power of this first artificial memory was incalculable: to restructure thought, to engender history. It is still incalculable, though one statistic gives a hint: whereas the total vocabulary of any oral language measures a few thousand words, the single language that has been written most widely, English, has a documented vocabulary of well over a million words, a corpus that grows by thousands of words a year. These words do not exist only in the present. Each word has a provenance and a history that melts into its present life.

With words we begin to leave traces behind us like breadcrumbs: memories in symbols for others to follow. Ants deploy their pheromones, trails of chemical information; Theseus unwound Ariadne's thread. Now people leave paper trails. Writing comes into being to retain information across time and across space. Before writing, communication is evanescent and local; sounds carry a few yards and fade to oblivion. The evanescence of the spoken word went without saying. So fleeting was speech that the rare phenomenon of the echo, a sound heard once and then again, seemed a sort of magic. "This miraculous rebounding of the voice, the Greeks have a pretty name for, and call it Echo," wrote Pliny. "The spoken symbol," as Samuel Butler observed, "perishes instantly without material trace, and if it lives at all does so only in the minds of those who heard it." Butler was able to formulate this truth just as it was being falsified for the first time, at the end of the nineteenth century, by the arrival of the electric technologies for capturing speech. It was precisely because it was no longer completely true that it could be clearly seen. Butler completed the distinction: "The written symbol extends infinitely, as regards time and space, the range within which one mind can communicate with another; it gives the writer's mind a life limited by the duration of ink, paper, and readers, as against that of his flesh and blood body."

But the new channel does more than extend the previous channel. It enables reuse and "re-collection"—new modes. It permits whole new architectures of information. Among them are history, law, business, mathematics, and logic. Apart from their content, these categories represent new techniques. The power lies not just in the knowledge, preserved and passed forward, valuable as it is, but in the methodology: encoded visual indications, the act of transference, substituting signs for things. And then, later, signs for signs.

Paleolithic people began at least 30,000 years ago to scratch and paint shapes that recalled to the eye images of horses, fishes, and hunters. These signs in clay and on cave walls served purposes of art or magic, and historians are loath to call them writing, but they began the recording of mental states in external media. In another way, knots in cords and notches in sticks served as aids to memory. These could be carried as messages. Marks in pottery and masonry could signify ownership. Marks, images, pictographs, petroglyphs—as these forms grew stylized, conventional, and thus increasingly abstract, they approached what we understand as writing, but one more transition was crucial, from the representation of things to the representation of spoken language: that is, representation twice removed. There is a progression from pictographic, *writing the picture;* to ideographic, *writing the idea;* and then logographic, *writing the word*.

Chinese script began this transition between 4,500 and 8,000 years ago: signs that began as pictures came to represent meaningful units of sound. Because the basic unit was the word, thousands of distinct symbols were required. This is efficient in one way, inefficient in another. Chinese unifies an array of distinct spoken languages: people who cannot speak to one another can write to one another. It employs at least fifty thousand symbols, about six thousand commonly used and known to most literate Chinese. In swift diagrammatic strokes they encode multidimensional semantic relationships. One device is simple repetition: *tree* + *tree* + *tree* = *forest;* more abstractly, *sun* + *moon* = *brightness* and *east* + *east* = *everywhere.* The process of compounding creates surprises: *grain* + *knife* = *profit; hand* + *eye* = *look.* Characters can be transformed in meaning by reorienting their elements: *child* to *childbirth* and *man* to *corpse.* Some elements are phonetic; some even punning. The entirety is the richest and most complex writing system that humanity has ever evolved. Considering scripts in terms of how many symbols are required and how much meaning each individual symbol conveys, Chinese thus became an extreme case: the largest set of symbols, and the most meaningful individually. Writing systems could take alternative paths: fewer symbols, each carrying less information. An intermediate stage is the syllabary, a phonetic writing system using individual characters to represent syllables, which may or may not be meaningful. A few hundred characters can serve a language.

The writing system at the opposite extreme took the longest to emerge: the alphabet, one symbol for one minimal sound. The alphabet is the most reductive, the most subversive of all scripts.

In all the languages of earth there is only one word for *alphabet* (*alfabet, alfabeto,* алфавит, αλφάβητο). The alphabet was invented only once. All known alphabets, used today or found buried on tablets and stone, descend from the same original ancestor, which arose near the eastern littoral of the Mediterranean Sea, sometime not much before 1500 BCE, in a region that became a politically unstable crossroads of culture, covering Palestine, Phoenicia, and Assyria. To the east lay the great civilization of Mesopotamia, with its cuneiform script already a millennium old; down the shoreline to the southwest lay Egypt, where hieroglyphics developed simultaneously and independently. Traders traveled, too, from Cyprus and Crete, bringing their own incompatible systems. With glyphs from Minoan, Hittite, and Anatolian, it made for a symbolic stew. The ruling priestly classes were invested in their writing systems. Whoever owned the scripts owned the laws and the rites. But self-preservation had to compete with the desire for rapid communication. The scripts were conservative; the new technology was pragmatic. A stripped-down symbol system, just twenty-two signs, was the innovation of Semitic peoples in or near Palestine. Scholars naturally look to Kiriath-sepher, translatable as "city of the book," and Byblos, "city of papyrus," but no one knows exactly, and no one can know. The paleographer has a unique bootstrap problem. It is only writing that makes its own history possible. The foremost twentieth-century authority on the alphabet, David Diringer, quoted an earlier scholar: "There never was a man who could sit down and say: 'Now I am going to be the first man to write.'"

The alphabet spread by contagion. The new technology was both the virus and the vector of transmission. It could not be monopolized, and it could not be suppressed. Even children could learn these few, lightweight, semantically empty letters. Divergent routes led to alphabets of the Arab world and of northern Africa; to Hebrew and Phoenician; across central Asia, to Brahmi and related Indian script; and to Greece. The new civilization arising there brought the alphabet to a high degree of perfection. Among others, the Latin and Cyrillic alphabets followed along.

Greece had not needed the alphabet to create literature—a fact that scholars realized only grudgingly, beginning in the 1930s. That was when Milman Parry, a structural linguist who studied the living tradition of oral epic poetry in Bosnia and Herzegovina, proposed that the *Iliad* and the *Odyssey* not only could have been but must have been composed and sung without benefit of writing. The meter, the formulaic redundancy, in effect the very poetry of the great works served first and foremost to aid memory. Its incantatory power made of the verse a time capsule, able to transmit a virtual encyclopedia of culture across generations. His argument was first controversial and then overwhelmingly persuasive—but only because the poems *were* written down, sometime in the sixth or seventh century BCE. This act—the transcribing of the Homeric epics—echoes through the ages. "It was something like a thunder-clap in human history, which the bias of familiarity has converted into the rustle of papers on a desk," said Eric

Havelock, a British classical scholar who followed Parry. "It constituted an intrusion into culture, with results that proved irreversible. It laid the basis for the destruction of the oral way of life and the oral modes of thought."

The transcription of Homer converted this great poetry into a new medium and made of it something unplanned: from a momentary string of words created every time anew by the rhapsode and fading again even as it echoed in the listener's ear, to a fixed but portable line on a papyrus sheet. Whether this alien, dry mode would suit the creation of poetry and song remained to be seen. In the meantime the written word helped more mundane forms of discourse: petitions to the gods, statements of law, and economic agreements. Writing also gave rise to discourse about discourse. Written texts became objects of a new sort of interest.

But how was one to speak about them? The words to describe the elements of this discourse did not exist in the lexicon of Homer. The language of an oral culture had to be wrenched into new forms; thus a new vocabulary emerged. Poems were seen to have *topics*—the word previously meaning "place." They possessed *structure*, by analogy with buildings. They were made of *plot* and *diction*. Aristotle could now see the works of the bards as "representations of life," born of the natural impulse toward imitation that begins in childhood. But he had also to account for other writing with other purposes—the Socratic dialogues, for example, and medical or scientific treatises—and this general type of work, including, presumably, his own, "happens, up to the present day, to have no name." Under construction was a whole realm of abstraction, forcibly divorced from the concrete. Havelock described it as cultural warfare, a new consciousness and a new language at war with the old consciousness and the old language: "Their conflict produced essential and permanent contributions to the vocabulary of all abstract thought. Body and space, matter and motion, permanence and change, quality and quantity, combination and separation, are among the counters of common currency now available."

Aristotle himself, son of the physician to the king of Macedonia and an avid, organized thinker, was attempting to systematize knowledge. The persistence of writing made it possible to impose structure on what was known about the world and, then, on what was known about knowing. As soon as one could set words down, examine them, look at them anew the next day, and consider their meaning, one became a philosopher, and the philosopher began with a clean slate and a vast project of definition to undertake. Knowledge could begin to pull itself up by the bootstraps. For Aristotle the most basic notions were worth recording and were necessary to record:

> A *beginning* is that which itself does not follow necessarily from anything else, but some second thing naturally exists or occurs after it. Conversely, an *end* is that which does itself naturally follow from something else, either necessarily or in general, but there is nothing else after it. A *middle* is that which itself comes after something else, and some other thing comes after it.

These are statements not about experience but about the uses of language to structure experience. In the same way, the Greeks created *categories* (this word originally meaning "accusations" or "predictions") as a means of classifying animal species, insects, and fishes. In turn, they could then classify ideas. This was a radical, alien mode of thought. Plato had warned that it would repel most people:

> The multitude cannot accept the idea of beauty in itself rather than many beautiful things, nor anything conceived in its essence instead of the many specific things. Thus the multitude cannot be philosophic.

For "the multitude" we may understand "the preliterate." They "lose themselves and wander amid the multiplicities of multifarious things," declared Plato, looking back on the oral culture that still surrounded him. They "have no vivid pattern in their souls."

And what vivid pattern was that? Havelock focused on the process of converting, mentally, from a "prose of narrative" to a "prose of ideas"; organizing experience in terms of categories rather than events; embracing the discipline of abstraction. He had a word in mind for this process, and the word was *thinking*. This was the discovery, not just of the self, but of the *thinking* self—in effect, the true beginning of consciousness.

In our world of ingrained literacy, thinking and writing seem scarcely related activities. We can imagine the latter depending on the former, but surely not the other way around: everyone thinks, whether or not they write. But Havelock was right. The written word—the persistent word—was a prerequisite for conscious thought as we understand it. It was the trigger for a wholesale, irreversible change in the human psyche—*psyche* being the word favored by Socrates/Plato as they struggled to understand. Plato, as Havelock puts it,

> is trying for the first time in history to identify this group of general mental qualities, and seeking for a term which will label them satisfactorily under a single type. . . . He it was who hailed the portent and correctly identified it. In so doing, he so to speak confirmed and clinched the guesses of a previous generation which had been feeling its way towards the *idea* that you could "think," and that thinking was a very special kind of psychic activity, very uncomfortable, but also very exciting, and one which required a very novel use of Greek.

Taking the next step on the road of abstraction, Aristotle deployed categories and relationships in a regimented order to develop a symbolism of reasoning: logic—from λόγος, *logos*, the not-quite-translatable word from which so much flows, meaning "speech" or "reason" or "discourse" or, ultimately, just "word."

Logic might be imagined to exist independent of writing—syllogisms can be spoken as well as written—but it did not. Speech is too fleeting to allow for analysis. Logic descended from the written word, in Greece as well as India and China, where it developed independently. Logic turns the act of abstraction into a tool for determining what is true and what is false: truth can be discovered in words alone, apart from concrete experience. Logic takes its form in chains: sequences whose members connect one to another. Conclusions follow from premises. These require a degree of constancy. They have no power unless people can examine and evaluate them. In contrast, an oral narrative proceeds by accretion, the words passing by in a line of parade past the viewing stand, briefly present and then gone, interacting with one another via memory and association. There are no syllogisms in Homer. Experience is arranged in terms of events, not categories. Only with writing does narrative structure come to embody sustained rational argument. Aristotle crossed another level, by seeing the study of such argument—not just the use of argument, but its study—as a tool. His logic expresses an ongoing self-consciousness about the words in which they are composed. When Aristotle unfurls premises and conclusions—*If it is possible for no man to be a horse, it is also admissible for no horse to be a man; and if it is admissible for no garment to be white, it is also admissible for nothing white to be a garment. For if any white thing must be a garment, then some garment will necessarily be white*—he neither requires nor implies any personal experience of horses, garments, or colors. He has departed that realm. Yet he claims through the manipulation of words to create knowledge anyway, and a superior brand of knowledge at that.

"We know that formal logic is the invention of Greek culture after it had interiorized the technology of alphabetic writing," Walter Ong says—it is true of India and China as well—"and so made a permanent part of its noetic resources the kind of thinking that alphabetic writing made possible." For evidence Ong turns to fieldwork of the Russian psychologist Aleksandr Romanovich Luria among illiterate peoples in remote Uzbekistan and Kyrgyzstan in Central Asia in the 1930s. Luria found striking differences between illiterate and even slightly literate subjects, not in what they knew, but in how they thought. Logic implicates symbolism directly: things are members of classes; they possess qualities, which are abstracted and generalized. Oral people lacked the categories that become second nature even to illiterate individuals in literate cultures: for example, for geometrical shapes. Shown drawings of circles and squares, they named them as "plate, sieve, bucket, watch, or moon" and "mirror, door, house, apricot drying board." They could not, or would not, accept logical syllogisms. A typical question:

In the Far North, where there is snow, all bears are white.
Novaya Zembla is in the Far North and there is always snow there.
What color are the bears?

Typical response: "I don't know. I've seen a black bear. I've never seen any others. . . . Each locality has its own animals."

By contrast, a man who has just learned to read and write responds, "To go by your words, they should all be white." To go by your words—in that phrase, a level is crossed. The information has been detached from any person, detached from the speaker's experience. Now it lives in the words, little life-support modules. Spoken words also transport information, but not with the self-consciousness that writing brings. Literate people take for granted their own awareness of words, along with the array of word-related machinery: classification, reference, definition. Before literacy, there is nothing obvious about such techniques. "Try to explain to me what a tree is," Luria says, and a peasant replies, "Why should I? Everyone knows what a tree is, they don't need me telling them."

"Basically the peasant was right," Ong comments. "There is no way to refute the world of primary orality. All you can do is walk away from it into literacy."

It is a twisting journey from things to words, from words to categories, from categories to metaphor and logic. Unnatural as it seemed to define *tree*, it was even trickier to define *word*, and helpful ancillary words like *define* were not at first available, the need never having existed. "In the infancy of logic, a form of thought has to be invented before the content can be filled up," said Benjamin Jowett, Aristotle's nineteenth-century translator. Spoken languages needed further evolution.

Language and reasoning fit so well that users could not always see the flaws and gaps. Still, as soon as any culture invented logic, paradoxes appeared. In China, nearly contemporaneously with Aristotle, the philosopher Gongsun Long captured some of these in the form of a dialogue, known as "When a White Horse Is Not a Horse." It was written on bamboo strips, tied with string, before the invention of paper. It begins:

> *Can it be that a white horse is not a horse?*
> It can.
> *How?*
> "Horse" is that by means of which one names the shape. "White" is that by means of which one names the color. What names the color is not what names the shape. Hence, I say that a white horse is not a horse.

On its face, this is unfathomable. It begins to come into focus as a statement about language and logic. Gongsun Long was a member of the Mingjia, the School of Names, and his delving into these paradoxes formed part of what Chinese historians call the "language crisis," a running debate over the nature of language. Names are not the things they name. Classes are not coextensive with subclasses. Thus innocent-seeming inferences get derailed: "a man dislikes white horses" does not imply "a man dislikes horses."

> You think that horses that are colored are not horses. In the world, it is not the case that there are horses with no color. Can it be that there are no horses in the world?

The philosopher shines his light on the process of abstracting into classes based on properties: whiteness; horsiness. Are these classes part of reality, or do they exist only in language?

Horses certainly have color. Hence, there are white horses. If it were the case that horses had no color, there would simply be horses, and then how could one select a white horse? A white horse is a horse and white. A horse and a white horse are different. Hence, I say that a white horse is not a horse.

Two millennia later, philosophers continue to struggle with these texts. The paths of logic into modern thought are roundabout, broken, and complex. Since the paradoxes seem to be in language, or about language, one way to banish them was to purify the medium: eliminate ambiguous words and woolly syntax, employ symbols that were rigorous and pure. To turn, that is, to mathematics. By the beginning of the twentieth century, it seemed that only a system of purpose-built symbols could make logic work properly—free of error and paradoxes. This dream was to prove illusory; the paradoxes would creep back in, but no one could hope to understand until the paths of logic and mathematics converged.

Mathematics, too, followed from the invention of writing. Greece is often thought of as the springhead for the river that becomes modern mathematics, with all its many tributaries down the centuries. But the Greeks themselves alluded to another tradition—to them, ancient— which they called Chaldean, and which we understand to be Babylonian. That tradition vanished into the sands, not to surface until the end of the nineteenth century, when tablets of clay were dug up from the mounds of lost cities.

First there were scores, then thousands of tablets, typically the size of a human hand, etched with a distinctive, edgy, angular writing called cuneiform, "wedge shaped." Mature cuneiform was neither pictographic (the symbols were spare and abstract) nor alphabetic (they were far too numerous). By 3000 BCE a system with about seven hundred symbols flourished in Uruk, the walled city, probably the largest in the world, home of the

A CUNEIFORM TABLET

hero-king Gilgamesh, in the alluvial marshes near the Euphrates River. German archeologists excavated Uruk in a series of digs all through the twentieth century. The materials for this most ancient of information technologies lay readily at hand. With damp clay held in one hand and a stylus of sharpened reed in the other, a scribe would imprint tiny characters in columns and rows.

The result: cryptic messages from an alien culture. They took generations to decipher. "Writing, like a theater curtain going up on these dazzling civilizations, lets us stare directly but imperfectly at them," writes the psychologist Julian Jaynes. Some Europeans took umbrage at first. "To the Assyrians, the Chaldeans, and Egyptians," wrote the seventeenth-century divine Thomas Sprat, "we owe the Invention" but also the "Corruption of knowledge," when they concealed it with their strange scripts. "It was the custom of their Wise men, to wrap up their Observations on Nature, and the Manners of Men, in the dark Shadows of *Hieroglyphicks*" (as though friendlier ancients would have used an alphabet more familiar to Sprat). The earliest examples of cuneiform baffled archeologists and paleolinguists the longest, because the first language to be written, Sumerian, left no other traces in culture or speech. Sumerian turned out to be a linguistic rarity, an isolate, with no known descendants. When scholars did learn to read the Uruk tablets, they found them to be, in their way, humdrum: civic memoranda, contracts and laws, and receipts and bills for barley, livestock, oil, reed mats, and pottery. Nothing like poetry or literature appeared in cuneiform for hundreds of years to come. The tablets were the quotidiana of nascent commerce and bureaucracy. The tablets not only recorded the commerce and the bureaucracy but, in the first place, made them possible.

Even then, cuneiform incorporated signs for counting and measurement. Different characters, used in different ways, could denote numbers and weights. A more systematic approach to the writing of numbers did not take shape until the time of Hammurabi, 1750 BCE, when Mesopotamia was unified around the great city of Babylon. Hammurabi himself was probably the first literate king, writing his own cuneiform rather than depending on scribes, and his empire building manifested the connection between writing and social control. "This process of conquest and influence is made possible by letters and tablets and stelae in an abundance that had never been known before," Jaynes declares. "Writing was a new method of civil direction, indeed the model that begins our own memo-communicating government."

The writing of numbers had evolved into an elaborate system. Numerals were composed of just two basic parts, a vertical wedge for 1 (Υ) and an angle wedge for 10 ($\triangleleft$). These were combined to form the standard characters, so that $\Uparrow\!\!\!\!\Uparrow$ represented 3 and $\triangleleft\!\!\!\Uparrow\!\!\!\!\Uparrow$ represented 16, and so on. But the Babylonian system was not decimal, base 10; it was sexagesimal, base 60. Each of the numerals from 1 to 60 had its own character. To form large numbers, the Babylonians used numerals in places: $\Upsilon \triangleleft$ was 70 (one 60 plus ten 1s); $\Upsilon \triangleleft\!\!\triangleleft\!\!\Uparrow\!\!\!\!\Uparrow$ was 616 (ten 60s plus sixteen 1s), and so on. None of this was clear when the tablets first began to surface. A basic theme with variations, encountered many times, proved to be multiplication tables. In a sexagesimal system these had to cover the numbers from 1 to 19 as well as 20, 30, 40, and 50. Even more difficult to unravel were tables of reciprocals, making possible division and fractional numbers: in the 60-based system, reciprocals were 2:30, 3:20, 4:15, 5:12 . . . and then, using extra places, 8:7,30, 9:6,40, and so on.*

---

* It is customary to transcribe a two-place sexagesimal cuneiform number with a comma—such as "7,30." But the scribes did not use such punctuation, and in fact their notation left the place values undefined; that is, their numbers were what we would call "floating point." A two-place number like 7,30 could be 450 (seven 60s + thirty 1s) or 7½ (seven 1s + thirty 1/60s).

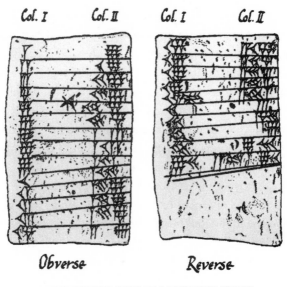

A MATHEMATICAL TABLE ON A CUNEIFORM TABLET
ANALYZED BY ASGER AABOE

These symbols were hardly words—or they were words of a peculiar, slender, rigid sort. They seemed to arrange themselves into visible patterns in the clay, repetitious, almost artistic, not like any prose or poetry archeologists had encountered. They were like maps of a mysterious city. This was the key to deciphering them, finally: the ordered chaos that seems to guarantee the presence of meaning. It seemed like a task for mathematicians, anyway, and finally it was. They recognized geometric progressions, tables of powers, and even instructions for computing square roots and cube roots. Familiar as they were with the rise of mathematics a millennium later in ancient Greece, these scholars were astounded at the breadth and depth of mathematical knowledge that existed before in Mesopotamia. "It was assumed that the Babylonians had had some sort of number mysticism or numerology," wrote Asger Aaboe in 1963, "but we now know how far short of the truth this assumption was." The Babylonians computed linear equations, quadratic equations, and Pythagorean numbers long before Pythagoras. In contrast to the Greek mathematics that followed, Babylonian mathematics did not emphasize geometry, except for practical problems; the Babylonians calculated areas and perimeters but did not prove theorems. Yet they could (in effect) reduce elaborate second-degree polynomials. Their mathematics seemed to value computational power above all.

That could not be appreciated until computational power began to mean something. By the time modern mathematicians turned their attention to Babylon, many important tablets had already been destroyed or scattered. Fragments retrieved from Uruk before 1914, for example, were dispersed to Berlin, Paris, and Chicago and only fifty years later were discovered to hold the beginning methods of astronomy. To demonstrate this, Otto Neugebauer, the leading twentieth-century historian of ancient mathematics, had to reassemble tablets whose fragments had made their way to opposite sides of the Atlantic Ocean. In 1949, when the number of cuneiform tablets housed in museums reached (at his rough guess) a half million, Neugebauer lamented, "Our task can therefore properly be compared with restoring the history of mathematics from a few torn pages which have accidentally survived the destruction of a great library."

In 1972, Donald Knuth, an early computer scientist at Stanford, looked at the remains of an Old Babylonian tablet the size of a paperback book, half lying in the British Museum in London, one-fourth in the Staatliche Museen in Berlin, and the rest missing, and saw what he could only describe, anachronistically, as an algorithm:

A cistern.
The height is 3,20, and a volume of 27,46,40 has been excavated.
The length exceeds the width by 50.
You should take the reciprocal of the height, 3,20, obtaining 18.
Multiply this by the volume, 27,46,40, obtaining 8,20.
Take half of 50 and square it, obtaining 10,25.
Add 8,20, and you get 8,30,25.
The square root is 2,55.

Make two copies of this, adding to the one and subtracting from the other.
You find that 3,20 is the length and 2,30 is the width.
This is the procedure.

"This is the procedure" was a standard closing, like a benediction, and for Knuth redolent with meaning. In the Louvre he found a "procedure" that reminded him of a stack program on a Burroughs B5500. "We can commend the Babylonians for developing a nice way to explain an algorithm by example as the algorithm itself was being defined," said Knuth. By then he himself was engrossed in the project of defining and explaining the algorithm; he was amazed by what he found on the ancient tablets. The scribes wrote instructions for placing numbers in certain locations—for making "copies" of a number, and for keeping a number "in your head." This idea, of abstract quantities occupying abstract places, would not come back to life till much later.

Where is a symbol? What is a symbol? Even to ask such questions required a self-consciousness that did not come naturally. Once asked, the questions continued to loom. *Look at these signs*, philosophers implored. *What are they?*

"Fundamentally letters are shapes indicating voices," explained John of Salisbury in medieval England. "Hence they represent things which they bring to mind through the windows of the eyes." John served as secretary and scribe to the Archbishop of Canterbury in the twelfth century. He served the cause of Aristotle as an advocate and salesman. His Metalogicon not only set forth the principles of Aristotelian logic but urged his contemporaries to convert, as though to a new religion. (He did not mince words: "Let him who is not come to logic be plagued with continuous and everlasting filth.") Putting pen to parchment in this time of barest literacy, he tried to examine the act of writing and the effect of words: "Frequently they speak voicelessly the utterances of the absent."

The idea of writing was still entangled with the idea of speaking. The mixing of the visual and the auditory continued to create puzzles, and so also did the mixing of past and future: utterances of the absent. Writing leapt across these levels.

Every user of this technology was a novice. Those composing formal legal documents, such as charters and deeds, often felt the need to express their sensation of speaking to an invisible audience: "Oh! all ye who shall have heard this and have seen!" (They found it awkward to keep tenses straight, like voicemail novices leaving their first messages circa 1980.) Many charters ended with the word "Goodbye." Before writing could feel natural in itself—could become second nature—these echoes of voices had to fade away. Writing in and of itself had to reshape human consciousness.

Among the many abilities gained by the written culture, not the least was the power of looking inward upon itself. Writers loved to discuss writing, far more than bards ever bothered to discuss speech. They could *see* the medium and its messages, hold them up to the mind's eye for study and analysis. And they could criticize it—for from the very start, the new abilities were accompanied by a nagging sense of loss. It was a form of nostalgia. Plato felt it:

> I cannot help feeling, Phaedrus, [says Socrates] that writing is unfortu-
> nately like painting; for the creations of the painter have the attitude of
> life, and yet if you ask them a question they preserve a solemn silence. . . .
> You would imagine that they had intelligence, but if you want to know
> anything and put a question to one of them, the speaker always gives one
> unvarying answer.

Unfortunately the written word stands still. It is stable and immobile. Plato's qualms were mostly set aside in the succeeding millennia, as the culture of literacy developed its many gifts: history and the law; the sciences and philosophy; the reflective explication of art and literature itself. None of that could have emerged from pure orality. Great poetry could and did, but it was expensive and rare. To make the epics of Homer, to let them be heard, to sustain them across the years and the miles required a considerable share of the available cultural energy.

Then the vanished world of primary orality was not much missed. Not until the twentieth century, amid a burgeoning of new media for communication, did the qualms and the nostalgia resurface. Marshall McLuhan, who became the most famous spokesman for the bygone oral culture, did so in the service of an argument for modernity. He hailed the new "electric age" not for its newness but for its return to the roots of human creativity. He saw it as a revival of the old orality. "We are in our century 'winding the tape backward,'" he declared, finding his metaphorical tape in one of the newest information technologies. He constructed a series of polemical contrasts: the printed word vs. the spoken word; cold/hot; static/fluid; neutral/magical; impoverished/rich; regimented/creative; mechanical/ organic; separatist/integrative. "The alphabet is a technology of visual fragmentation and specialism," he wrote. It leads to "a desert of classified data." One way of framing McLuhan's critique of print would be to say that print offers only a narrow channel of communication. The channel is linear and even fragmented. By contrast, speech—in the primal case, face-to-face human intercourse, alive with gesture and touch—engages all the senses, not just hearing. If the ideal of communication is a meeting of souls, then writing is a sad shadow of the ideal.

The same criticism was made of other constrained channels, created by later technologies—the telegraph, the telephone, radio, and e-mail. Jonathan Miller rephrases McLuhan's argument in quasi-technical terms of information: "The larger the number of senses involved, the better the chance of transmitting a reliable copy of the sender's mental state."* In the stream of words past the ear or eye, we sense not just the items one by one but their rhythms and tones, which is to say their music. We, the

---

* Not that Miller agrees. On the contrary: "It is hard to overestimate the subtle reflexive effects of literacy upon the creative imagination, providing as it does a cumulative deposit of ideas, images, and idioms upon whose rich and appreciating funds every artist enjoys an unlimited right of withdrawal."

listener or the reader, do not hear, or read, one word at a time; we get messages in groupings small and large. Human memory being what it is, larger patterns can be grasped in writing than in sound. The eye can glance back. McLuhan considered this damaging, or at least diminishing. "Acoustic space is organic and integral," he said, "perceived through the simultaneous interplay of all the senses; whereas 'rational' or pictorial space is uniform, sequential and continuous and creates a closed world with none of the rich resonance of the tribal echoland." For McLuhan, the tribal echoland is Eden.

> By their dependence on the spoken word for information, people were drawn together into a tribal mesh . . . the spoken word is more emotionally laden than the written. . . . Audile-tactile tribal man partook of the collective unconscious, lived in a magical integral world patterned by myth and ritual, its values divine.*

Up to a point, maybe. Yet three centuries earlier, Thomas Hobbes, looking from a vantage where literacy was new, had taken a less rosy view. He could see the preliterate culture more clearly: "Men lived upon gross experience," he wrote. "There was no method; that is to say, no sowing nor planting of knowledge by itself, apart from the weeds and common plants of error and conjecture." A sorry place, neither magical nor divine.

Was McLuhan right, or was Hobbes? If we are ambivalent, the ambivalence began with Plato. He witnessed writing's rising dominion; he asserted its force and feared its lifelessness. The writer-philosopher embodied a paradox. The same paradox was destined to reappear in different guises, each technology of information bringing its own powers and its own fears. It turns out that the "forgetfulness" Plato feared does not arise. It does not arise because Plato himself, with his mentor

---

* The interviewer asked plaintively, "But aren't there corresponding gains in insight, understanding and cultural diversity to compensate detribalized man?" McLuhan responded, "Your question reflects all the institutionalized biases of literate man."

Socrates and his disciple Aristotle, designed a vocabulary of ideas, organized them into categories, set down rules of logic, and so fulfilled the promise of the technology of writing. All this made knowledge more durable stuff than before.

And the atom of knowledge was the word. Or was it? For some time to come, the word continued to elude its pursuers, whether it was a fleeting burst of sound or a fixed cluster of marks. "Most literate persons, when you say, 'Think of a word,' at least in some vague fashion think of something before their eyes," Ong says, "where a real word can never be at all." Where do we look for the words, then? In the dictionary, of course. Ong also said: "It is demoralizing to remind oneself that there is no dictionary in the mind, that lexicographical apparatus is a very late accretion to language."

3 | TWO WORDBOOKS

(The Uncertainty in Our Writing, the Inconstancy in Our Letters)

In such busie, and active times, there arise more new thoughts of men, which must be signifi'd, and varied by new expressions.
—Thomas Sprat (1667)

A VILLAGE SCHOOLMASTER AND PRIEST made a book in 1604 with a rambling title that began "A Table Alphabeticall, conteyning and teaching the true writing, and understanding of hard usuall English wordes," and went on with more hints to its purpose, which was unusual and needed explanation:

> With the interpretation thereof by plaine English words, gathered for the benefit & helpe of Ladies, Gentlewomen, or any other unskilfull persons.
>   Whereby they may the more easily and better understand many hard English wordes, which they shall heare or read in Scriptures, Sermons, or elsewhere, and also be made able to use the same aptly themselves.

The title page omitted the name of the author, Robert Cawdrey, but included a motto from Latin—"As good not read, as not to understand"—and situated the publisher with as much formality and exactness as could be expected in a time when the *address*, as a specification of place, did not yet exist:

> At London, Printed by I. R. for Edmund Weaver, & are to be sold at his shop at the great North doore of Paules Church.

CAWDREY'S TITLE PAGE

Even in London's densely packed streets, shops and homes were seldom to be found by number. The alphabet, however, had a definite order—the first and second letters providing its very name—and that order had been maintained since the early Phoenician times, through all the borrowing and evolution that followed.

Cawdrey lived in a time of information poverty. He would not have thought so, even had he possessed the concept. On the contrary, he would have considered himself to be in the midst of an information explosion, which he himself was trying to abet and organize. But four centuries later, his own life is shrouded in the obscurity of missing knowledge.

His *Table Alphabeticall* appears as a milestone in the history of information, yet of its entire first edition, just one worn copy survived into the future. When and where he was born remain unknown—probably in the late 1530s; probably in the Midlands. Parish registers notwithstanding, people's lives were almost wholly undocumented. No one has even a definitive spelling for Cawdrey's name (Cowdrey, Cawdry). But then, no one agreed on the spelling of most names: they were spoken, seldom written.

In fact, few had any concept of "spelling"—the idea that each word, when written, should take a particular predetermined form of letters. The word *cony* (rabbit) appeared variously as *conny, conye, conie, connie, coni, cuny, cunny*, and *cunnie* in a single 1591 pamphlet. Others spelled it differently. And for that matter Cawdrey himself, on the title page of his book for "teaching the true writing," wrote *wordes* in one sentence and *words* in the next. Language did not function as a storehouse of words, from which users could summon the correct items, preformed. On the contrary, words were fugitive, on the fly, expected to vanish again thereafter. When spoken, they were not available to be compared with, or measured against, other instantiations of themselves. Every time people dipped quill in ink to form a word on paper they made a fresh choice of whatever letters seemed to suit the task. But this was changing. The availability—the solidity—of the printed book inspired a sense that the written word *should be* a certain way, that one form was right and others wrong. First this sense was unconscious; then it began to rise toward general awareness. Printers themselves made it their business.

To spell (from an old Germanic word) first meant to speak or to utter. Then it meant to read, slowly, letter by letter. Then, by extension, just around Cawdrey's time, it meant to write words letter by letter. The last was a somewhat poetic usage. "Spell Eva back and Ave shall you find," wrote the Jesuit poet Robert Southwell (shortly before being hanged and quartered in 1595). When certain educators did begin to consider the idea of spelling, they would say "right writing"—or, to borrow from

Greek, "*orthography*." Few bothered, but one who did was a school headmaster in London, Richard Mulcaster. He assembled a primer, titled "The first part [a second part was not to be] of the Elementarie which entreateth chefelie of the right writing of our English tung." He published it in 1582 ("at London by Thomas Vautroullier dwelling in the blak-friers by Lud-gate"), including his own list of about eight thousand words and a plea for the idea of a dictionary:

> It were a thing verie praiseworthie in my opinion, and no lesse profitable than praise worthie, if some one well learned and as laborious a man, wold gather all the words which we use in our English tung . . . into one dictionarie, and besides the right writing, which is incident to the Alphabete, wold open unto us therein, both their naturall force, and their proper use.

He recognized another motivating factor: the quickening pace of commerce and transportation made other languages a palpable presence, forcing an awareness of the English language as just one among many. "Forenners and strangers do wonder at us," Mulcaster wrote, "both for the uncertaintie in our writing, and the inconstancie in our letters." Language was no longer invisible like the air.

Barely 5 million people on earth spoke English (a rough estimate; no one tried to count the population of England, Scotland, or Ireland until 1801). Barely a million of those could write. Of all the world's languages English was already the most checkered, the most mottled, the most polygenetic. Its history showed continual corruption and enrichment from without. Its oldest core words, the words that felt most basic, came from the language spoken by the Angles, Saxons, and Jutes, Germanic tribes that crossed the North Sea into England in the fifth century, pushing aside the Celtic inhabitants. Not much of Celtic penetrated the Anglo-Saxon speech, but Viking invaders brought more words from Norse and Danish: *egg, sky, anger, give, get*. Latin came by way of Christian missionaries; they wrote in the alphabet of the Romans, which replaced the runic scripts that spread in central and northern Europe early in the first millennium. Then came the influence of French.

Influence, to Robert Cawdrey, meant "a flowing in." The Norman Conquest was more like a deluge, linguistically. English peasants of the lower classes continued to breed *cows*, *pigs*, and *oxen* (Germanic words), but in the second millennium the upper classes dined on *beef, pork*, and *mutton* (French). By medieval times French and Latin roots accounted for more than half of the common vocabulary. More alien words came when intellectuals began consciously to borrow from Latin and Greek to express concepts the language had not before needed. Cawdrey found this habit irritating. "Some men seek so far for outlandish English, that they forget altogether their mothers language, so that if some of their mothers were alive, they were not able to tell, or understand what they say," he complained. "One might well charge them, for counterfeyting the Kings English."

Four hundred years after Cawdrey published his book of words, John Simpson retraced Cawdrey's path. Simpson was in certain respects his natural heir: the editor of a grander book of words, the *Oxford English Dictionary*. Simpson, a pale, soft-spoken man, saw Cawdrey as obstinate, uncompromising, and even pugnacious. The schoolteacher was ordained a deacon and then a priest of the Church of England in a restless time, when Puritanism was on the rise. Nonconformity led him into trouble. He seems to have been guilty of "not Conforming himself" to some of the sacraments, such as "the Cross in Baptism, and the Ring in Marriage." As a village priest he did not care to bow down to bishops and archbishops. He preached a form of equality unwelcome to church authorities. "There was preferred secretly an Information against him for speaking diverse Words in the Pulpit, tending to the depraving of the Book of Common Prayer. . . . And so being judged a dangerous Person, if he should continue preaching, but infecting the People with Principles different from the Religion established." Cawdrey was degraded from the priesthood and deprived of his benefice. He continued to fight the case for years, to no avail.

All that time, he collected words ("**collect**, gather"). He published two instructional treatises, one on catechism ("**catechiser**, that teacheth the principles of Christian religion") and one on *A godlie forme of householde government for the ordering of private families*, and in 1604 he produced a different sort of book: nothing more than a list of words, with brief definitions.

Why? Simpson says, "We have already seen that he was committed to simplicity in language, and that he was strong-minded to the point of obstinacy." He was still preaching—now, to preachers. "Such as by their place and calling (but especially Preachers) as have occasion to speak publiquely before the ignorant people," Cawdrey declared in his introductory note, "are to bee admonished." He admonishes them. "Never affect any strange ynckhorne termes." (An *inkhorn* was an inkpot; by *inkhorn term* he meant a bookish word.) "Labour to speake so as is commonly received, and so as the most ignorant may well understand them." And above all do not affect to speak like a foreigner:

> Some far journied gentlemen, at their returne home, like as they love to go in forraine apparrell, so they will pouder their talke with over-sea language. He that commeth lately out of France, will talk French English, and never blush at the matter.

Cawdrey had no idea of listing *all* the words—whatever that would mean. By 1604 William Shakespeare had written most of his plays, employing a vocabulary of nearly 30,000, but these words were not available to Cawdrey or anyone else. Cawdrey did not bother with the most common words, nor the most inkhorn and Frenchified words; he listed only the "hard usual" words, words difficult enough to need some explanation but still "proper unto the tongue wherein we speake" and "plaine for all men to perceive." He compiled 2,500. He knew that many were derived from Greek, French, and Latin ("**derive**, fetch from"), and he marked these accordingly. The book Cawdrey made was the first English dictionary. The word *dictionary* was not in it.

Although Cawdrey cited no authorities, he had relied on some. He copied the remarks about inkhorn terms and the far-journeyed gentlemen in their foreign apparel from Thomas Wilson's successful book *The Arte of Rhetorique*. For the words themselves he found several sources ("**source**, wave, or issuing foorth of water"). He found about half his words in a primer for teaching reading, called *The English Schoolemaister*, by Edmund Coote, first published in 1596 and widely reprinted thereafter. Coote claimed that a schoolmaster could teach a hundred students more quickly with his text than forty without it. He found it worthwhile to explain the benefits of teaching people to read: "So more knowledge will be brought into this Land, and moe bookes bought, than otherwise would have been." Coote included a long glossary, which Cawdrey plundered.

That Cawdrey should arrange his words in alphabetical order, to make his *Table Alphabeticall*, was not self-evident. He knew he could not count on even his educated readers to be versed in alphabetical order, so he tried to produce a small how-to manual. He struggled with this: whether to describe the ordering in logical, schematic terms or in terms of a step-by-step procedure, an algorithm. "Gentle reader," he wrote—again adapting freely from Coote—

> thou must learne the Alphabet, to wit, the order of the Letters as they stand, perfectly without booke, and where every Letter standeth: as *b* neere the beginning, *n* about the middest, and *t* toward the end. Nowe if the word, which thou art desirous to finde, begin with *a* then looke in the beginning of this Table, but if with *v* looke towards the end. Againe, if thy word beginne with *ca* looke in the beginning of the letter *c* but if with *cu* then looke toward the end of that letter. And so of all the rest. &c.

It was not easy to explain. Friar Johannes Balbus of Genoa tried in his 1286 *Catholicon*. Balbus thought he was inventing alphabetical order for the first time, and his instructions were painstaking: "For example I intend to discuss *amo* and *bibo*. I will discuss *amo* before *bibo* because *a* is the first letter of *amo* and *b* is the first letter of *bibo* and *a* is before *b* in the alphabet. Similarly . . ." He rehearsed a long list of examples and concluded: "I beg of you, therefore, good reader, do not scorn this great labor of mine and this order as something worthless."

In the ancient world, alphabetical lists scarcely appeared until around 250 BCE, in papyrus texts from Alexandria. The great library there seems to have used at least some alphabetization in organizing its books. The need for such an artificial ordering scheme arises only with large collections of data, not otherwise ordered. And the possibility of alphabetical order arises only in languages possessing an alphabet: a discrete small symbol set with its own conventional sequence ("**abecedarie**, the order of the Letters, or hee that useth them"). Even then the system is unnatural. It forces the user to detach information from meaning; to treat words strictly as character strings; to focus abstractly on the configuration of the word. Furthermore, alphabetical ordering comprises a pair of procedures, one the inverse of the other: organizing a list and looking up items; sorting and searching. In either direction the procedure is recursive ("**recourse**, a running backe againe"). The basic operation is a binary decision: greater than or less than. This operation is performed first on one letter; then, nested as a subroutine, on the next letter; and (as Cawdrey put it, struggling with the awkwardness) "so of all the rest. &c." This makes for astounding efficiency. The system scales easily to any size, the macrostructure being identical to the microstructure. A person who understands alphabetical order homes in on any one item in a list of a thousand or a million, unerringly, with perfect confidence. And without knowing anything about the meaning.

Not until 1613 was the first alphabetical catalogue made—not printed, but written in two small handbooks—for the Bodleian Library at Oxford.

The first catalogue of a university library, made at Leiden, Holland, two decades earlier, was arranged by subject matter, as a shelf list (about 450 books), with no alphabetical index. Of one thing Cawdrey could be sure: his typical reader, a literate, book-buying Englishman at the turn of the seventeenth century, could live a lifetime without ever encountering a set of data ordered alphabetically.

More sensible ways of ordering words came first and lingered for a long time. In China the closest thing to a dictionary for many centuries was the *Erya*, author unknown, date unknown but probably around the third century BCE. It arranged its two thousand entries by meaning, in topical categories: kinship, building, tools and weapons, the heavens, the earth, plants and animals. Egyptian had word lists organized on philosophical or educational principles; so did Arabic. These lists were arranging not the words themselves, mainly, but rather the world: the things for which the words stood. In Germany, a century after Cawdrey, the philosopher and mathematician Gottfried Wilhelm Leibniz made this distinction explicit:

> Let me mention that the words or names of all things and actions can be brought into a list in two different ways, according to the alphabet and according to nature. . . . The former go from the word to the thing, the latter from the thing to the word.

Topical lists were thought provoking, imperfect, and creative. Alphabetical lists were mechanical, effective, and automatic. Considered alphabetically, words are no more than tokens, each placed in a slot. In effect they may as well be numbers.

Meaning comes into the dictionary in its definitions, of course. Cawdrey's crucial models were dictionaries for translation, especially a 1587 Latin-English *Dictionarium* by Thomas Thomas. A bilingual dictionary had a clearer purpose than a dictionary of one language alone: mapping

Latin onto English made a kind of sense that translating English to English did not. Yet definitions were the point, Cawdrey's stated purpose being after all to help people understand and use hard words. He approached the task of definition with a trepidation that remains palpable. Even as he defined his words, Cawdrey still did not quite believe in their solidity. Meanings were even more fluid than spellings. *Define*, to Cawdrey, was for things, not for words: "**define**, to shew clearly what a thing is." It was reality, in all its richness, that needed defining. *Interpret* meant "open, make plaine, to shewe the sence and meaning of a thing." For him the relationship between the thing and the word was like the relationship between an object and its shadow.

The relevant concepts had not reached maturity:

figurate, to shadowe, or represent, or to counterfaite
type, figure, example, shadowe of any thing
represent, expresse, beare shew of a thing

An earlier contemporary of Cawdrey's, Ralph Lever, made up his own word: "**saywhat**, corruptly called a definition: but it is a saying which telleth what a thing is, it may more aptly be called a saywhat." This did not catch on. It took almost another century—and the examples of Cawdrey and his successors—for the modern sense to come into focus: "Definition," John Locke finally writes in 1690, "being nothing but making another understand by Words, what Idea the Term defin'd stands for." And Locke still takes an operational view. Definition is communication: making another understand; sending a message.

Cawdrey borrows definitions from his sources, combines them, and adapts them. In many case he simply maps one word onto another:

orifice, mouth
baud, whore
helmet, head peece

For a small class of words he uses a special designation, the letter *k:* "stand-eth for a kind of." He does not consider it his job to say *what* kind. Thus:

> **crocodile**, *k* beast
> **alablaster**, *k* stone
> **citron**, *k* fruit

But linking pairs of words, either as synonyms or as members of a class, can carry a lexicographer only so far. The relationships among the words of a language are far too complex for so linear an approach ("**chaos**, a confused heap of mingle-mangle"). Sometimes Cawdrey tries to cope by adding one or more extra synonyms, definition by triangulation:

> **specke**, spot, or marke
> **cynicall**, doggish, froward
> **vapor**, moisture, ayre, hote breath, or reaking

For other words, representing concepts and abstractions, further removed from the concrete realm of the senses, Cawdrey needs to find another style altogether. He makes it up as he goes along. He must speak to his reader, in prose but not quite in sentences, and we can hear him struggle, both to understand certain words and to express his understanding.

> **gargarise**, to wash the mouth, and throate within, by stirring some
>   liquor up and downe in the mouth
> **hipocrite**, such a one as in his outward apparrell, countenaunce,
>   & behaviour, pretendeth to be another man, then he is indeede,
>   or a deceiver
> **buggerie**, coniunction with one of the same kinde, or of men with beasts
> **theologie**, divinitie, the science of living blessedly for ever

Among the most troublesome were technical terms from new sciences:

cypher, a circle in numbering, of no value of it selfe, but serveth to
    make up the number, and to make other figures of more value
horizon, a circle, deviding the halfe of the firmament, from the other
    halfe which we see not
zodiack, a circle in the heaven, wherein be placed the 12 signes, and in
    which the Sunne is mooved

Not just the words but the knowledge was in flux. The language was examining itself. Even when Cawdrey is copying from Coote or Thomas, he is fundamentally alone, with no authority to consult.

One of Cawdrey's hard usual words was *science* ("knowledge, or skill"). Science did not yet exist as an institution responsible for learning about the material universe and its laws. Natural philosophers were beginning to have a special interest in the nature of words and their meaning. They needed better than they had. When Galileo pointed his first telescope skyward and discovered sunspots in 1611, he immediately anticipated controversy—traditionally the sun was an epitome of purity—and he sensed that science could not proceed without first solving a problem of language:

> So long as men were in fact obliged to call the sun "most pure and most lucid," no shadows or impurities whatever had been perceived in it; but now that it shows itself to us as partly impure and spotty; why should we not call it "spotted and not pure"? For names and attributes must be accommodated to the essence of things, and not the essence to the names, since things come first and names afterwards.

When Isaac Newton embarked on his great program, he encountered a fundamental lack of definition where it was most needed. He began with a semantic sleight of hand: "I do not define time, space, place, and motion, as being well known to all," he wrote deceptively. Defining these words was his very purpose. There were no agreed standards for weights and measures. *Weight* and *measure* were themselves vague terms. Latin seemed more reliable than English, precisely because it was less worn by everyday use, but the Romans had not possessed the necessary words either. Newton's raw notes reveal a struggle hidden in the finished product. He tried expressions like *quantitas materiae*. Too hard for Cawdrey: "**materiall**, of some matter, or importance." Newton suggested (to himself) "that which arises from its density and bulk conjointly." He considered more words: "This quantity I designate under the name of body or mass." Without the right words he could not proceed. *Velocity, force, gravity*—none of these were yet suitable. They could not be defined in terms of one another; there was nothing in visible nature at which anyone could point a finger; and there was no book in which to look them up.

As for Robert Cawdrey, his mark on history ends with the publication of his *Table Alphabeticall* in 1604. No one knows when he died. No one knows how many copies the printer made. There are no records ("**records**, writings layde up for remembrance"). A single copy made its way to the Bodleian Library in Oxford, which has preserved it. All the others disappeared. A second edition appeared in 1609, slightly expanded ("much inlarged," the title page claims falsely) by Cawdrey's son, Thomas, and a third and fourth appeared in 1613 and 1617, and there the life of this book ended.

It was overshadowed by a new dictionary, twice as comprehensive, *An English Expositour: Teaching the Interpretation of the hardest Words used in our Language, with sundry Explications, Descriptions, and Discourses.* Its compiler, John Bullokar, otherwise left as faint a mark on the historical record as Cawdrey did. He was doctor of physic; he lived for some time in Chichester; his dates of birth and death are uncertain; he is said to have visited London in 1611 and there to have seen a dead crocodile; and little else is known. His *Expositour* appeared in 1616 and went through several editions in the succeeding decades. Then in 1656 a London barrister, Thomas Blount, published his *Glossographia: or a Dictionary, Interpreting all such Hard Words of Whatsoever Language, now used in our refined English Tongue.* Blount's dictionary listed more than eleven thousand words, many of which, he recognized, were new, reaching London in the hurly-burly of trade and commerce—

> **coffa** or **cauphe**, a kind of drink among the Turks and Persians, (and of late introduced among us) which is black, thick and bitter, destrained from Berries of that nature, and name, thought good and very wholesom: they say it expels melancholy.

—or home-grown, such as "**tom-boy**, a girle or wench that leaps up and down like a boy." He seems to have known he was aiming at a moving target. The dictionary maker's "labor," he wrote in his preface, "would find no end, since our English tongue daily changes habit." Blount's definitions were much more elaborate than Cawdrey's, and he tried to provide information about the origins of words as well.

Neither Bullokar nor Blount so much as mentioned Cawdrey. He was already forgotten. But in 1933, upon the publication of the greatest word book of all, the first editors of the *Oxford English Dictionary* did pay their respects to his "slim, small volume." They called it "the original acorn" from which their oak had grown. (Cawdrey: "**akecorne**, *k* fruit.")

Four hundred and two years after the *Table Alphabeticall*, the International Astronomical Union voted to declare Pluto a nonplanet, and John Simpson had to make a quick decision. He and his band of lexicographers in Oxford were working on the *P*'s. *Pletzel, plish, pod person, point-and-shoot,* and *polyamorous* were among the new words entering the *OED*. The entry for Pluto was itself relatively new. The planet had been discovered only in 1930, too late for the *OED*'s first edition. The name Minerva was first proposed and then rejected because there was already an asteroid Minerva. In terms of names, the heavens were beginning to fill up. Then "Pluto" was suggested by Venetia Burney, an eleven-year-old resident of Oxford. The

OED caught up by adding an entry for Pluto in its second edition: "1. A small planet of the solar system lying beyond the orbit of Neptune . . . 2. The name of a cartoon dog that made its first appearance in Walt Disney's *Moose Hunt*, released in April 1931."

"We really don't like being pushed into megachanges," Simpson said, but he had little choice. The Disney meaning of *Pluto* had proved more stable than the astronomical sense, which was downgraded to "small planetary body." Consequences rippled through the *OED*. *Pluto* was removed from the list under *planet n.* 3a. *Plutonian* was revised (not to be confused with *pluton*, *plutey*, or *plutonyl*).

Simpson was the sixth in a distinguished line, the editors of the *Oxford English Dictionary*, whose names rolled fluently off his tongue—"Murray, Bradley, Craigie, Onions, Burchfield, so however many fingers that is"—and saw himself as a steward of their traditions, as well as traditions of English lexicography extending back to Cawdrey by way of Samuel Johnson. James Murray in the nineteenth century established a working method based on index cards, slips of paper 6 inches by 4 inches. At any given moment a thousand such slips sat on Simpson's desk, and within a stone's throw were millions more, filling metal files and wooden boxes with the ink of two centuries. But the word-slips had gone obsolete. They had become treeware. *Treeware* had just entered the *OED* as "computing slang, freq. humorous"; *blog* was recognized in 2003, *dot-commer* in 2004, *cyberpet* in 2005, and the verb *to Google* in 2006. Simpson himself Googled often. Beside the word-slips his desk held conduits into the nervous system of the language: instantaneous connection to a worldwide network of proxy amateur lexicographers and access to a vast, interlocking set of databases growing asymptotically toward the ideal of All Previous Text. The dictionary had met cyberspace, and neither would be the same thereafter. However much Simpson loved the *OED*'s roots and legacy, he was leading a revolution, willy-nilly—in what it was, what it knew, what it saw. Where Cawdrey had been isolated, Simpson was connected.

The English language, spoken now by more than a billion people globally, has entered a period of ferment, and the perspective available in these venerable Oxford offices is both intimate and sweeping. The language upon which the lexicographers eavesdrop has become wild and amorphous: a great, swirling, expanding cloud of messaging and speech; newspapers, magazines, pamphlets; menus and business memos; Internet news groups and chat-room conversations; television and radio broadcasts and phonograph records. By contrast, the dictionary itself has acquired the status of a monument, definitive and towering. It exerts an influence on the language it tries to observe. It wears its authoritative role reluctantly. The lexicographers may recall Ambrose Bierce's sardonic century-old definition: "**dictionary**, a malevolent literary device for cramping the growth of a language and making it hard and inelastic." Nowadays they stress that they do not presume (or deign) to disapprove any particular usage or spelling. But they cannot disavow a strong ambition: the goal of completeness. They want every word, all the lingo: idioms and euphemisms, sacred or profane, dead or alive, the King's English or the street's. It is an ideal only: the constraints of space and time are ever present and, at the margins, the question of what qualifies as a word can become impossible to answer. Still, to the extent possible, the *OED* is meant to be a perfect record, perfect mirror of the language.

The dictionary ratifies the persistence of the word. It declares that the meanings of words come from other words. It implies that all words, taken together, form an interlocking structure: interlocking, because all words are defined in terms of other words. This could never have been an issue in an oral culture, where language was barely visible. Only when printing—and the dictionary—put the language into separate relief, as an object to be scrutinized, could anyone develop a sense of word meaning as interdependent and even circular. Words had to be considered as words, representing other words, apart from things. In the twentieth century, when the technologies of logic advanced to high levels, the potential for circularity became a problem. "In giving explanations I already have to use language full blown," complained Ludwig Wittgenstein. He echoed Newton's frustration three centuries earlier, but with an extra twist, because where Newton wanted words for nature's laws, Wittgenstein wanted words for words: "When I talk about language (words, sentences, etc.) I must speak the language of every day. Is this language somehow too coarse and material for what we want to say?" Yes. And the language was always in flux.

James Murray was speaking of the language as well as the book when he said, in 1900, "The English Dictionary, like the English Constitution, is the creation of no one man, and of no one age; it is a growth that has slowly developed itself adown the ages." The first edition of what became the *OED* was one of the largest books that had ever been made: *A New English Dictionary on Historical Principles*, 414,825 words in ten weighty volumes, presented to King George V and President Calvin Coolidge in 1928. The work had taken decades; Murray himself was dead; and the dictionary was understood to be out of date even as the volumes were bound and sewn. Several supplements followed, but not till 1989 did the second edition appear: twenty volumes, totaling 22,000 pages. It weighed 138 pounds. The third edition is different. It is weightless, taking its shape in the digital realm. It may never again involve paper and ink. Beginning in the year 2000, a revision of the entire contents began to appear online in quarterly installments, each comprising several thousand revised entries and hundreds of new words.

Cawdrey had begun work naturally enough with the letter *A*, and so had James Murray in 1879, but Simpson chose to begin with *M*. He was wary of the *A*'s. To insiders it had long been clear that the *OED* as printed was not a seamless masterpiece. The early letters still bore scars of the immaturity of the uncertain work in Murray's first days. "Basically he got here, sorted his suitcases out and started setting up text," Simpson said. "It just took them a long time to sort out their policy and things, so if we started at A, then we'd be making our job doubly difficult. I think they'd sorted themselves out by . . . well, I was going to say D, but Murray always said that E was the worst letter, because his assistant, Henry Bradley, started E, and Murray always said that he did that rather badly. So then we thought, maybe it's safe to start with G, H. But you get to G and H and there's I, J, K, and you know, you think, well, start after that."

The first thousand entries from *M* to *mahurat* went online in the spring of 2000. A year later, the lexicographers reached words starting with *me: me-ism* (a creed for modern times), *meds* (colloq. for drugs), *medspeak* (doctors' jargon), *meet-and-greet* (a N. Amer. type of social occasion), and an assortment of combined forms under *media* (baron, circus, darling, hype, savvy) and *mega-* (pixel, bitch, dose, hit, trend). This was no longer a language spoken by 5 million mostly illiterate inhabitants of a small island. As the *OED* revised the entries letter by letter, it also began adding neologisms wherever they arose; waiting for the alphabetical sequence became impractical. Thus one installment in 2001 saw the arrival of *acid jazz, Bollywood, channel surfing, double-click, emoticon, feel-good, gangsta, hyperlink*, and many more. *Kool-Aid* was recognized as a new word, not because the *OED* feels obliged to list proprietary names (the original Kool-Ade powdered drink had been patented in the United States in 1927) but because a special usage could no longer be ignored: "to drink the Kool-Aid: to demonstrate unquestioning obedience or loyalty." The growth of this peculiar expression since the use of a powdered beverage in a mass poisoning in Guyana in 1978 bespoke a certain density of global communication.

But they were no slaves to fashion, these Oxford lexicographers. As a rule a neologism needs five years of solid evidence for admission to the canon. Every proposed word undergoes intense scrutiny. The approval of a new word is a solemn matter. It must be in general use, beyond any particular place of origin; the *OED* is global, recognizing words from everywhere English is spoken, but it does not want to capture local quirks. Once added, a word cannot come out. A word can go obsolete or rare, but the most ancient and forgotten words have a way of reappearing—rediscovered or spontaneously reinvented—and in any case they are part of the language's history. All 2,500 of Cawdrey's words are in the *OED*, perforce. For thirty-one of them Cawdrey's little book was the first known usage. For a few

Cawdrey is all alone. This is troublesome. The *OED* is irrevocably committed. Cawdrey, for example, has "**onust**, loaden, overcharged"; so the *OED* has "loaded, burdened," but it is an outlier, a one-off. Did Cawdrey make it up? "I'm tending towards the view that he was attempting to reproduce vocabulary he had heard or seen," Simpson said. "But I can't be absolutely sure." Cawdrey has "**hallucinate**, to deceive, or blind"; the *OED* duly gave "to deceive" as the first sense of the word, though it never found anyone else who used it that way. In cases like these, the editors can add their double caveat "*Obs. rare.*" But there it is.

For the twenty-first-century *OED* a single source is never enough. Strangely, considering the vastness of the enterprise and its constituency, individual men and women strive to have their own nonce-words ratified by the *OED*. *Nonce-word*, in fact, was coined by James Murray himself. He got it in. An American psychologist, Sondra Smalley, coined the word *codependency* in 1979 and began lobbying for it in the eighties; the editors finally drafted an entry in the nineties, when they judged the word to have become established. W. H. Auden declared that he wanted to be recognized as an *OED* word coiner—and he was, at long last, for *motted*, *metalogue*, *spitzy*, and others. The dictionary had thus become engaged in a feedback loop. It inspired a twisty self-consciousness in the language's users and creators. Anthony Burgess whinged in print about his inability to break through: "I invented some years ago the word *amation*, for the art or act of making love, and still think it useful. But I have to persuade others to use it *in print* before it is eligible for lexicographicizing (if that word exists)"—he knew it did not. "T. S. Eliot's large authority got the shameful (in my view) *juvescence* into the previous volume of the Supplement." Burgess was quite sure that Eliot simply misspelled *juvenescence*. If so, the misspelling was either copied or reprised twenty-eight years later by Stephen Spender, so *juvescence* has two citations, not one. The *OED* admits that it is rare.

As hard as the *OED* tries to embody the language's fluidity, it cannot help but serve as an agent of its crystallization. The problem of spelling poses characteristic difficulties. "*Every* form in which a word has occurred throughout its history" is meant to be included. So for *mackerel* ("a well-known sea-fish, *Scomber scombrus*, much used for food") the second edition in 1989 listed nineteen alternative spellings. The unearthing of sources never ends, though, so the third edition revised entry in 2002 listed no fewer than thirty: *maccarel, mackaral, mackarel, mackarell, mackerell, mackeril, mackreel, mackrel, mackrell, mackril, macquerel, macquerell, macrel, macrell, macrelle, macril, macrill, makarell, makcaral, makerel, makerell, makerelle, makral, makrall, makreill, makrel, makrell, makyrelle, maquerel,* and *maycril*. As lexicographers, the editors would never declare these alternatives to be wrong: misspellings. They do not wish to declare their choice of spelling for the headword, *mackerel*, to be "correct." They emphasize that they examine the evidence and choose "the most common current spelling." Even so, arbitrary considerations come into play: "Oxford's house style occasionally takes precedence, as with verbs which can end -ize or -ise, where the -ize spelling is always used." They know that no matter how often and how firmly they disclaim a prescriptive authority, a reader will turn to the dictionary to find out how a word should be spelled. They cannot escape inconsistencies. They feel obliged to include words that make purists wince. A new entry as of December 2003 memorialized *nucular*: "= nuclear *a.* (in various senses)." Yet they refuse to count evident misprints found by way of Internet searches. They do not recognize *straight-laced*, even though statistical evidence finds that bastardized form outnumbering *strait-laced*. For the crystallization of spelling, the *OED* offers a conventional explanation: "Since the invention of the printing press, spelling has become much less variable, partly because printers wanted uniformity and partly because of a growing interest in language study during the Renaissance." This is true. But it omits the role of the dictionary itself, arbitrator and exemplar.

For Cawdrey the dictionary was a snapshot; he could not see past his moment in time. Samuel Johnson was more explicitly aware of the dictionary's historical dimension. He justified his ambitious program in part as a means of bringing a wild thing under control—the wild thing being the language, "which, while it was employed in the cultivation of every species of literature, has itself been hitherto neglected; suffered to spread, under the direction of chance, into wild exuberance; resigned to the tyranny of time and fashion; and exposed to the corruptions of ignorance, and caprices of innovation." Not until the *OED*, though, did lexicography attempt to reveal the whole shape of a language across time. The *OED* becomes a historical panorama. The project gains poignancy if the electronic age is seen as a new age of orality, the word breaking free from the bonds of cold print. No publishing institution better embodies those bonds, but the *OED*, too, tries to throw them off. The editors feel they can no longer wait for a new word to appear in print, let alone in a respectably bound book, before they must take note. For *tighty-whities* (men's underwear), new in 2007, they cite a typescript of North Carolina campus slang. For *kitesurfer*, they cite a posting to the Usenet newsgroup alt.kite and later a New Zealand newspaper found via an online database. Bits in the ether.

When Murray began work on the new dictionary, the idea was to find the words, and with them the signposts to their history. No one had any idea how many words were there to be found. By then the best and most comprehensive dictionary of English was American: Noah Webster's, seventy thousand words. That was a baseline. Where were the rest to be discovered? For the first editors of what became the *OED*, it went almost without saying that the source, the wellspring, should be the literature of the language—particularly the books of distinction and quality. The dictionary's first readers combed Milton and Shakespeare (still the single most quoted author, with more than thirty thousand references), Fielding and Swift, histories and sermons, philosophers and poets. Murray announced in a famous public appeal in 1879:

> A thousand readers are wanted. The later sixteenth-century literature is very fairly done; yet here several books remain to be read. The seventeenth century, with so many more writers, naturally shows still more unexplored territory.

He considered the territory to be large but bounded. The founders of the dictionary explicitly meant to find every word, however many that would ultimately be. They planned a complete inventory. Why should they not? The number of books was unknown but not unlimited, and the number of words in those books was countable. The task seemed formidable but finite.

It no longer seems finite. Lexicographers are accepting the language's boundlessness. They know by heart Murray's famous remark: "The circle of the English language has a well-defined centre but no discernable circumference." In the center are the words everyone knows. At the edges, where Murray placed slang and cant and scientific jargon and foreign border crossers, everyone's sense of the language differs and no one's can be called "standard."

Murray called the center "well defined," but infinitude and fuzziness can be seen there. The easiest, most common words—the words Cawdrey had no thought of including—require, in the *OED*, the most extensive entries. The entry for *make* alone would fill a book: it teases apart ninety-eight distinct senses of the verb, and some of these senses have a dozen or more subsenses. Samuel Johnson saw the problem with these words and settled on a solution: he threw up his hands.

> My labor has likewise been much increased by a class of verbs too frequent in the English language, of which the signification is so loose and general, the use so vague and indeterminate, and the senses detorted so widely from the first idea, that it is hard to trace them through the maze of variation, to catch them on the brink of utter inanity, to circumscribe them by any limitations, or interpret them by any words of distinct and settled meaning; such are *bear, break, come, cast, full, get, give, do, put, set, go, run, make, take, turn, throw.* If of these the whole power is not accurately delivered, it must be remembered, that while our language is yet living, and variable by the caprice of every one that speaks it, these words are hourly shifting their relations, and can no more be ascertained in a dictionary, than a grove, in the agitation of a storm, can be accurately delineated from its picture in the water.

Johnson had a point. These are words that any speaker of English can press into new service at any time, on any occasion, alone or in combination, inventively or not, with hopes of being understood. In every revision, the *OED*'s entry for a word like *make* subdivides further and thus grows larger. The task is unbounded in an inward-facing direction.

The more obvious kind of unboundedness appears at the edges. Neologism never ceases. Words are coined by committee: *transistor*, Bell Laboratories, 1948. Or by wags: *booboisie*, H. L. Mencken, 1922. Most arise through spontaneous generation, organisms appearing in a petri dish, like *blog* (c. 1999). One batch of arrivals includes *agroterrorism*, *bada-bing*, *bahookie* (a body part), *beer pong* (a drinking game), *bippy* (as in, you bet your ————), *chucklesome*, *cypherpunk*, *tuneage*, and *wonky*. None are what Cawdrey would have seen as "hard, usual words," and none are anywhere near Murray's well-defined center, but they now belong to the common language. Even *bada-bing*: "Suggesting something happening suddenly, emphatically, or easily and predictably; 'Just like that!', 'Presto!'" The historical citations begin with a 1965 audio recording of a comedy routine by Pat Cooper and continue with newspaper clippings, a television news transcript, and a line of dialogue from the first *Godfather* movie: "You've gotta get up close like this and bada-bing! you blow their brains all over your nice Ivy League suit." The lexicographers also provide an etymology, an exquisite piece of guesswork: "Origin uncertain. Perh. imitative of the sound of a drum roll and cymbal clash. Perh. cf. Italian *bada bene* mark well."

The English language no longer has such a thing as a geographic center, if it ever did. The universe of human discourse always has backwaters. The language spoken in one valley diverges from the language of the next valley, and so on. There are more valleys now than ever, even if the valleys are not so isolated. "We are listening to the language," said Peter Gilliver, an *OED* lexicographer and resident historian. "When you are listening to the language by collecting pieces of paper, that's fine, but now it's as if we can hear everything said anywhere. Take an expatriate community living in a non-English-speaking part of the world, expatriates who live at Buenos Aires or something. Their English, the English that they speak to one another every day, is full of borrowings from local Spanish. And so they would regard those words as part of their idiolect, their personal vocabulary." Only now they may also speak in chat rooms and on blogs. When they coin a word, anyone may hear. Then it may or may not become part of the language.

If there is an ultimate limit to the sensitivity of lexicographers' ears, no one has yet found it. Spontaneous coinages can have an audience of one. They can be as ephemeral as atomic particles in a bubble chamber. But many neologisms require a level of shared cultural knowledge. Perhaps *bada-bing* would not truly have become part of twenty-first-century English had it not been for the common experience of viewers of a particular American television program (though it is not cited by the *OED*).

The whole word hoard—the lexis—constitutes a symbol set of the language. It is the fundamental symbol set, in one way: words are the first units of meaning any language recognizes. They are recognized universally. But in another way it is far from fundamental: as communication evolves, messages in a language can be broken down and composed and transmitted in much smaller sets of symbols: the alphabet; dots and dashes; drumbeats high and low. These symbol sets are discrete. The lexis is not. It is messier. It keeps on growing. Lexicography turns out to be a science poorly suited to exact measurement. English, the largest and most widely shared language, can be said very roughly to possess a number of units of meaning that approaches a million. Linguists have no special yardsticks of their own; when they try to quantify the pace of neologism, they tend to look to the dictionary for guidance, and even the best dictionary runs from that responsibility. The edges always blur. A clear line cannot be drawn between word and unword.

So we count as we can. Robert Cawdrey's little book, making no pretense to completeness, contained a vocabulary of only 2,500. We possess now a more complete dictionary of English as it was circa 1600: the subset of the *OED* comprising words then current. That vocabulary numbers 60,000 and keeps growing, because the discovery of sixteenth-century sources never ends. Even so, it is a tiny fraction of the words used four centuries later. The explanation for this explosive growth, from 60,000 to a million, is not simple. Much of what now needs naming did not yet exist, of course. And much of what existed was not recognized. There was no call for *transistor* in 1600, nor *nanobacterium*, nor *webcam*, nor *fen-phen*. Some of the growth comes from mitosis. The guitar divides into the electric and the acoustic; other words divide in reflection of delicate nuances (as of March 2007 the *OED* assigned a new entry to *prevert* as a form of *pervert*, taking the view that *prevert* was not just an error but a deliberately humorous effect). Other new words appear without any corresponding innovation in the world of real things. They crystallize in the solvent of universal information.

What, in the world, is a *mondegreen*? It is a misheard lyric, as when, for example, the Christian hymn is heard as "Lead on, O kinky turtle . . ."). In sifting the evidence, the *OED* first cites a 1954 essay in *Harper's Magazine* by Sylvia Wright: "What I shall hereafter call mondegreens, since no one else has thought up a word for them." She explained the idea and the word this way:

> When I was a child, my mother used to read aloud to me from Percy's Reliques, and one of my favorite poems began, as I remember:
>
> *Ye Highlands and ye Lowlands,*
> *Oh, where hae ye been?*
> *They hae slain the Earl Amurray,*
> *And Lady Mondegreen.*

There the word lay, for some time. A quarter-century later, William Safire discussed the word in a column about language in *The New York Times Magazine*. Fifteen years after that, Steven Pinker, in his book *The Language Instinct*, offered a brace of examples, from "A girl with colitis goes by" to "Gladly the cross-eyed bear," and observed, "The interesting thing about mondegreens is that the mishearings are generally *less* plausible than the intended lyrics." But it was not books or magazines that gave the word its life; it was Internet sites, compiling mondegreens by the thousands. The *OED* recognized the word in June 2004.

A mondegreen is not a transistor, inherently modern. Its modernity is harder to explain. The ingredients—songs, words, and imperfect understanding—are all as old as civilization. Yet for mondegreens to arise in the culture, and for *mondegreen* to exist in the lexis, required something new: a modern level of linguistic self-consciousness and interconnectedness. People needed to mishear lyrics not just once, not just several times, but often enough to become aware of the mishearing as a thing worth discussing. They needed to have other such people with whom to share the recognition. Until the most modern times, mondegreens, like countless other cultural or psychological phenomena, simply did not need to be named. Songs themselves were not so common; not heard, anyway, on elevators and mobile phones. The word *lyrics*, meaning the words of a song, did not exist until the nineteenth century. The conditions for mondegreens took a long time to ripen. Similarly, the verb *to gaslight* now means "to manipulate a person by psychological means into questioning his or her own sanity"; it exists only because enough people saw the 1944 film of that title and could assume that their listeners had seen it, too. Might not the language Cawdrey spoke—which was, after all, the abounding and fertile language of Shakespeare—have found use for such a word? No matter: the technology for *gaslight* had not been invented. Nor had the technology for motion pictures.

The lexis is a measure of shared experience, which comes from interconnectedness. The number of users of the language forms only the first part of the equation: jumping in four centuries from 5 million English speakers to a billion. The driving factor is the number of connections between and among those speakers. A mathematician might say that messaging grows not geometrically, but combinatorially, which is much,

much faster. "I think of it as a saucepan under which the temperature has been turned up," Gilliver said. "Any word, because of the interconnectedness of the English-speaking world, can spring from the backwater. And they are still backwaters, but they have this instant connection to ordinary, everyday discourse." Like the printing press, the telegraph, and the telephone before it, the Internet is transforming the language simply by transmitting information differently. What makes cyberspace different from all previous information technologies is its intermixing of scales from the largest to the smallest without prejudice, broadcasting to the millions, narrowcasting to groups, instant messaging one to one.

This comes as quite an unexpected consequence of the invention of computing machinery. At first, that had seemed to be about numbers.

4 | TO THROW THE POWERS
OF THOUGHT INTO WHEEL-WORK

(Lo, the Raptured Arithmetician)

*Light almost solar has been extracted from the refuse of fish; fire
has been sifted by the lamp of Davy; and machinery has been
taught arithmetic instead of poetry.*

—Charles Babbage (1832)

NO ONE DOUBTED THAT Charles Babbage was brilliant. Nor did anyone
quite understand the nature of his genius, which remained out of focus
for a long time. What did he hope to achieve? For that matter, what,
exactly, was his vocation? On his death in London in 1871 the *Times*
obituarist declared him "one of the most active and original of original
thinkers" but seemed to feel he was best known for his long, cranky
crusade against street musicians and organ-grinders. He might not have
minded. He was multifarious and took pride in it. "He showed great
desire to inquire into the causes of things that astonish childish minds,"
said an American eulogist. "He eviscerated toys to ascertain their manner
of working." Babbage did not quite belong in his time, which called itself
the Steam Age or the Machine Age. He did revel in the uses of steam and
machinery and considered himself a thoroughly modern man, but he
also pursued an assortment of hobbies and obsessions—cipher cracking, lock picking, lighthouses, tree rings, the post—whose logic became
clearer a century later. Examining the economics of the mail, he pursued

a counterintuitive insight, that the significant cost comes not from the physical transport of paper packets but from their "verification"—the calculation of distances and the collection of correct fees—and thus he invented the modern idea of standardized postal rates. He loved boating, by which he meant not "the manual labor of rowing but the more intellectual art of sailing." He was a train buff. He devised a railroad recording device that used inking pens to trace curves on sheets of paper a thousand feet long: a combination seismograph and speedometer, inscribing the history of a train's velocity and all the bumps and shakes along the way.

As a young man, stopping at an inn in the north of England, he was amused to hear that his fellow travelers had been debating his trade:

> "The tall gentleman in the corner," said my informant, "maintained you were in the hardware line; whilst the fat gentleman who sat next to you at supper was quite sure that you were in the spirit trade. Another of the party declared that they were both mistaken: he said you were travelling for a great iron-master."
>
> "Well," said I, "you, I presume, knew my vocation better than our friends."
>
> "Yes," said my informant, "I knew perfectly well that you were in the Nottingham lace trade."

He might have been described as a professional mathematician, yet here he was touring the country's workshops and manufactories, trying to discover the state of the art in machine tools. He noted, "Those who enjoy leisure can scarcely find a more interesting and instructive pursuit than the examination of the workshops of their own country, which contain within them a rich mine of knowledge, too generally neglected by the wealthier classes." He himself neglected no vein of knowledge. He did become expert on the manufacture of Nottingham lace; also the use of gunpowder in quarrying limestone; precision glass cutting with diamonds; and all known uses of machinery to produce power, save time, and communicate signals. He analyzed hydraulic presses, air pumps, gas meters, and screw cutters. By the end of his tour he knew as much as anyone in England about the making of pins. His knowledge was practical and methodical. He estimated that a pound of pins required the work of ten men and women for at least seven and a half hours, drawing wire, straightening wire, pointing the wire, twisting and cutting heads from the spiral coils, tinning or whitening, and finally papering. He computed the cost of each phase in millionths of a penny. And he noted that this process, when finally perfected, had reached its last days: an American had invented an automatic machine to accomplish the same task, faster.

Babbage invented his own machine, a great, gleaming engine of brass and pewter, comprising thousands of cranks and rotors, cogs and gearwheels, all tooled with the utmost precision. He spent his long life improving it, first in one and then in another incarnation, but all, mainly, in his mind. It never came to fruition anywhere else. It thus occupies an extreme and peculiar place in the annals of invention: a failure, and also one of humanity's grandest intellectual achievements. It failed on a colossal scale, as a scientific-industrial project "at the expense of the nation, to be held as national property," financed by the Treasury for almost twenty years, beginning in 1823 with a Parliamentary appropriation of £1,500 and ending in 1842, when the prime minister shut it down. Later, Babbage's engine was forgotten. It vanished from the lineage of invention. Later still, however, it was rediscovered, and it became influential in retrospect, to shine as a beacon from the past.

Like the looms, forges, naileries, and glassworks he studied in his travels across northern England, Babbage's machine was designed to manufacture vast quantities of a certain commodity. The commodity was numbers. The engine opened a channel from the corporeal world of matter to a world of pure abstraction. The engine consumed no raw materials—input and output being weightless—but needed a considerable force to turn the gears. All that wheel-work would fill a room and weigh several tons. Producing numbers, as Babbage conceived it, required a degree of mechanical complexity at the very limit of available technology. Pins were easy, compared with numbers.

It was not natural to think of numbers as a manufactured commodity. They existed in the mind, or in ideal abstraction, in their perfect infinitude. No machine could add to the world's supply. The numbers produced by Babbage's engine were meant to be those with significance: numbers with a meaning. For example, 2.096910013 has a meaning, as the logarithm of 125. (Whether *every* number has a meaning would be a conundrum for the next century.) The meaning of a number could be expressed as a relationship to other numbers, or as the answer to a certain question of arithmetic. Babbage himself did not speak in terms of meaning; he tried to explain his engine pragmatically, in terms of putting numbers into the machine and seeing other numbers come out, or, a bit more fancifully, in terms of posing questions to the machine and expecting an answer. Either way, he had trouble getting the point across. He grumbled:

> On two occasions I have been asked,—"Pray, Mr. Babbage, if you put into the machine wrong figures, will the right answers come out?" In one case a member of the Upper, and in the other a member of the Lower, House put this question. I am not able rightly to apprehend the kind of confusion of ideas that could provoke such a question.

Anyway, the machine was not meant to be a sort of oracle, to be consulted by individuals who would travel from far and wide for mathematical answers. The engine's chief mission was to print out numbers en masse. For portability, the facts of arithmetic could be expressed in tables and bound in books.

To Babbage the world seemed made of such facts. They were the "constants of Nature and Art." He collected them everywhere. He compiled a Table of Constants of the Class Mammalia: wherever he went he timed the breaths and heartbeats of pigs and cows. He invented a statistical methodology with tables of life expectancy for the somewhat shady business of life insurance. He drew up a table of the weight in Troy grains per square yard of various fabrics: cambric, calico, nankeen, muslins, silk gauze, and "caterpillar veils." Another table revealed the relative frequencies of all the double-letter combinations in English, French, Italian, German, and Latin. He researched, computed, and published a Table of the Relative Frequency of the Causes of Breaking of Plate Glass Windows, distinguishing 464 different causes, no less than fourteen of which involved "drunken men, women, or boys." But the tables closest to his heart were the purest: tables of numbers and only numbers, marching neatly across and down the pages in stately rows and columns, patterns for abstract appreciation.

A book of numbers: amid all the species of information technology, how peculiar and powerful an object this is. "Lo! the raptured arithmetician!" wrote Élie de Joncourt in 1762. "Easily satisfied, he asks no Brussels lace, nor a coach and six." Joncourt's own contribution was a small quarto volume registering the first 19,999 triangular numbers. It was a treasure box of exactitude, perfection, and close reckoning. These numbers were so simple, just the sums of the first n whole numbers: 1, 3 (1+2), 6 (1+2+3), 10 (1+2+3+4), 15, 21, 28, and so on. They had interested number theorists since Pythagoras. They offered little in the way of utility, but Joncourt rhapsodized about his pleasure in compiling them and Babbage quoted him with heartfelt sympathy: "Numbers have many charms, unseen by vulgar eyes, and only discovered to the unwearied and respectful sons of Art. Sweet joy may arise from such contemplations."

Tables of numbers had been part of the book business even before the beginning of the print era. Working in Baghdad in the ninth century, Abu Abdullah Mohammad Ibn Musa al-Khwarizmi, whose name survives in the word *algorithm*, devised tables of trigonometric functions that spread west across Europe and east to China, made by hand and copied by hand, for hundreds of years. Printing brought number tables into their own: they were a natural first application for the mass production of data in the raw. For people in need of arithmetic, multiplication tables covered more and more territory: 10 × 1,000, then 10 × 10,000, and later as far as 1,000 × 1,000. There were tables of squares and cubes, roots and reciprocals. An early form of table was the ephemeris or almanac, listing positions of the sun, moon, and planets for sky-gazers. Tradespeople found uses for number books. In 1582 Simon Stevin produced *Tafelen van Interest*, a compendium of interest tables for bankers and moneylenders. He promoted the new decimal arithmetic "to astrologers, land-measurers, measurers of tapestry and wine casks and stereometricians, in general, mint masters and merchants all." He might have added sailors. When Christopher Columbus set off for the Indies, he carried as an aid to navigation a book of tables by Regiomontanus printed in Nuremberg two decades after the invention of moveable type in Europe.

Joncourt's book of triangular numbers was purer than any of these—which is also to say useless. Any arbitrary triangular number can be found (or made) by an algorithm: multiply *n* by *n* + 1 and divide by 2. So Joncourt's whole compendium, as a bundle of information to be stored and transmitted, collapses in a puff to a one-line formula. The formula contains all the information. With it, anyone capable of simple multiplication (not many were) could generate any triangular number on demand. Joncourt knew this. Still he and his publisher, M. Husson, at the Hague, found it worthwhile to set the tables in metal type, three pairs of columns to a page, each pair listing thirty natural numbers alongside their corresponding triangular numbers, from 1(1) to 19,999(199,990,000), every numeral chosen individually by the compositor from his cases of metal type and lined up in a galley frame and wedged into an iron chase to be placed upon the press.

Why? Besides the obsession and the ebullience, the creators of number tables had a sense of their economic worth. Consciously or not, they reckoned the price of these special data by weighing the difficulty of computing them versus looking them up in a book. Precomputation plus data storage plus data transmission usually came out cheaper than ad hoc computation. "Computers" and "calculators" existed: they were people with special skills, and all in all, computing was costly.

Beginning in 1767, England's Board of Longitude ordered published a yearly *Nautical Almanac*, with position tables for the sun, moon, stars, planets, and moons of Jupiter. Over the next half century a network of computers did the work—thirty-four men and one woman, Mary Edwards of Ludlow, Shropshire, all working from their homes. Their painstaking labor paid £70 a year. Computing was a cottage industry. Some mathematical sense was required but no particular genius; rules were laid out in steps for each type of calculation. In any case the computers, being human, made errors, so the same work was often farmed out twice for the sake of redundancy. (Unfortunately, being human, computers were sometimes caught saving themselves labor by copying from one other.) To manage the information flow the project employed a Comparer of the Ephemeris and Corrector of the Proofs. Communication between the computers and comparer went by post, men on foot or on horseback, a few days per message.

A seventeenth-century invention had catalyzed the whole enterprise. This invention was itself a species of number, given the name *logarithm*. It was number as tool. Henry Briggs explained:

> Logarithmes are Numbers invented for the more easie working of questions in Arithmetike and Geometrie. The name is derived of *Logos*, which signifies *Reason*, and *Arithmos*, signifying *Numbers*. By them all troublesome Multiplications and Divisions in Arithmetike are avoided, and performed onely by Addition in stead of Multiplication, and by Subtraction in stead of Division.

In 1614 Briggs was a professor of geometry—the first professor of geometry—at Gresham College, London, later to be the birthplace of the Royal Society. Without logarithms he had already created two books of tables, *A Table to find the Height of the Pole, the Magnetic Declination being given* and *Tables for the Improvement of Navigation*, when a book came from Edinburgh promising to "take away all the difficultie that heretofore hath beene in mathematical calculations."

> There is nothing (right well beloved Students in the Mathematickes) that is so troublesome to Mathematicall practice, not that doth more molest and hinder Calculators, then the Multiplications, Divisions, square and cubicall Extractions of great numbers, which besides the tedious expence of time, are for the most part subject to many slippery errors.

This new book proposed a method that would do away with most of the expense and the errors. It was like an electric flashlight sent to a lightless world. The author was a wealthy Scotsman, John Napier (or Napper, Nepair, Naper, or Neper), the eighth laird of Merchiston Castle, a theologian and well-known astrologer who also made a hobby of mathematics. Briggs was agog. "Naper, lord of Markinston, hath set my head and hands a work," he wrote. "I hope to see him this summer, if it please God, for I never saw book, which pleased me better, and made me more wonder." He made his pilgrimage to Scotland and their first meeting, as he reported later, began with a quarter hour of silence: "spent, each beholding other almost with admiration before one word was spoke."

Briggs broke the trance: "My Lord, I have undertaken this long journey purposely to see your person, and to know by what engine of wit or ingenuity you came first to think of this most excellent help unto astronomy, viz. the Logarithms; but, my Lord, being by you found out, I wonder nobody else found it out before, when now known it is so easy." He stayed with the laird for several weeks, studying.

In modern terms a logarithm is an exponent. A student learns that the logarithm of 100, using 10 as the base, is 2, because $100 = 10^2$. The logarithm of 1,000,000 is 6, because 6 is the exponent in the expression $1,000,000 = 10^6$. To multiply two numbers, a calculator could just look up their logarithms and add those. For example:

$$100 \times 1,000,000 = 10^2 \times 10^6 = 10^{(2+6)}$$

Looking up and adding are easier than multiplying.

But Napier did not express his idea this way, in terms of exponents. He grasped the thing viscerally: he was thinking in terms of a relationship between differences and ratios. A series of numbers with a fixed difference is an arithmetic progression: 0, 1, 2, 3, 4, 5 . . . When the numbers are separated by a fixed ratio, the progression is geometric: 1, 2, 4, 8, 16, 32 . . . Set these progressions side by side,

0	1	2	3	4	5	. . .	(base 2 logarithms)
1	2	4	8	16	32	. . .	(natural numbers)

and the result is a crude table of logarithms—crude, because the whole-number exponents are the easy ones. A useful table of logarithms had to fill in the gaps, with many decimal places of accuracy.

In Napier's mind was an analogy: differences are to ratios as addition is to multiplication. His thinking crossed over from one plane to another, from spatial relationships to pure numbers. Aligning these scales side by side, he gave a calculator a practical means of converting multiplication into addition—downshifting, in effect, from the difficult task to the easier one. In a way, the method is a kind of translation, or encoding. The natural numbers are encoded as logarithms. The calculator looks them up in a table, the code book. In this new language, calculation is easy: addition instead of multiplication, or multiplication instead of exponentiation. When the work is done, the result is translated back into the language of natural numbers. Napier, of course, could not think in terms of encoding.

Briggs revised and extended the necessary number sequences and published a book of his own, *Logarithmicall Arithmetike*, full of pragmatic applications. Besides the logarithms he presented tables of latitude of the sun's declination year by year; showed how to find the distance between any two places, given their latitudes and longitudes; and laid out a star guide with declinations, distance to the pole, and right ascension.

Some of this represented knowledge never compiled and some was oral knowledge making the transition to print, as could be seen in the not-quite-formal names of the stars: the *Pole Starre, girdle of Andromeda, Whales Bellie, the brightest in the harpe,* and *the first in the great Beares taile next her rump.* Briggs also considered matters of finance, offering rules for computing with interest, backward and forward in time. The new technology was a watershed: "It may be here also noted that the use of a 100 pound for a day at the rate of 8, 9, 10, or the like for a yeare hath beene scarcely known, till by Logarithms it was found out: for otherwise it requires so many laborious extractions of roots, as will cost more paines than the knowledge of the thing is accompted to be worth." Knowledge has a value and a discovery cost, each to be counted and weighed.

Even this exciting discovery took several years to travel as far as Johannes Kepler, who employed it in perfecting his celestial tables in 1627, based on the laboriously acquired data of Tycho Brahe. "A Scottish baron has appeared on the scene (his name I have forgotten) who has done an excellent thing," Kepler wrote a friend, "transforming all multiplication and division into addition and subtraction." Kepler's tables were far more accurate—perhaps thirty times more—than any of his medieval predecessors, and the accuracy made possible an entirely new thing, his harmonious heliocentric system, with planets orbiting the sun in ellipses. From that time until the arrival of electronic machines, the majority of human computation was performed by means of logarithms. A teacher of Kepler's sniffed, "It is not fitting for a professor of mathematics to manifest

NATURAL NUMBERS	LOGARITHMS BASE 2
1	0
2	1
3	1.5850
4	2
5	2.3219
6	2.5850
7	2.8074
8	3
9	3.1699
10	3.3219
11	3.4594
12	3.5850
13	3.7004
14	3.8074
15	3.9069
16	4
17	4.0875
18	4.1699
19	4.2479
20	4.3219
21	4.3923
22	4.4594
23	4.5236
24	4.5850
25	4.6439
26	4.7004
27	4.7549
28	4.8074
29	4.8580
30	4.9069
31	4.9542
32	5
33	5.0444
34	5.0875
35	5.1293
36	5.1699
37	5.2095
38	5.2479
39	5.2854
40	5.3219
41	5.3576
42	5.3923
43	5.4263
44	5.4594
45	5.4919
46	5.5236
47	5.5546
48	5.5850
49	5.6147
50	5.6439

childish joy just because reckoning is made easier." But why not? Across the centuries they all felt that joy in reckoning: Napier and Briggs, Kepler and Babbage, making their lists, building their towers of ratio and proportion, perfecting their mechanisms for transforming numbers into numbers. And then the world's commerce validated their pleasure.

Charles Babbage was born on Boxing Day 1791, near the end of the century that began with Newton. His home was on the south side of the River Thames in Walworth, Surrey, still a rural hamlet, though the London Bridge was scarcely a half hour's walk even for a small boy. He was the son of a banker, who was himself the son and grandson of goldsmiths. In the London of Babbage's childhood, the Machine Age made itself felt everywhere. A new breed of impresario was showing off machinery in exhibitions. The shows that drew the biggest crowds featured automata—mechanical dolls, ingenious and delicate, with wheels and pinions mimicking life itself. Charles Babbage went with his mother to John Merlin's Mechanical Museum in Hanover Square, full of clockwork and music boxes and, most interesting, simulacra of living things. A metal swan bent its neck to catch a metal fish, moved by hidden motors and cams. In the artist's attic workshop Charles saw a pair of naked dancing women, gliding and bowing, crafted in silver at one-fifth life size. Merlin himself, their elderly creator, said he had devoted years to these machines, his favorites, still unfinished. One of the figurines especially impressed Charles with its (or her) grace and seeming liveliness. "This lady attitudinized in a most fascinating manner," he recalled. "Her eyes were full of imagination, and irresistible." Indeed, when he was a man in his forties he found Merlin's silver dancer at an auction, bought it for £35, installed it on a pedestal in his home, and dressed its nude form in custom finery.

The boy also loved mathematics—an interest far removed from the mechanical arts, as it seemed. He taught himself in bits and pieces from such books as he could find. In 1810 he entered Trinity College, Cambridge—Isaac Newton's domain and still the moral center of mathematics in England. Babbage was immediately disappointed: he discovered that he already knew more of the modern subject than his tutors, and the further knowledge he sought was not to be found there, maybe not anywhere in England. He began to acquire foreign books—especially books from Napoleon's France, with which England was at war. From a specialty bookseller in London he got Lagrange's *Théorie des fonctions analytiques* and "the great work of Lacroix, on the *Differential and Integral Calculus*."

He was right: at Cambridge mathematics was stagnating. A century earlier Newton had been only the second professor of mathematics the university ever had; all the subject's power and prestige came from his legacy. Now his great shadow lay across English mathematics as a curse. The most advanced students learned his brilliant and esoteric "fluxions" and the geometrical proofs of his *Principia*. In the hands of anyone but Newton, the old methods of geometry brought little but frustration. His peculiar formulations of the calculus did his heirs little good. They were increasingly isolated. The English professoriate "regarded any attempt at innovation as a sin against the memory of Newton," one nineteenth-century mathematician said. For the running river of modern mathematics a student had to look elsewhere, to the Continent, to "analysis" and the language of differentiation as invented by Newton's rival and nemesis, Gottfried Wilhelm Leibniz. Fundamentally, there was only one calculus. Newton and Leibniz knew how similar their work was—enough that each accused the other of plagiarism. But they had devised incompatible systems of notation—different languages—and in practice these surface differences mattered more than the underlying sameness. Symbols and operators were what a mathematician had to work with, after all. Babbage, unlike most students, made himself fluent in both—"the dots of Newton, the *d*'s of Leibnitz"—and felt he had seen the light. "It is always difficult to think and reason in a new language."

Indeed, language itself struck him as a fit subject for philosophical study—a subject into which he found himself sidetracked from time to time. Thinking about language, while thinking *in* language, leads to puzzles and paradoxes. Babbage tried for a while to invent, or construct, a universal language, a symbol system that would be free of local idiosyncrasies and imperfections. He was not the first to try. Leibniz himself had claimed to be on the verge of a *characteristica universalis* that would give humanity "a new kind of an instrument increasing the powers of reason far more than any optical instrument has ever aided the power of vision." As philosophers came face to face with the multiplicity of the world's dialects, they so often saw language not as a perfect vessel for truth but as a leaky sieve. Confusion about the meanings of words led to contradictions. Ambiguities and false metaphors were surely not inherent in the nature of things, but arose from a poor choice of signs. If only one could find a proper mental technology, a true philosophical language! Its symbols, properly chosen, must be universal, transparent, and immutable, Babbage argued. Working systematically, he managed to create a grammar and began to write down a lexicon but ran aground on a problem of storage and retrieval—stopped "by the apparent impossibility of arranging signs in any consecutive order, so as to find, as in a dictionary, the meaning of each when wanted." Nevertheless he felt that language was a thing a person could invent. Ideally, language should be rationalized, made predictable and mechanical. The gears should mesh.

Still an undergraduate, he aimed at a new revival of English mathematics—a suitable cause for founding an advocacy group and launching a crusade. He joined with two other promising students, John Herschel and George Peacock, to form what they named the Analytical Society, "for the propagation of *d*'s" and against "the heresy of dots," or as Babbage said, "the Dot-age of the University." (He was pleased with his own "wicked pun.") In their campaign to free the calculus from English dotage, Babbage lamented "the cloud of dispute and national acrimony, which has been thrown over its origin." Never mind if it seemed French. He declared, "We have now to re-import the exotic, with nearly a century of foreign improvement, and to render it once more indigenous among us." They were rebels against Newton in the heart of Newton-land. They met over breakfast every Sunday after chapel.

"Of course we were much ridiculed by the Dons," Babbage recalled. "It was darkly hinted that we were young infidels, and that no good would come of us." Yet their evangelism worked: the new methods spread from the bottom up, students learning faster than their teachers. "The brows of many a Cambridge moderator were elevated, half in ire, half in admiration, at the unusual answers which began to appear in examination papers," wrote Herschel. The dots of Newton faded from the scene, his fluxions replaced by the notation and language of Leibniz.

Meanwhile Babbage never lacked companions with whom he could quaff wine or play whist for six-penny points. With one set of friends he formed a Ghost Club, dedicated to collecting evidence for and against occult spirits. With another set he founded a club called the Extractors, meant to sort out issues of sanity and insanity according to a set of procedures:

1. Every member shall communicate his address to the Secretary once in six months.
2. If this communication is delayed beyond twelve months, it shall be taken for granted that his relatives had shut him up as insane.
3. Every effort legal and illegal shall be made to get him out of the madhouse [hence the name "Extractors"].
4. Every candidate for admission as a member shall produce six certificates. Three that he is sane and three others that he is insane.

But the Analytical Society was serious. It was with no irony, all earnestness, that these mathematical friends, Babbage and Herschel and Peacock, resolved to "do their best to leave the world a wiser place than they found it." They rented rooms and read papers to one another and published their "Transactions." And in those rooms, as Babbage nodded over a book of logarithms, one of them interrupted: "Well, Babbage, what are you dreaming about?"

"I am thinking that all these Tables might be calculated by machinery," he replied.

Anyway that was how Babbage reported the conversation fifty years later. Every good invention needs a eureka story, and he had another in reserve. He and Herschel were laboring together to produce a manuscript of logarithm tables for the Cambridge Astronomical Society. These very logarithms had been computed before; logarithms must always be computed and recomputed and compared and mistrusted. No wonder Babbage and Herschel, laboring over their own manuscript at Cambridge, found the work tedious. "I wish to God these calculations had been executed by steam," cried Babbage, and Herschel replied simply, "It is quite possible."

Steam was the driver of all engines, the enabler of industry. If only for these few decades, the word stood for power and force and all that was vigorous and modern. Formerly, water or wind drove the mills, and most of the world's work still depended on the brawn of people and horses and livestock. But hot steam, generated by burning coal and brought under control by ingenious inventors, had portability and versatility. It replaced muscles everywhere. It became a watchword: people on the go would now "steam up" or "get more steam on" or "blow off steam." Benjamin Disraeli hailed "your moral steam which can work the world." Steam became the most powerful transmitter of energy known to humanity.

It was odd even so that Babbage thought to exert this potent force in a weightless realm—applying steam to thought and arithmetic. Numbers were the grist for his mill. Racks would slide, pinions would turn, and the mind's work would be done.

It should be done automatically, Babbage declared. What did it mean to call a machine "automatic"? For him it was not just a matter of semantics but a principle for judging a machine's usefulness. Calculating devices, such as they were, could be divided into two classes: the first requiring human intervention, the second truly self-acting. To decide whether a machine qualified as automatic, he needed to ask a question that would have been simpler if the words *input* and *output* had been invented: "Whether, when the numbers on which it is to operate are placed in the instrument, it is capable of arriving at its result by the mere motion of a spring, a descending weight, or any other constant force." This was a farsighted standard. It eliminated virtually all the devices ever used or conceived as tools for arithmetic—and there had been many, from the beginning of recorded history. Pebbles in bags, knotted strings, and tally sticks of wood or bone served as short-term memory aids. Abacuses and slide rules applied more complex hardware to abstract reckoning. Then, in the seventeenth century, a few mathematicians conceived the first calculating devices worthy of the name *machine*, for adding and—through repetition of the adding—multiplying. Blaise Pascal made an adding machine in 1642 with a row of revolving disks, one for each decimal digit. Three decades later Leibniz improved on Pascal by using a cylindrical drum with protruding teeth to manage "carrying" from one digit to the next.* Fundamentally, however, the prototypes of Pascal and Leibniz remained closer to the abacus—a passive register of memory states—than to a kinetic machine. As Babbage saw, they were not automatic.

It would not occur to him to use a device for a one-time calculation, no matter how difficult. Machinery excelled at repetition—"intolerable labour and fatiguing monotony." The demand for computation, he foresaw, would grow as the uses of commerce, industry, and science came together. "I will yet venture to predict, that a time will arrive, when the accumulating labour which arises from the arithmetical application of mathematical formulae, acting as a constantly retarding force, shall ultimately impede the useful progress of the science, unless this or some

---

* Leibniz dreamed grandly of mechanizing algebra and even reason itself. "We may give final praise to the machine," he wrote. "It will be desirable to all who are engaged in computations . . . the managers of financial affairs, the administrators of others' estates, merchants, surveyors, geographers, navigators, astronomers. . . . For it is unworthy of excellent men to lose hours like slaves in the labor of calculation."

equivalent method is devised for relieving it from the overwhelming incumbrance of numerical detail."

In the information-poor world, where any table of numbers was a rarity, centuries went by before people began systematically to gather different printed tables in order to check one against another. When they did, they found unexpected flaws. For example, Taylor's *Logarithms*, the standard quarto printed in London in 1792, contained (it eventually transpired) nineteen errors of either one or two digits. These were itemized in the *Nautical Almanac*, for, as the Admiralty knew well, every error was a potential shipwreck.

Unfortunately, one of the nineteen corrections proved erroneous, so the next year's *Nautical Almanac* printed an "erratum of the errata." This in turn introduced yet another error. "Confusion is worse confounded," declared *The Edinburgh Review*. The next almanac would have to put forth an "Erratum of the Erratum of the Errata in Taylor's *Logarithms*."

Particular mistakes had their own private histories. When Ireland established its Ordnance Survey, to map the entire country on a finer scale than any nation had ever accomplished, the first order of business was to ensure that the surveyors—teams of sappers and miners—had 250 sets of logarithmic tables, relatively portable and accurate to seven places. The survey office compared thirteen tables published in London over the preceding two hundred years, as well as tables from Paris, Avignon, Berlin, Leipzig, Gouda, Florence, and China. Six errors were discovered in almost every volume—and they were the *same* six errors. The conclusion was inescapable: these tables had been copied, one from another, at least in part.

Errors arose from mistakes in carrying. Errors arose from the inversion of digits, sometimes by the computers themselves and sometimes by the printer. Printers were liable to transpose digits in successive lines of type. What a mysterious, fallible thing the human mind seemed to be! All these errors, one commentator mused, "would afford a curious subject of metaphysical speculation respecting the operation of the faculty of memory."

Human computers had no future, he saw: "It is only by the *mechanical fabrication of tables* that such errors can be rendered impossible."

Babbage proceeded by exposing mechanical principles within the numbers. He saw that some of the structure could be revealed by computing differences between one sequence and another. The "calculus of finite differences" had been explored by mathematicians (especially the French) for a hundred years. Its power was to reduce high-level calculations to simple addition, ready to be routinized. For Babbage the method was so crucial that he named his machine from its first conception the Difference Engine.

By way of example (for he felt the need to publicize and explain his conception many times as the years passed) Babbage offered the Table of Triangular Numbers. Like many of the sequences of concern, this was a ladder, starting on the ground and rising ever higher:

1, 3, 6, 10, 15, 21 . . .

He illustrated the idea by imagining a child placing groups of marbles on the sand:

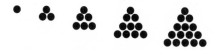

Suppose the child wants to know "how many marbles the thirtieth or any other distant group might contain." (It is a child after Babbage's own heart.) "Perhaps he might go to papa to obtain this information; but I much fear papa would snub him, and would tell him that it was nonsense—that it was useless—that nobody knew the number, and so forth." Understandably papa knows nothing of the Table of Triangular Numbers published at the Hague by É. de Joncourt, professor of philosophy. "If papa fail to inform him, let him go to mamma, who will not fail to find means to satisfy her darling's curiosity." Meanwhile, Babbage answers the question by means of a table of differences. The first column contains the number sequence in question. The next columns are derived by repeated subtractions, until a constant is reached—a column made up entirely of a single number.

Number of the Group	Number of Marbles in Each Group	1st Difference Difference Between Each Group and the Next	2nd Difference
1	1	1	1
2	3	2	1
3	6	3	1
4	10	4	1
5	15	5	1
6	21	6	1
7	28	7	1

Any polynomial function can be reduced by the method of differences, and all well-behaved functions, including logarithms, can be effectively approximated. Equations of higher degree require higher-order differences. Babbage offered another concrete geometrical example that requires a table of third differences: piles of cannonballs in the form of a triangular pyramid—the triangular numbers translated to three dimensions.

Number	Table	1st Difference	2nd Difference	3rd Difference
1	1	3	3	1
2	4	6	4	1
3	10	10	5	1
4	20	15	6	1
5	35	21	7	1
6	56	28	8	1

The Difference Engine would run this process in reverse: instead of repeated subtraction to find the differences, it would generate sequences of numbers by a cascade of additions. To accomplish this, Babbage conceived a system of figure wheels, marked with the numerals 0 to 9, placed along an axis to represent the decimal digits of a number: the units, the tens, the hundreds, and so on. The wheels would have gears. The gears along each axis would mesh with the gears of the next, to add the successive digits. As the machinery transmitted motion, wheel to wheel, it would be transmitting information, in tiny increments, the numbers summing across the axes. A mechanical complication arose, of course, when any sum passed 9. Then a unit had to be carried to the next decimal place. To manage this, Babbage placed a projecting tooth on each wheel, between the 9 and 0. The tooth would push a lever, which would in turn transmit its motion to the next wheel above.

At this point in the history of computing machinery, a new theme appears: the obsession with time. It occurred to Babbage that his machine had to compute faster than the human mind and as fast as possible. He had an idea for parallel processing: number wheels arrayed along an axis could add a row of the digits all at once. "If this could be accomplished," he noted, "it would render additions and subtractions with numbers

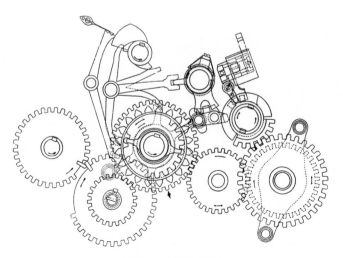

BABBAGE'S WHEEL-WORK

having ten, twenty, fifty, or any number of figures, as rapid as those operations are with single figures." He could see a problem, however. The digits of a single addition could not be managed with complete independence because of the carrying. The carries could overflow and cascade through a whole set of wheels. If the carries were known in advance, then the additions could proceed in parallel. But that knowledge did not become available in timely fashion. "Unfortunately," he wrote, "there are multitudes of cases in which the carriages that become due are only known in successive periods of time." He counted up the time, assuming one second per operation: to add two fifty-digit numbers might take only nine seconds in itself, but the carrying, in the worst case, could require fifty seconds more. Bad news indeed. "Multitudes of contrivances were designed, and almost endless drawings made, for the purpose of economizing the time," Babbage wrote ruefully. By 1820 he had settled on a design. He acquired his own lathe, used it himself and hired metalworkers, and in 1822 managed to present the Royal Society with a small working model, gleaming and futuristic.

He was living in London near the Regent's Park as a sort of gentleman philosopher, publishing mathematical papers and occasionally lecturing to the public on astronomy. He married a wealthy young woman from Shropshire, Georgiana Whitmore, the youngest of eight sisters. Beyond what money she had, he was supported mainly by a £300 allowance from his father—whom he resented as a tyrannical, ungenerous, and above all close-minded old man. "It is scarcely too much to assert that he *believes* nothing he *hears*, and only half of what he sees," Babbage wrote his friend Herschel. When his father died, in 1827, Babbage inherited a fortune of £100,000. He briefly became an actuary for a new Protector Life Assurance Company and computed statistical tables rationalizing life expectancies. He tried to get a university professorship, so far unsuccessfully, but he had an increasingly lively social life, and in scholarly circles people were beginning to know his name. With Herschel's help he was elected a fellow of the Royal Society.

Even his misfires kindled his reputation. On behalf of *The Edinburgh Journal of Science* Sir David Brewster sent him a classic in the annals of rejection letters: "It is with no inconsiderable degree of reluctance that I decline the offer of any Paper from you. I think, however, you will upon reconsideration of the subject be of opinion that I have no other alternative. The subjects you propose for a series of Mathematical and Metaphysical Essays are so very profound, that there is perhaps not a single subscriber to our Journal who could follow them." On behalf of his nascent invention, Babbage began a campaign of demonstrations and letters. By 1823 the Treasury and the Exchequer had grown interested. He promised them "logarithmic tables as cheap as potatoes"—how could they resist? Logarithms saved ships. The Lords of the Treasury authorized a first appropriation of £1,500.

As an abstract conception the Difference Engine generated excitement that did not need to wait for anything so mundane as the machine's actual construction. The idea was landing in fertile soil. Dionysius Lardner, a popular lecturer on technical subjects, devoted a series of public talks to Babbage, hailing his "proposition to reduce arithmetic to the dominion of mechanism,—to substitute an automaton for a compositor,—to throw the powers of thought into wheel-work." The engine "must, when completed," he said, "produce important effects, not only on the progress of science, but on that of civilization." It would be the *rational* machine. It would be a junction point for two roads—mechanism and thought. Its admirers sometimes struggled with their explanations of this intersection: "The question is set to the instrument," Henry Colebrooke told the Astronomical Society, "or the instrument is set to the question." Either way, he said, "by simply giving motion the solution is wrought."

But the engine made slower progress in the realm of brass and wrought iron. Babbage tore out the stables in back of his London house and replaced them with a forge, foundry, and fireproofed workshop. He engaged Joseph Clement, a draftsman and inventor, self-educated, the son of a village weaver who had made himself into England's preeminent mechanical engineer. Babbage and Clement realized that they would have to make new tools. Inside a colossal iron frame the design called for the most intricate and precise parts—axles, gears, springs, and pins, and above all figure wheels by the hundreds and then thousands. Hand tools could never produce the components with the needed precision. Before Babbage could have a manufactory of number tables, he would have to build new manufactories of parts. The rest of the Industrial Revolution, too, needed standardization in its parts: interchangeable screws of uniform thread count and pitch; screws as fundamental units. The lathes of Clement and his journeymen began to produce them.

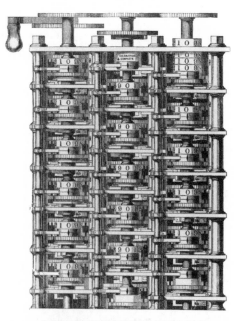

A WOODCUT IMPRESSION (1853) OF A
SMALL PORTION OF THE DIFFERENCE ENGINE

As the difficulties grew, so did Babbage's ambitions. After ten years, the engine stood twenty-four inches high, with six vertical axles and dozens of wheels, capable of computing six-figure results. Ten years after that, the scale—on paper—had reached 160 cubic feet, 15 tons, and 25,000 parts, and the paper had spread, too, the drawings covering more than 400 square feet. The level of complexity was confounding. Babbage solved the problem of adding many digits at once by separating the "adding motions" from the "carrying motions" and then staggering the timing of the carries. The addition would begin with a rush of grinding gears, first the odd-numbered columns of dials, then the even columns. Then the carries would recoil across the rows. To keep the motion synchronized, parts of the machine would need to "know" at critical times that a carry was pending. The information was conveyed by the state of a latch. For the first time, but not the last, a device was invested with memory. "It is in effect a memorandum taken by the machine," wrote his publicizer, Dionysius Lardner. Babbage himself was self-conscious about anthropomorphizing but could not resist. "The mechanical means I employed to make these carriages," he suggested, "bears some slight analogy to the operation of the faculty of memory."

In ordinary language, to describe even this basic process of addition required a great effulgence of words, naming the metal parts, accounting for their interactions, and sorting out interdependencies that multiplied to form a long chain of causality. Lardner's own explanation of "carrying," for example, was epic. A single isolated instant of the action involved a dial, an index, a thumb, an axis, a trigger, a notch, a hook, a claw, a spring, a tooth, and a ratchet wheel:

Now, at the moment that the division between 9 and 0 on the dial B^2 passes under the index, a thumb placed on the axis of this dial touches a trigger which raises out of the notch of the hook which sustains the claw just mentioned, and allows it to fall back by the recoil of the spring, and drop into the next tooth of the ratchet wheel.

Hundreds of words later, summing up, Lardner resorted to a metaphor suggesting fluid dynamics:

> There are two systems of waves of mechanical action continually flowing from the bottom to the top; and two streams of similar action constantly passing from the right to the left. The crests of the first system of adding waves fall upon the last difference, and upon every alternate one proceeding upwards. . . . The first stream of carrying action passes from right to left along the highest row and every alternate row.

This was one way of abstracting from the particular—the particulars being so intricate. And then he surrendered. "Its wonders, however, are still greater in its details," he wrote. "We despair of doing it justice."

Nor were ordinary draftsman's plans sufficient for describing this machine that was more than a machine. It was a dynamical system, its many parts each capable of several modes or states, sometimes at rest and sometimes in motion, propagating their influence along convoluted channels. Could it ever be specified completely, on paper? Babbage, for his own purposes, devised a new formal tool, a system of "mechanical notation" (his term). This was a language of signs meant to represent not just the physical form of a machine but its more elusive properties: its timing and its logic. It was an extraordinary ambition, as Babbage himself appreciated. In 1826 he proudly reported to the Royal Society "On a Method of Expressing by Signs the Action of Machinery." In part it was an exercise in classification. He analyzed the different ways in which something—motion, or power—could be "communicated" through a system. There were many ways. A part could receive its influence simply by being attached to another part, "as a pin on a wheel, or a wheel and pinion on the same axis." Or transmission could occur "by stiff friction." A part might be driven constantly by another part "as happens when a wheel is driven by a pinion"—or *not* constantly, "as is the case when a stud lifts a bolt once in the course of a revolution." Here a vision of logical branching entered the scheme: the path of communication would vary depending on the alternative states of some part of the machine.

Babbage's mechanical notation followed naturally from his work on symbolic notation in mathematical analysis. Machinery, like mathematics, needed rigor and definition for progress. "The forms of ordinary language were far too diffuse," he wrote. "The signs, if they have been properly chosen, and if they should be generally adopted, will form as it were an universal language." Language was never a side issue for Babbage.

He finally won a university post, at Cambridge: the prestigious Lucasian Professorship of Mathematics, formerly occupied by Newton. As in Newton's time, the work was not onerous. Babbage did not have to teach students, deliver lectures, or even live in Cambridge, and this was just as well, because he was also becoming a popular fixture of London social life. At home at One Dorset Street he hosted a regular Saturday soirée that drew a glittering crowd—politicians, artists, dukes and duchesses, and the greatest English scientists of the age: Charles Darwin, Michael Faraday, and Charles Lyell, among others.* They marveled at his calculating machine and, on display nearby, the dancing automaton of his youth. (In invitations he would write, "I hope you intend to patronise the 'Silver Lady.' She is to appear in new dresses and decorations.") He was a mathematical raconteur—that was no contradiction, in this time and place. Lyell reported approvingly that he "jokes and reasons in high mathematics." He published a much-quoted treatise applying probability theory to the theological question of miracles. With tongue in cheek he wrote Alfred, Lord Tennyson, to suggest a correction for the poet's couplet: "Every minute dies a man, / Every minute one is born."

I need hardly point out to you that this calculation would tend to keep the sum total of the world's population in a state of perpetual equipoise,

---

* Another guest, Charles Dickens, put something of Babbage into the character of Daniel Doyce in *Little Dorrit*. Doyce is an inventor mistreated by the government he tries to serve: "He is well known as a very ingenious man. . . . He perfects an invention (involving a very curious secret process) of great importance to his country and his fellow-creatures. I won't say how much money it cost him, or how many years of his life he had been about it, but he brought it to perfection." Dickens added: "A composed and unobtrusive self-sustainment was noticeable in Daniel Doyce—a calm knowledge that what was true must remain true."

whereas it is a well-known fact that the said sum total is constantly on the increase. I would therefore take the liberty of suggesting that in the next edition of your excellent poem the erroneous calculation to which I refer should be corrected as follows: "Every moment dies a man / And one and a sixteenth is born." I may add that the exact figures are 1.167, but something must, of course, be conceded to the laws of metre.

Fascinated with his own celebrity, he kept a scrapbook—"the pros and cons in parallel columns, from which he obtained a sort of balance," as one visitor described it. "I was told repeatedly that he spent all his days in gloating and grumbling over what people said of him."

But progress on the engine, the main source of his fame, was faltering. In 1832 he and his engineer Clement produced a working demonstration piece. Babbage displayed it at his parties to guests who found it miraculous or merely puzzling. The Difference Engine stands—for a replica works today, in the Science Museum in London—as a milestone of what could be achieved in precision engineering. In the composition of its alloys, the exactness of its dimensions, the interchangeability of its parts, nothing surpassed this segment of an unfinished machine. Still, it was a curio. And it was as far as Babbage could go.

He and his engineer fell into disputes. Clement demanded more and more money from Babbage and from the Treasury, which began to suspect profiteering. He withheld parts and drawings and fought over control of the specialized machine tools in their workshops. The government, after more than a decade and £17,000, was losing faith in Babbage, and he in the government. In his dealing with lords and ministers Babbage could be imperious. He was developing a sour view of the Englishman's attitude toward technological innovation: "If you speak to him of a machine for peeling a potato, he will pronounce it impossible: if you peel a potato with it before his eyes, he will declare it useless, because it will not slice a pineapple." They no longer saw the point.

"What shall we do to get rid of Mr. Babbage and his calculating machine?" Prime Minister Robert Peel wrote one of his advisers in

August 1842. "Surely if completed it would be worthless as far as science is concerned. . . . It will be in my opinion a very costly toy." He had no trouble finding voices inimical to Babbage in the civil service. Perhaps the most damning was George Biddell Airy, the Astronomer Royal, a starched and methodical figure, who with no equivocation told Peel precisely what he wanted to hear: that the engine was useless. He added this personal note: "I think it likely he

CHARLES BABBAGE
(1860)

lives in a sort of dream as to its utility." Peel's government terminated the project. As for Babbage's dream, it continued. It had already taken another turn. The engine in his mind had advanced into a new dimension. And he had met Ada Byron.

In the Strand, at the north end of the Lowther shopping arcade, visitors thronged to the National Gallery of Practical Science, "Blending Instruction with Amusement," a combination toy store and technology show set up by an American entrepreneur. For the admission price of a shilling, a visitor could touch the "electrical eel," listen to lectures on the newest science, and watch a model steamboat cruising a seventy-foot trough and the Perkins steam gun emitting a spray of bullets. For a guinea, she could sit for a "daguerreotype" or "photographic" portrait, by which a faithful and pleasing likeness could be obtained in "less than One Second." Or she could watch, as young Augusta Ada Byron did, a weaver demonstrating the automated Jacquard loom, in which the patterns to be woven in cloth were encoded as holes punched into pasteboard cards.

Ada was "the child of love," her father had written, "—though born in bitterness, and nurtured in convulsion." Her father was a poet. When she was barely a month old, in 1816, the already notorious Lord Byron, twenty-seven, and the bright, wealthy, and mathematically knowledge-able Anne Isabella Milbanke (Annabella), twenty-three, separated after a year of marriage. Byron left England and never saw his daughter again. Her mother refused to tell her who her father was until she was eight and he died in Greece, an international celebrity. The poet had begged for any news of his daughter: "Is the Girl imaginative?—at *her* present age I have an idea that I had many feelings & notions which people would not believe if I stated them *now*." Yes, she was imaginative.

She was a prodigy, clever at mathematics, encouraged by tutors, tal-ented in drawing and music, fantastically inventive and profoundly lonely. When she was twelve, she set about inventing a means of fly-ing. "I am going to begin my paper wings tomorrow," she wrote to her mother. She hoped "to bring the art of flying to very great perfection. I think of writing a book of *Flyology* illustrated with plates." For a while she signed her letters "your very affectionate Carrier Pigeon." She asked her mother to find a book illustrating bird anatomy, because she was reluctant "to dissect even a bird." She analyzed her daily situation with a care for logic.

> Miss Stamp desires me to say that at present she is not particularly pleased with me on account of some very foolish conduct yesterday about a sim-ple thing, and which she said was not only foolish but showed a spirit of inattention, and though today she has not had reason to be dissatisfied with me on the whole yet she says that she can not directly efface the recollection of the past.

She was growing up in a well-kept cloister of her mother's arranging. She had years of sickliness, a severe bout of measles, and episodes of what was called neurasthenia or hysteria. ("When I am weak," she wrote, "I am always so exceedingly terrified, at *nobody knows what*, that I can hardly help having an agitated look & manner.") Green drapery enclosed the portrait of her father that hung in one room. In her teens she developed a romantic interest in her tutor, which led to a certain amount of sneaking about the house and gardens and to lovemaking as intimate as possible without, she said, actual "connection." The tutor was dismissed. Then, in the spring, wearing white satin and tulle, the seventeen-year-old made her ritual debut at court, where she met the king and queen, the most important dukes, and the French diplomat Talleyrand, whom she described as an "old monkey."

A month later she met Charles Babbage. With her mother, she went to see what Lady Byron called his "thinking machine," the portion of the Difference Engine in his salon. Babbage saw a sparkling, self-possessed young woman with porcelain features and a notorious name, who managed to reveal that she knew more mathematics than most men graduating from university. She saw an imposing forty-one-year-old, authoritative eyebrows anchoring his strong-boned face, who possessed wit and charm and did not wear these qualities lightly. He seemed a kind of visionary— just what she was seeking. She admired the machine, too. An onlooker reported: "While other visitors gazed at the working of this beautiful instrument with the sort of expression, and I dare say the sort of feeling, that some savages are said to have shown on first seeing a looking-glass or hearing a gun, Miss Byron, young as she was, understood its working, and saw the great beauty of the invention." Her feeling for the beauty and abstractions of mathematics, fed only in morsels from her succession of tutors, was overflowing. It had no outlet. A woman could not attend university in England, nor join a scientific society (with two exceptions: the botanical and horticultural).

Ada became a tutor for the young daughters of one of her mother's friends. When writing to them, she signed herself, "your affectionate & untenable Instructress." On her own she studied Euclid. Forms burgeoned in her mind. "I do not consider that I know a proposition," she wrote another tutor, "until I can imagine to myself a figure in the air, and go through the construction & demonstration without any book or assistance whatever." She could not forget Babbage, either, or his "gem of

AUGUSTA ADA BYRON KING, COUNTESS OF LOVELACE,
AS PAINTED IN 1836 BY MARGARET CARPENTER.
"I CONCLUDE SHE IS BENT ON DISPLAYING THE
WHOLE EXPANSE OF MY CAPACIOUS JAW BONE,
UPON WHICH I THINK THE WORD MATHEMATICS
SHOULD BE WRITTEN."

all mechanism." To another friend she reported her "great anxiety about the machine." Her gaze turned inward, often. She liked to think about herself thinking.

Babbage himself had moved far beyond the machine on display in his drawing room; he was planning a new machine, still an engine of computation but transmuted into another species. He called this the Analytical Engine. Motivating him was a quiet awareness of the Difference Engine's limitations: it could not, merely by adding differences, compute every sort of number or solve any mathematical problem. Inspiring him, as well, was the loom on display in the Strand, invented by Joseph-Marie Jacquard, controlled by instructions encoded and stored as holes punched in cards.

What caught Babbage's fancy was not the weaving, but rather the encoding, from one medium to another, of patterns. The patterns would appear in damask, eventually, but first were "sent to a peculiar artist." This specialist, as he said,

> punches holes in a set of pasteboard cards in such a manner that when those cards are placed in a Jacquard loom, it will then weave upon its produce the exact pattern designed by the artist.

The notion of abstracting information away from its physical substrate required careful emphasis. Babbage explained, for example, that the weaver might choose different threads and different colors—"but in all these cases the *form* of the pattern will be precisely the same." As Babbage conceived his machine now, it raised this very process of abstraction to higher and higher degrees. He meant the cogs and wheels to handle not just numbers but variables standing in for numbers. Variables were to be filled or determined by the outcomes of prior calculations, and, further, the very operations—such as addition or multiplication—were to be changeable, depending on prior outcomes. He imagined these abstract information quantities being stored in cards: variable cards and operation cards. He thought of the machine as embodying laws and of the cards as communicating these laws. Lacking a ready-made vocabulary, he found it awkward to express his fundamental working concepts; for example,

> how the machine could perform the act of judgment sometimes required during an analytical inquiry, when two or more different courses presented themselves, especially as the proper course to be adopted could not be known in many cases until all the previous portion had been gone through.

He made clear, though, that information—representations of number and process—would course through the machinery. It would pass to and from certain special physical locations, which Babbage named a *store*, for storage, and a *mill*, for action.

In all this he had an intellectual companion now in Ada, first his acolyte and then his muse. She married a sensible and promising aristocrat, William King, her senior by a decade and a favorite of her mother. In the space of a few years he was elevated to the peerage as earl of Lovelace—making Ada, therefore, a countess—and, still in her early twenties, she bore three children. She managed their homes, in Surrey and London, practiced the harp for hours daily ("I am at present a condemned slave to *my harp*, no easy Task master"), danced at balls, met the new queen, Victoria, and sat for her portrait, self-consciously ("I conclude [the artist] is bent on displaying the whole expanse of my capacious jaw bone, upon which I think the word Mathematics should be written"). She suffered terrible dark moods and bouts of illness, including cholera. Her interests and behavior still set her apart. One morning she went alone in her carriage, dressed plainly, to see a model of Edward Davy's "electrical telegraph" at Exeter Hall

> & the only other person was a middle-aged gentleman who chose to behave as if *I* were the show [she wrote to her mother] which of course I thought was the most impudent and unpardonable.—I am sure he took me for a very young (& I suppose he thought rather handsome) governess. . . . He stopped as long as I did, & then followed me out.— I took care to look as aristocratic & *as like a Countess* as possible. . . . I must try & add a little age to my appearance. . . . I would go & see something everyday & I am sure London would never be exhausted.

Lady Lovelace adored her husband but reserved much of her mental life for Babbage. She had dreams, waking dreams, of something she could not be and something she could not achieve, except by proxy, through his genius. "I have a peculiar *way* of *learning*," she wrote to him, "& I think it must be a peculiar man to teach me successfully." Her growing desperation went side by side with a powerful confidence in her untried abilities. "I hope you are bearing me in mind," she wrote some months later, "I mean my mathematical interests. You know this is the greatest favour any one can do me.—Perhaps, none of us can estimate *how* great. . . ."

> You know I am by nature a bit of a philosopher, & a very great speculator,
> —so that I look on through a very immeasurable vista, and though I see
> nothing but vague & cloudy uncertainty in the foreground of our being,
> yet I fancy I discern a very bright light a good way further on, and this
> makes me care much less about the cloudiness & indistinctness which is
> near.—Am I too imaginative for you? I think not.

The mathematician and logician Augustus De Morgan, a friend of Babbage and of Lady Byron, became Ada's teacher by post. He sent her exercises. She sent him questions and musings and doubts ("I could wish I went on quicker"; "I am sorry to say I am sadly obstinate about the Term at which Convergence begins"; "I have enclosed my Demonstration of *my* view of the case"; "functional Equations are complete Will-o-the-wisps to me"; "However I try to keep my metaphysical head in order"). Despite her naïveté, or because of it, he recognized a "power of thinking . . . so utterly out of the common way for any beginner, man or woman." She had rapidly mastered trigonometry and integral and differential calculus, and he told her mother privately that if he had encountered "such power" in a Cambridge student he would have anticipated "an original mathematical investigator, perhaps of first rate eminence." She was fearless about drilling down to first principles. Where she felt difficulties, real difficulties lay.

One winter she grew obsessed with a fashionable puzzle known as Solitaire, the Rubik's Cube of its day. Thirty-two pegs were arranged on a board with thirty-three holes, and the rules were simple: Any peg may

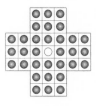

jump over another immediately adjacent, and the peg jumped over is removed, until no more jumps are possible. The object is to finish with only one peg remaining. "People may try thousands of times, and not succeed in this," she wrote Babbage excitedly.

> I *have* done it by trying & observation & can now do it at any time, but I want to know if the problem admits of being put into a mathematical Formula, & solved in this manner. . . . There must be a definite principle, a compound I imagine of numerical & geometrical properties, on which the solution depends, & which can be put into symbolic language.

A formal solution to a game—the very idea of such a thing was original. The desire to create a language of symbols, in which the solution could be encoded—this way of thinking was Babbage's, as she well knew.

She pondered her growing powers of mind. They were not strictly mathematical, as she saw it. She saw mathematics as merely a part of a greater imaginative world. Mathematical transformations reminded her "of certain sprites & fairies one reads of, who are at one's elbows in *one* shape now, & the next minute in a form most dissimilar; and uncommonly deceptive, troublesome & tantalizing are the mathematical sprites & fairies sometimes; like the types I have found for them in the world of Fiction." *Imagination*—the cherished quality. She mused on it; it was her heritage from her never-present father.

> We talk *much* of Imagination. We talk of the Imagination of Poets, the Imagination of Artists &c; I am inclined to think that in general we don't know very exactly *what* we are talking about. . . .
>
> It is that which penetrates into the unseen worlds around us, the worlds of Science. It is that which feels & discovers what *is*, the *real* which we see not, which *exists* not for our *senses*. Those who have learned to walk on the threshold of the unknown worlds . . . may then with the fair white

wings of Imagination hope to soar further into the unexplored amidst which we live.

She began to believe she had a divine mission to fulfill. She used that word, *mission*. "I have on my mind most strongly the impression that Heaven has allotted me some peculiar *intellectual-moral* mission to perform." She had powers. She confided in her mother:

> I believe myself to possess a most singular combination of qualities exactly fitted to make me *pre-eminently* a discoverer of the *hidden realities* of nature. . . . The belief has been *forced* upon me, & most slow have I been to admit it even.

She listed her qualities:

> Firstly: Owing to some peculiarity in my nervous system, I have *perceptions* of some things, which no one else has; or at least very few, if any. . . . Some might say an *intuitive* perception of hidden things;—that is of things hidden from eyes, ears & the ordinary senses. . . .
>
> Secondly;—my immense reasoning faculties;
>
> Thirdly; . . . the power not only of throwing my whole energy & existence into whatever I choose, but also bring to bear on any one subject or idea, a vast apparatus from all sorts of apparently irrelevant & extraneous sources. I can throw *rays* from every quarter of the universe into *one* vast focus.

She admitted that this sounded mad but insisted she was being logical and cool. She knew her life's course now, she told her mother. "*What* a mountain I have to climb! It is enough to frighten anyone who had not all that most insatiable & restless energy, which from my babyhood has been the plague of your life & my own. However it has found food I believe at last." She had found it in the Analytical Engine.

———

Babbage meanwhile, restless and omnivorous, was diverting his energies to another burgeoning technology, steam's most powerful expression, the railroad. The newly formed Great Western Railway was laying down track and preparing trial runs of locomotive engines from Bristol to London under the supervision of Isambard Kingdom Brunel, the brilliant engineer, then just twenty-seven years old. Brunel asked Babbage for help, and Babbage decided to begin with an information-gathering program—characteristically ingenious and grandiose. He outfitted an entire railway carriage. On a specially built, independently suspended table, rollers unwound sheets of paper a thousand feet long, while pens drew lines to "express" (as Babbage put it) measurements of the vibrations and forces felt by the carriage in every direction. A chronometer marked the passage of time in half seconds. He covered two miles of paper this way.

As he traversed the rails, he realized that a peculiar danger of steam locomotion lay in its outracing every previous means of communication. Trains lost track of one another. Until the most regular and disciplined scheduling was imposed, hazard ran with every movement. One Sunday Babbage and Brunel, operating in different engines, barely avoided smashing into each other. Other people, too, worried about this new gap between the speeds of travel and messaging. An important London banker told Babbage he disapproved: "It will enable our clerks to plunder us, and then be off to Liverpool on their way to America at the rate of twenty miles an hour." Babbage could only express the hope that science might yet find a remedy for the problem it had created. ("Possibly we might send lightning to outstrip the culprit.")

As for his own engine—the one that would travel nowhere—he had found a fine new metaphor. It would be, he said, "a locomotive that lays down its own railway."

Bitter as he was about England's waning interest in his visionary plans, Babbage found admirers on the continent, particularly in Italy—"the country of Archimedes and Galileo," as he put it to his new friends. In the summer of 1840 he gathered up his sheaves of drawings and journeyed by way of Paris and Lyon, where he watched the great Jacquard loom at Manufacture d'Étoffes pour Ameublements et Ornements d'Église, to Turin, the capital of Sardinia, for an assembly of mathematicians and engineers. There he made his first (and last) public presentation of the Analytical Engine. "The discovery of the Analytical Engine is so much in advance of my own country, and I fear even of the age," he said. He met the Sardinian king, Charles Albert, and, more significantly, an ambitious young mathematician named Luigi Menabrea. Later Menabrea was to become a general, a diplomat, and the prime minister of Italy; now he prepared a scientific report, "*Notions sur la machine analytique*," to introduce Babbage's plan to a broader community of European philosophers.

As soon as this reached Ada Lovelace, she began translating it into English, correcting errors on the basis of her own knowledge. She did that on her own, without telling either Menabrea or Babbage.

When she finally did show Babbage her draft, in 1843, he responded enthusiastically, urging her to write on her own behalf, and their extraordinary collaboration began in earnest. They sent letters by messenger back and forth across London at a ferocious pace—"My Dear Babbage" and "My Dear Lady Lovelace"—and met whenever they could at her home in St. James's Square. The pace was almost frantic. Though he was the eminence, fifty-one years old to her twenty-seven, she took charge, mixing stern command with banter. "I want you to answer me the following question by return of post"; "Be kind enough to write this out properly for me"; "You were a little harum-scarum and inaccurate"; "I wish you were as accurate and as much to be relied on as myself." She proposed to sign her work with her initials—nothing so forward as her name—not to "*proclaim* who has written it," merely to "*individualize* and *identify* it with other productions of the said A.A.L."

Her exposition took the form of notes lettered A through G, extending to nearly three times the length of Menabrea's essay. They offered a vision of the future more general and more prescient than any expressed by Babbage himself. How general? The engine did not just calculate;

it performed *operations*, she said, defining an operation as "any process which alters the mutual relation of two or more things," and declaring: "This is the most general definition, and would include all subjects in the universe." The science of operations, as she conceived it,

> is a science of itself, and has its own abstract truth and value; just as logic has its own peculiar truth and value, independently of the subjects to which we may apply its reasonings and processes. . . . One main reason why the separate nature of the science of operations has been little felt, and in general little dwelt on, is the *shifting* meaning of many of the symbols used.

Symbols and *meaning:* she was emphatically not speaking of mathematics alone. The engine "might act upon other things besides *number*." Babbage had inscribed numerals on those thousands of dials, but their working could represent symbols more abstractly. The engine might process any meaningful relationships. It might manipulate language. It might create music. "Supposing, for instance, that the fundamental relations of pitched sounds in the science of harmony and of musical composition were susceptible of such expression and adaptations, the engine might compose elaborate and scientific pieces of music of any degree of complexity or extent."

It had been an engine of numbers; now it became an engine of information. A.A.L. perceived that more distinctly and more imaginatively than Babbage himself. She explained his prospective, notional, virtual creation as though it already existed:

> The Analytical Engine does not occupy common ground with mere "calculating machines." It holds a position wholly its own. . . . A new, a vast, and a powerful language is developed . . . in which to wield its truths so that these may become of more speedy and accurate practical application for the purposes of mankind than the means hitherto in our possession have rendered possible. Thus not only the mental and the material, but the theoretical and the practical in the mathematical world, are brought into more intimate and effective connexion with each other.

> . . . We may say most aptly, that the Analytical Engine *weaves algebraical patterns* just as the Jacquard-loom weaves flowers and leaves.

For this flight of fancy she took full responsibility. "Whether the inventor of this engine had any such views in his mind while working out the invention, or whether he may subsequently ever have regarded it under this phase, we do not know; but it is one that forcibly occurred to ourselves."

She proceeded from the poetic to the practical. She set forth on a virtuoso excursion through a hypothetical program by which this hypothetical machine might compute a famously deep-seated infinite series, the Bernoulli numbers. These numbers arise in the summing of numbers from 1 to n raised to integral powers, and they occur in various guises all through number theory. No direct formula generates them, but they can be worked out methodically, by expanding certain formulas further and further and looking at the coefficients each time. She began with examples; the simplest, she wrote, would be the expansion of

$$\frac{x}{\epsilon^x - 1} = \frac{1}{1 + \frac{x}{2} + \frac{x^2}{2 \cdot 3} + \frac{x^3}{2 \cdot 3 \cdot 4} + \&c.}$$

and another approach would be via

$$B_{2n-1} = \frac{\pm 2^n}{(2^{2n}-1)2^{n-1}} \left\{ \begin{array}{l} \frac{1}{2} n^{2n-1} \\[1mm] - (n-1)^{2n-1} \left\{ 1 + \frac{1}{2} \cdot \frac{2n}{1} \right\} \\[1mm] + (n-2)^{2n-1} \left\{ 1 + \frac{1}{2} + \frac{1}{2} \cdot \frac{2n \cdot (2n-1)}{1 \cdot 2} \right\} \\[1mm] - (n-3)^{2n-1} \left\{ 1 + \frac{2n}{1} + \frac{2n \cdot (2n-1)}{1 \cdot 2} + \right. \\[1mm] \left. + \frac{1}{2} \cdot \frac{2n \cdot (2n-1) \cdot (2n-2)}{1 \cdot 2 \cdot 3} \right\} \\[1mm] + \; \ldots \qquad \ldots \qquad \ldots \qquad \ldots \end{array} \right\}$$

but she would take a more challenging path, because "our object is not simplicity . . . but the illustration of the powers of the engine."

She devised a process, a set of rules, a sequence of operations. In another century this would be called an algorithm, later a computer program, but for now the concept demanded painstaking explanation. The trickiest point was that her algorithm was recursive. It ran in a loop. The result of one iteration became food for the next. Babbage had alluded to this approach as "the Engine eating its own tail." A.A.L. explained: "We easily perceive that since every successive function is arranged in a series following the same law, there would be a cycle of a cycle of a cycle, &c. . . . The question is so exceedingly complicated, that perhaps few persons can be expected to follow. . . . Still it is a very important case as regards the engine, and suggests ideas peculiar to itself, which we should regret to pass wholly without allusion."

A core idea was the entity she and Babbage called the *variable*. Variables were, in hardware terms, the machine's columns of number dials. But there were "Variable cards," too. In software terms they were a sort of receptacle or envelope, capable of representing, or storing, a number of many decimal digits. ("What is there in a name?" Babbage wrote. "It is merely an empty basket until you put something in it.") Variables were the machine's units of information. This was quite distinct from the algebraic variable. As A.A.L. explained, "The origin of this appellation is, that the values on the columns are destined to change, that is to vary, in every conceivable manner." Numbers *traveled*, in effect, from variable cards to variables, from variables to the mill (for operations), from the mill to the store. To solve the problem of generating Bernoulli numbers, she choreographed an intricate dance. She worked days and sometimes through the night, messaging Babbage across London, struggling with sickness and ominous pains, her mind soaring:

> That *brain* of mine is something more than merely *mortal;* as time will show; (if only my *breathing* & some other et-ceteras do not make too rapid a progress *towards* instead of *from* mortality).
>
> Before ten years are over, the Devil's in it if I have not sucked out some of the life-blood from the mysteries of this universe, in a way that no purely mortal lips or brains could do.

No one knows what almost *awful* energy & power lie yet undeveloped in that *wiry* little system of mine. I say *awful*, because you may imagine what it *might* be under certain circumstances. . . .

I am doggedly attacking & sifting to the very bottom, all the ways of deducing the Bernoulli Numbers. . . . I am grappling with this subject, & *connecting* it with others.

She was programming the machine. She programmed it in her mind, because the machine did not exist. The complexities she encountered for the first time became familiar to programmers of the next century:

How multifarious and how mutually complicated are the considerations which the working of such an engine involve. There are frequently several distinct sets of effects going on simultaneously; all in a manner independent of each other, and yet to a greater or less degree exercising a mutual influence. To adjust each to every other, and indeed even to perceive and trace them out with perfect correctness and success, entails difficulties whose nature partakes to a certain extent of those involved in every question where conditions are very numerous and inter-complicated.

She reported her feelings to Babbage: "I am in much dismay at having got into so amazing a quagmire & botheration." And nine days later: "I find that my plans & ideas keep gaining in clearness, & assuming more of the *crystalline* & less & less of the *nebulous* form." She knew she had achieved something utterly new. Ten days later still, struggling over the final proofs with "Mr Taylors Printing Office" in Fleet Street, she declared: "I do not think you possess half *my* forethought, & power of foreseeing all *possible* contingencies (*probable* & *improbable*, just alike).— . . . I do *not* believe that my father was (or ever could have been) such a *Poet* as *I shall* be an *Analyst*; (& Metaphysician); for with me the two go together indissolubly."

Who would have used this machine? Not clerks or shopkeepers, said Babbage's son, many years later. Common arithmetic was never the purpose—"It would be like using the steam hammer to crush the nut." He paraphrased Leibniz: "It is not made for those who sell vegetables or little fishes, but for observatories, or the private rooms of calculators, or for others who can easily bear the expense, and need a good deal of calculation." Babbage's engine had not been well understood, not by his government and not by the many friends who passed through his salon, but in its time its influence traveled far.

In America, a country bursting with invention and scientific optimism, Edgar Allan Poe wrote, "What shall we think of the calculating machine of Mr. Babbage? What shall we think of an engine of wood and metal which can . . . render the exactitude of its operations mathematically certain through its power of correcting its possible errors?" Ralph Waldo Emerson had met Babbage in London and declared in 1870, "Steam is an apt scholar and a strong-shouldered fellow, but it has not yet done all its work."

It already walks about the field like a man, and will do anything required of it. It irrigates crops, and drags away a mountain. It must sew our shirts, it must drive our gigs; taught by Mr. Babbage, it must calculate interest and logarithms. . . . It is yet coming to render many higher services of a mechanico-intellectual kind.

Its wonders met disapproval, too. Some critics feared a rivalry between mechanism and mind. "What a satire is that machine on the mere mathematician!" said Oliver Wendell Holmes Sr. "A Frankenstein-monster, a thing without brains and without heart, too stupid to make a blunder; which turns out results like a corn-sheller, and never grows any wiser or better, though it grind a thousand bushels of them!" They all spoke as though the engine were real, but it never was. It remained poised before its own future.

Midway between his time and ours, the *Dictionary of National Biography* granted Charles Babbage a brief entry—almost entirely devoid of relevance or consequence:

mathematician and scientific mechanician; . . . obtained government grant for making a calculating machine . . . but the work of construction ceased, owning to disagreements with the engineer; offered the government an improved design, which was refused on grounds of expense; . . . Lucasian professor of mathematics, Cambridge, but delivered no lectures.

Babbage's interests, straying so far from mathematics, seeming so miscellaneous, did possess a common thread that neither he nor his contemporaries could perceive. His obsessions belonged to no category—that is, no category yet existing. His true subject was information: messaging, encoding, processing.

He took up two quirky and apparently unphilosophical challenges, which he himself noted had a deep connection one to the other: picking locks and deciphering codes. Deciphering, he said, was "one of the most fascinating of arts, and I fear I have wasted upon it more time than it deserves." To rationalize the process, he set out to perform a "complete analysis" of the English language. He created sets of special dictionaries: lists of the words of one letter, two letters, three letters, and so on; and lists of words alphabetized by their initial letter, second letter, third letter, and so on. With these at hand he designed methodologies for solving anagram puzzles and word squares.

In tree rings he saw nature encoding messages about the past. A profound lesson: that a tree records a whole complex of information in its solid substance. "Every shower that falls, every change of temperature that occurs, and every wind that blows, leaves on the vegetable world the traces of its passage; slight, indeed, and imperceptible, perhaps, to us, but not the less permanently recorded in the depths of those woody fabrics."

In London workshops he had observed speaking tubes, made of tin, "by which the directions of the superintendent are instantly conveyed to the remotest parts." He classified this technology as a contribution to the "economy of time" and suggested that no one had yet discovered a limit on the distance over which spoken messages might travel. He made a quick calculation: "Admitting it to be possible between London and Liverpool, about seventeen minutes would elapse before the words spoken at one end would reach the other extremity of the pipe." In the 1820s he had an idea for transmitting written messages, "enclosed in small cylinders along wires suspended from posts, and from towers, or from church steeples," and he built a working model in his London house. He grew obsessed with other variations on the theme of sending messages over the greatest possible distances. The post bag dispatched nightly from Bristol, he noted, weighed less than one hundred pounds. To send these messages 120 miles, "a coach and apparatus, weighing above thirty hundred weight, are put in motion, and also conveyed over the same space." What a waste! Suppose, instead, he suggested, post towns were linked by a series of high pillars erected every hundred feet or so. Steel wires would stretch from pillar to pillar. Within cities, church steeples might serve as the pillars. Tin cases with wheels would roll along the wires and carry batches of letters. The expense would be "comparatively trifling," he said, "nor is it impossible that the stretched wire might itself be available for a species of telegraphic communication yet more rapid."

During the Great Exhibition of 1851, when England showcased its industrial achievement in a Crystal Palace, Babbage placed an oil lamp with a moveable shutter in an upstairs window at Dorset Street to create an "occulting light" apparatus that blinked coded signals to passersby. He drew up a standardized system for lighthouses to use in sending numerical signals and posted twelve copies to, as he said, "the proper authorities of the great maritime countries." In the United States, the Congress approved $5,000 for a trial program of Babbage's system. He studied sun signals and "zenith-light signals" flashed by mirrors, and Greenwich time signals for transmission to mariners. For communicating between stranded ships and rescuers on shore, he proposed that all nations adopt a standard list of a hundred questions and answers, assigned numbers, "to be printed on cards, and nailed up on several parts of every vessel." Similar signals, he suggested, could help the military, the police, the railways, or even, "for various social purposes," neighbors in the country.

These purposes were far from obvious. "For what purposes will the electric telegraph become useful?" the king of Sardinia, Charles Albert, asked Babbage in 1840. Babbage searched his mind for an illustration, "and at last I pointed out the probability that, by means of the electric telegraphs, his Majesty's fleet might receive warning of coming storms. . . ."

> This led to a new theory of storms, about which the king was very curious. By degrees I endeavoured to make it clear. I cited, as an illustration, a storm which had occurred but a short time before I left England. The damage done by it at Liverpool was very great, and at Glasgow immense. . . . I added that if there had been electric communication between Genoa and a few other places the people of Glasgow might have had information of one of those storms twenty-four hours previously to its arrival.

As for the engine, it had to be forgotten before it was remembered. It had no obvious progeny. It rematerialized like buried treasure and inspired a sense of puzzled wonder. With the computer era in full swing, the historian Jenny Uglow felt in Babbage's engines "a different sense of anachronism." Such failed inventions, she wrote, contain "ideas that lie like yellowing blueprints in dark cupboards, to be stumbled on afresh by later generations."

Meant first to generate number tables, the engine in its modern form instead rendered number tables obsolete. Did Babbage anticipate that? He did wonder how the future would make use of his vision. He guessed that a half century would pass before anyone would try again to create a general-purpose computing machine. In fact, it took most of a century for the necessary substrate of technology to be laid down. "If, unwarned by my example," he wrote in 1864, "any man shall undertake and shall succeed in really constructing an engine embodying in itself the whole of the executive department of mathematical analysis upon different principles or by simpler mechanical means, I have no fear of leaving my reputation in his charge, for he alone will be fully able to appreciate the nature of my efforts and the value of their results."

As he looked to the future, he saw a special role for one truth above all: "the maxim, that knowledge is power." He understood that literally. Knowledge "is itself the generator of physical force," he declared. Science gave the world steam, and soon, he suspected, would turn to the less tangible power of electricity: "Already it has nearly chained the ethereal fluid." And he looked further:

> It is the science of *calculation*—which becomes continually more necessary at each step of our progress, and which must ultimately govern the whole of the applications of science to the arts of life.

Some years before his death, he told a friend that he would gladly give up whatever time he had left, if only he could be allowed to live for three days, five centuries in the future.

As for his young friend Ada, countess of Lovelace, she died many years before him—a protracted, torturous death from cancer of the womb, her agony barely lessened by laudanum and cannabis. For a long time her family kept from her the truth of her illness. In the end she knew she was dying. "They say that '*coming events cast their shadows before*,'" she wrote to her mother. "May they not sometimes cast their *lights* before?" They buried her next to her father.

She, too, had a last dream of the future: "my being *in time* an *Autocrat*, in my own way." She would have regiments, marshaled before her. The iron rulers of the earth would have to give way. And of what would her regiments consist? "I do not at present divulge. I have however the hope that they will be most *harmoniously* disciplined troops;—consisting of vast *numbers*, & marching in irresistible power to the sound of *Music*. Is not this very mysterious? Certainly *my* troops must consist of *numbers*, or they can have no existence at all. . . . But then, *what* are these *Numbers*? There is a riddle—"

A NERVOUS SYSTEM FOR THE EARTH

(What Can One Expect of a Few Wretched Wires?)

*Is it a fact—or have I dreamt it—that, by means of electricity, the
world of matter has become a great nerve, vibrating thousands of
miles in a breathless point of time? Rather, the round globe is a vast
head, a brain, instinct with intelligence! Or, shall we say, it is itself
a thought, nothing but thought, and no longer the substance which
we deemed it!*

—Nathaniel Hawthorne (1851)

THREE CLERKS IN A SMALL ROOM UPSTAIRS in the Ferry House of
Jersey City handled the entire telegraph traffic of the city of New York
in 1846 and did not have to work very hard. They administered one end
of a single pair of wires leading to Baltimore and Washington. Incoming
messages were written down by hand, relayed by ferry across the Hudson
River to the Liberty Street pier, and delivered to the first office of the
Magnetic Telegraph Company at 16 Wall Street.

In London, where the river caused less difficulty, capitalists formed
the Electric Telegraph Company and began to lay their first copper wires,
twisted into cables, covered with gutta-percha, and drawn through iron
pipes, mainly alongside new railroad tracks. To house the central office
the company rented Founders' Hall, Lothbury, opposite the Bank of
England, and advertised its presence by installing an electric clock—
modern and apt, for already railroad time was telegraphic time. By 1849

the telegraph office boasted eight instruments, operated day and night. Four hundred battery cells provided the power. "We see before us a stuccoed wall, ornamented with an electric illuminated clock," reported Andrew Wynter, a journalist, in 1854. "Who would think that behind this narrow forehead lay the great brain—if we may so term it—of the nervous system of Britain?" He was neither the first nor the last to liken the electric telegraph to biological wiring: comparing cables to nerves; the nation, or the whole earth, to the human body.

The analogy linked one perplexing phenomenon with another. Electricity was an enigma wrapped in mystery verging on magic, and no one understood nerves, either. Nerves were at least known to conduct a form of electricity and thus, perhaps, to serve as conduits for the brain's control of the body. Anatomists examining nerve fibers wondered whether they might be insulated with the body's own version of gutta-percha. Maybe nerves were not just *like* wires; maybe they *were* wires, carrying messages from the nether regions to the sensorium. Alfred Smee, in his 1849 *Elements of Electro-Biology*, likened the brain to a battery and the nerves to "bio-telegraphs." Like any overused metaphor, this one soon grew ripe for satire. A newspaper reporter in Menlo Park, discovering Thomas A. Edison in the grip of a head cold, wrote: "The doctor came and looked at him, explained the relations of the trigeminal nerves and their analogy to an electric telegraph with three wires, and observed incidentally that in facial neuralgia each tooth might be regarded as a telegraph station with an operator." When the telephone arrived, it reinforced the analogy. "The time is close at hand," declared *Scientific American* in 1880, "when the scattered members of civilized communities will be as closely united, so far as instant telephonic communication is concerned, as the various members of the body now are by the nervous system." Considering how speculative the analogy was, it turned out well. Nerves really do transmit messages, and the telegraph and telephone did begin to turn human society, for the first time, into something like a coherent organism.

In their earliest days these inventions inspired exhilaration without precedent in the annals of technology. The excitement passed from place to place in daily newspapers and monthly magazines and, more to the point, along the wires themselves. A new sense of futurity arose: a sense that the world was in a state of change, that life for one's children and grandchildren would be very different, all because of this force and its uses. "Electricity is the poetry of science," an American historian declared in 1852.

Not that anyone knew what electricity was. "An invisible, intangible, imponderable agent," said one authority. Everyone agreed that it involved a "peculiar condition" either of molecules or of the ether (itself a nebulous, and ultimately doomed, conception). Thomas Browne, in the seventeenth century, described electrical effluvia as "threads of syrup, which elongate and contract." In the eighteenth, the kite-flying Benjamin Franklin proved "the sameness of lightning with electricity"—identifying those fearsome bolts from the sky with the odd terrestrial sparks and currents. Franklin followed the Abbé Jean-Antoine Nollet, a natural philosopher and a bit of a showman, who said in 1748, "Electricity in our hands is the same as thunder in the hands of nature" and to prove it organized an experiment employing a Leyden jar and iron wire to send a shock through two hundred Carthusian monks arranged in a circle one mile around. From the monks' almost simultaneous hops, starts, jerks, and cries, onlookers judged that the message—its information content small but not zero—sped round the circle at fantastic speed.

Later, it was Michael Faraday in England who did more than anyone to turn electricity from magic to science, but even so, in 1854, when Faraday was at the height of his investigations, Dionysius Lardner, the scientific writer who so admired Babbage, could quite accurately declare, "The World of Science is not agreed as to the physical character of Electricity." Some believed it to be a fluid "lighter and more subtle" than any gas; others suspected a compound of two fluids "having antagonistic properties"; and still others thought electricity was not a fluid at all, but something analogous to sound: "a series of undulations or vibrations." *Harper's Magazine* warned that "current" was just a metaphor and added mysteriously, "We are not to conceive of the electricity as carrying the message that we write, but rather as enabling the operator at the other end of the line to write a similar one."

Whatever its nature, electricity was appreciated as a natural force placed under human control. A young New York newspaper, *The Times*, explained it by way of contrast with steam:

> Both of them are powerful and even formidable agents wrested from nature, by the skill and power of man. But electricity is by far the subtlest energy of the two. It is an original natural element, while steam is an artificial production. . . . Electricity combined with magnetism, is a more subjective agent, and when evolved for transmission is ready to go forth, a safe and expeditious messenger to the ends of the habitable globe.

Looking back, rhapsodists found the modern age foretold in a verse from the book of Job: "Canst thou send lightnings, that they may go and say unto thee, Here we are?"

But lightning did not *say* anything—it dazzled, cracked, and burned, but to convey a message would require some ingenuity. In human hands, electricity could hardly accomplish anything, at first. It could not make a light brighter than a spark. It was silent. But it could be sent along wires to great distances—this was discovered early—and it seemed to turn wires into faint magnets. Those wires could be long: no one had found any limit to the range of the electric current. It took no time at all to see what this meant for the ancient dream of long-distance communication. It meant sympathetic needles.

Practical problems had to be solved: making wires, insulating them, storing currents, measuring them. A whole realm of engineering had to be invented. Apart from the engineering was a separate problem: the problem of the message itself. This was more a logic puzzle than a technical one. It was a problem of crossing levels, from kinetics to meaning.

What form would the message take? How would the telegraph convert this fluid into words? By virtue of magnetism, the influence propagated across a distance could perform work upon physical objects, such as needles, or iron filings, or even small levers. People had different ideas: the electromagnet might sound an alarum-bell; might govern the motion of wheel-work; might turn a handle, which might carry a pencil (but nineteenth-century engineering was not up to robotic handwriting). Or the current might discharge a cannon. Imagine discharging a cannon by sending a signal from miles away! Would-be inventors naturally looked to previous communications technologies, but the precedents were mostly the wrong sort.

Before there were electric telegraphs, there were just telegraphs: *les télégraphes*, invented and named by Claude Chappe in France during the Revolution.* They were optical; a "telegraph" was a tower for sending signals to other towers in line of sight. The task was to devise a signaling system more efficient and flexible than, say, bonfires. Working with his messaging partner, his brother Ignace, Claude tried out a series of different schemes, evolving over a period of years.

The first was peculiar and ingenious. The Chappe brothers set a pair of pendulum clocks to beat in synchrony, each with its pointer turning around a dial at relatively high speed. They experimented with this in their hometown, Brûlon, about one hundred miles west of Paris. Ignace, the sender, would wait till the pointer reached an agreed number and at that instant signal by ringing a bell or firing a gun or, more often, banging upon a *casserole*. Upon hearing the sound, Claude, stationed a quarter mile away, would read the appropriate number off his own clock. He could convert number to words by looking them up in a prearranged

---

* But Count Miot de Melito claimed in his memoirs that Chappe submitted his idea to the War Office with the name *tachygraphe* ("swift writer") and that he, Miot, proposed *télégraphe* instead—which "has become, so to speak, a household word."

list. This notion of communication via synchronized clocks reappeared in the twentieth century, in physicists' thought experiments and in electronic devices, but in 1791 it led nowhere. One drawback was that the two stations had to be linked both by sight and by sound—and if they were, the clocks had little to add. Another was the problem of getting the clocks synchronized in the first place and keeping them synchronized. Ultimately, fast long-distance messaging was what made synchronization possible—not the reverse. The scheme collapsed under the weight of its own cleverness.

Meanwhile the Chappes managed to draw more of their brothers, Pierre and René, into the project, with a corps of municipal officers and royal notaries to bear witness. The next attempt dispensed with clockwork and sound. The Chappes constructed a large wooden frame with five sliding shutters, to be raised and lowered with pulleys. By using each possible combination, this "telegraph" could transmit an alphabet of thirty-two symbols—2^5, another binary code, though the details do not survive. Claude was pleading for money from the newly formed Legislative Assembly, so he tried this hopeful message from Brûlon: "*L'Assembleé nationale récompensera les experiences utiles au public*" ("The National Assembly will reward experiments useful to the public"). The eight words took 6 minutes, 20 seconds to transmit, and they failed to come true.

Revolutionary France was both a good and a bad place for modernistic experimentation. When Claude erected a prototype telegraph in the parc Saint-Fargeau, in the northeast of Paris, a suspicious mob burned it to the ground, fearful of secret messaging. Citizen Chappe continued looking for a technology as swift and reliable as that other new device, the guillotine. He designed an apparatus with a great crossbeam supporting two giant arms manipulated by ropes. Like so many early machines, this was somewhat anthropomorphic in form. The arms could take any of seven angles, at 45-degree increments (not eight, because one would leave the arm hidden behind the beam), and the beam, too, could rotate, all under the control of an operator down below, manipulating a system of cranks and pulleys. To perfect this complex mechanism Chappe enlisted Abraham-Louis Breguet, the well-known watchmaker.

As intricate as the control problem was, the question of devising a suitable code proved even more difficult. From a strictly mechanical point of view, the arms and the beam could take any angle at all—the possibilities were infinite—but for efficient signaling Chappe had to limit the possibilities. The fewer meaningful positions, the less likelihood of confusion. He chose only two for the crossbeam, on top of the seven for each arm, giving a symbol space of 98 possible arrangements (7 × 7 × 2). Rather than just use these for letters and numerals, Chappe set out to devise an elaborate code. Certain signals were reserved for error correction and control: start and stop, acknowledgment, delay, conflict (a tower could not send messages in both directions at once), and failure. Others were used in pairs, pointing the operator to pages and line numbers in special code books with more than eight thousand potential entries: words and syllables as well as proper names of people and places. All this remained a carefully guarded secret. After all, the messages were to be broadcast in the sky, for anyone to see. Chappe took it for granted that the telegraph network of which he dreamed would be a department of the state, government owned and operated. He saw it not as an instrument of knowledge or of riches, but as an instrument of power. "The day will come," he wrote, "when the Government will be able to achieve the grandest idea we can possibly have of power, by using the telegraph system in order to spread directly, every day, every hour, and simultaneously, its influence over the whole republic."

With the country at war and authority now residing with the National Convention, Chappe managed to gain the attention of some influential legislators. "Citizen Chappe offers an ingenious method to write in the air, using a small number of symbols, simply formed from straight line segments," reported one of them, Gilbert Romme, in 1793. He persuaded the Convention to appropriate six thousand francs for the construction of three telegraph towers in a line north of Paris, seven to

A CHAPPE TELEGRAPH

nine miles apart. The Chappe brothers moved rapidly now and by the end of summer arranged a triumphant demonstration for the watching deputies. The deputies liked what they saw: a means of receiving news from the military frontier and transmitting their orders and decrees. They gave Chappe a salary, the use of a government horse, and an official appointment to the post of *ingénieur télégraphe*. He began work on a line of stations 120 miles long, from the Louvre in Paris to Lille, on the northern border. In less than a year he had eighteen in operation, and the first messages arrived from Lille: happily, news of victories over the Prussians and Austrians. The Convention was ecstatic. One deputy named a pantheon of four great human inventions: printing, gunpowder, the compass, and "the language of telegraph signs." He was right to focus on the language. In terms of hardware—ropes, levers, and wooden beams—the Chappes had invented nothing new.

Construction began on stations in branches extending east to Strasbourg, west to Brest, and south to Lyon. When Napoleon Bonaparte seized power in 1799, he ordered a message sent in every direction—*"Paris est tranquille et les bons citoyens sont contents"* ("Paris is quiet and the good citizens are happy")—and soon commissioned a line of new stations all the way to Milan. The telegraph system was setting a new standard for speed of communication, since the only real competition was a rider on horseback. But speed could be measured in two ways: in terms of distance or in terms of symbols and words. Chappe once claimed that a signal could go from Toulon to Paris—a line of 120 stations across 475 miles—in just ten or twelve minutes. But he could not make that claim for a full message, even a relatively short one. Three signals per minute was the most that could be expected of even the fastest telegraph

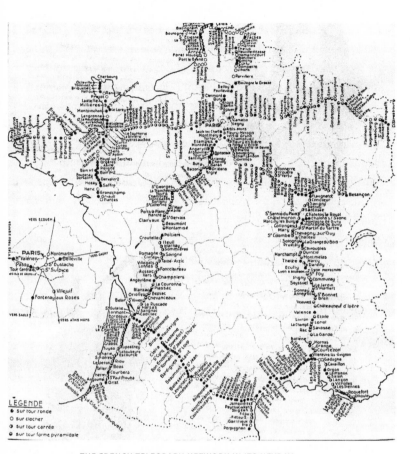

THE FRENCH TELEGRAPH NETWORK IN ITS HEYDAY

operator. The next operator in the chain, watching through a telescope, had to log each signal by hand in a notebook, reproduce it by turning his own cranks and pulleys, and watch to make sure it was received correctly by the next station. The signal chain was vulnerable and delicate: rain, fog, or an inattentive operator would break any message. When success rates were measured in the 1840s, only two out of three messages were found to arrive within a day during the warm months, and in winter the rate dropped to one in three. Coding and decoding took time, too, but only at the beginning and end of the line. Operators at intermediate stations were meant to relay signals without understanding them. Indeed, many *stationaires* were illiterate.

When messages did arrive, they could not always be trusted. Many relay stations meant many chances for error. Children everywhere know this, from playing the messaging game known in Britain as Chinese Whispers, in China as 以讹传讹, in Turkey as From Ear to Ear, and in the modern United States simply as Telephone. When his colleagues disregarded the problem of error correction, Ignace Chappe complained, "They have probably never performed experiments with more than two or three stations."

Today the old telegraphs are forgotten, but they were a sensation in their time. In London, a Drury Lane entertainer and songwriter named Charles Dibdin put the invention into a 1794 musical show and foresaw a marvelous future:

> If you'll only just promise you'll none of you laugh,
> I'll be after explaining the French telegraph!
> A machine that's endow'd with such wonderful pow'r,
> It writes, reads, and sends news fifty miles in an hour.
> . . .
> Oh! the dabblers in lott'ries will grow rich as Jews:
> 'Stead of flying of pigeons, to bring them the news,
> They'll a telegraph place upon Old Ormond Quay;
> Put another 'board ship, in the midst of the sea.
> . . .
> Adieu, penny-posts! mails and coaches, adieu;
> Your occupation's gone, 'tis all over wid you:
> In your place, telegraphs on our houses we'll see,
> To tell time, conduct lightning, dry shirts, and send news.

The telegraph towers spread across Europe and beyond, and their ruins dot the countrysides today. Telegraph Hill, Telegrafberget, Telegraphen-Berg are vestigial place names. Sweden, Denmark, and Belgium were early to develop systems on the French model. Germany soon followed. A line between Calcutta and Chunar began operating in 1823; between Alexandria and Cairo in 1824; and in Russia, Nicholas I organized 220 stations from Warsaw to St. Petersburg and Moscow. They held dominion over the world's communication and then, faster than they had arisen, went obsolete. Colonel Taliaferro Shaffner, a Kentucky inventor and historian, traveled to Russia in 1859 and was struck by the towers' height and their beauty, the care taken with their painting and landscaping with flowers, and by their sudden, universal death.

> These stations are now silent. No movements of the indicators are to be seen. They are still upon their high positions, fast yielding to the wasting hand of time. The electric wire, though less grand in its appearance, traverses the empire, and with burning flames inscribes in the distance the will of the emperor to sixty-six millions of human beings scattered over his wide-spread dominions.

In Shaffner's mind this was a one-way conversation. The sixty-six millions were not talking back to the emperor, nor to one another.

What was to be said, when writing in the air? Claude Chappe had proposed, "Anything that could be the subject of a correspondence." But his example—"Lukner has left for Mons to besiege that city, Bender is advancing for its defense"—made clear what he meant: dispatches of military and state import. Later Chappe proposed sending other types of information: shipping news, and financial quotations from bourses and stock exchanges. Napoleon would not allow it, though he did use the telegraph to proclaim the birth of his son, Napoleon II, in 1811. A communications infrastructure built with enormous government investment and capable of transmitting some hundreds of total words per day could hardly be used for private messaging. That was unimaginable—and when, in the next century, it became imaginable, some governments found it undesirable. No sooner did entrepreneurs begin to organize private telegraphy than France banned it outright: an 1837 law mandated

THE TELEGRAPH AT MONTMARTRE

imprisonment and fines for "anyone performing unauthorized transmissions of signals from one place to another, with the aid of telegraphic machines or by any other means." The idea of a global nervous system had to arise elsewhere. In the next year, 1838, the French authorities received a visit from an American with a proposal for a "telegraph" utilizing electrical wires: Samuel F. B. Morse. They turned him down flat. Compared to the majestic semaphore, electricity seemed gimcrack and insecure. No one could interfere with telegraph signals in the sky, but wire could be cut by saboteurs. Jules Guyot, a physician and scientist assigned to assess the technology, sniffed, "What can one expect of a few wretched wires?" What indeed.

The care and feeding of the delicate galvanic impulse presented a harsh set of technical challenges, and a different set appeared where electricity met language: where words had to be transmuted into a twinkling in the wire. The crossing point between electricity and language—also the interface between device and human—required new ingenuity. Many different schemes occurred to inventors. Virtually all were based in one way or another on the written alphabet, employing letters as an intermediate layer. This seemed so natural as to be not worth remarking. *Telegraph* meant "far writing," after all. So in 1774 Georges-Louis Le Sage of Geneva arranged twenty-four separate wires to designate twenty-four letters, each wire conveying just enough current to stir a piece of gold leaf or a pith ball suspended in a glass jar or "other bodies that can be as easily attracted, and are, at the same time, easily visible." That was too many wires to be practicable. A Frenchman named Lomond in 1787 ran a single wire across his apartment and claimed to be able to signal different letters by making a pith ball dance in different directions. "It appears that he has formed an alphabet of motions," reported a witness, but apparently only Lomond's wife could understand the code. In 1809 a German, Samuel Thomas von Sömmerring, made a bubble telegraph. Current passing through wires in a vessel of water produced bubbles of hydrogen; each wire, and thus each jet of bubbles, could indicate a single letter. While he was at it, von Sömmerring managed to make electricity ring a bell: he balanced a spoon in the water, upside down, so that enough bubbles would make it tilt, releasing a weight, driving a lever, and ringing the bell. "This secondary object, the alarum," he wrote in his diary, "cost me a great deal of reflection and many useless trials with wheelwork." Across the Atlantic, an American named Harrison Gray Dyer tried sending signals by making electric sparks form nitric acid that discolored litmus paper. He strung a wire on trees and stakes around a Long Island race track. The litmus paper had to be moved by hand.

Then came needles. The physicist André-Marie Ampère, a developer of the galvanometer, proposed using that as a signaling device: it was a needle deflected by electromagnetism—a compass pointing to a momentary artificial north. He, too, thought in terms of one needle for every letter. In Russia, Baron Pavel Schilling demonstrated a system with five needles and later reduced that to one: he assigned combinations of right and left signals to the letters and numerals. At Göttingen in 1833 the mathematician Carl Friedrich Gauss, working with a physicist, Wilhelm Weber, organized a similar scheme with one needle. The first deflection of the needle gave two possible signals, left or right. Two deflections combined gave four more possibilities (right + right, right + left, left + right, and left + left). Three deflections gave eight combinations, and four gave sixteen, for a total of thirty distinct signals. An operator would use pauses to separate the signals. Gauss and Weber organized their alphabet of deflections logically, beginning with the vowels and otherwise taking letters and digits in order:

right	= a
left	= e
right, right	= i
right, left	= o
left, right	= u
left, left	= b
right, right, right	= c (and k)
right, right, left	= d
etc.	

This scheme for encoding letters was binary, in a way. Each minimal unit, each little piece of signal, amounted to a choice between two possibilities, left or right. Each letter required a number of such choices, and that number was not predetermined. It could be one, as in right for a and left for e. It could be more, so the scheme was open-ended, allowing an alphabet of as many letters as needed. Gauss and Weber strung a doubled wire over a mile of houses and steeples between the Göttingen observatory and the physics institute. What they managed to say to each other has not been preserved.

Far away from these inventors' workrooms, the *telegraph* still meant towers, semaphores, shutters, and flags, but enthusiasm for new possibilities was beginning to build. Lecturing to the Boston Marine Society in 1833, a lawyer and philologist, John Pickering, declared, "It must be evident to the most common observer, that no means of conveying intelligence can ever be devised, that shall exceed or even equal the rapidity of the Telegraph, for, with the exception of the scarcely perceptible relay at each station, its rapidity may be compared with that of light itself." He was thinking particularly of the Telegraph on Central Wharf, a Chappe-like tower communicating shipping news with three other stations in a twelve-mile line across Boston Harbor. Meanwhile, dozens of young newspapers around the nation were modernistically calling themselves "*The Telegraph*." They, too, were in the far-writing business.

"Telegraphy is an element of power and order," Abraham Chappe had said, but the rising financial and mercantile classes were the next to grasp the value of information leaping across distance. Only two hundred miles separated the Stock Exchange on Threadneedle Street in London from the Bourse at the Palais Brongniart, but two hundred miles meant days. Fortunes could be made by bridging that gap. For speculators a private telegraph would be as useful as a time machine. The Rothschild banking family was using pigeons as postal carriers and, more reliably, a small fleet of boats to carry messengers across the Channel. The phenomenon of fast information from a distance, having been discovered, generated a cascade of excitement. Pickering in Boston did the math: "If there are now essential advantages to business in obtaining intelligence from New York in two days, or less, or at the rate of eight or ten miles an hour, any man can perceive that there may be a proportionate benefit, when we can transmit the same information for that distance by telegraph at the rate of four miles in a minute, or in the space of a single hour, from New York to Boston." The interest of governments in receiving military bulletins and projecting authority was surpassed by the desires of capitalists and newspapers, railroads and shipping companies. Still, in the sprawling United States, even the pressure of commerce was not enough to make optical telegraphy a reality. Only one prototype succeeded in linking two cities: New York and Philadelphia, in 1840. It transmitted stock prices and then lottery numbers and then was obsolete.

All the would-be inventors of the electrical telegraph—and there were many—worked from the same toolkit. They had their wires, and they had magnetic needles. They had batteries: galvanic cells, linked together, producing electricity from the reaction of metal strips immersed in acid baths. They did not have lights. They did not have motors. They had whatever mechanisms they could construct from wood and brass: pins, screws, wheels, springs, and levers. In the end they had the shared target at which they all aimed: the letters of the alphabet. (Edward Davy thought it was necessary to explain, in 1836, how and why the letters would suffice: "A single letter may be indicated at a time, each letter being taken down by the attendant as it arrives, so as to form words and sentences; but it will be easy to see that, from the infinite changes upon a number of letters, a great number of ordinary communications may be conveyed.") Along with this common stock list, in Vienna, Paris, London, Göttingen, St. Petersburg, and the United States, these pioneers shared a sense of their excited, competitive landscape, but no one knew clearly what anyone else was doing. They could not keep up with the relevant science; crucial advances in the science of electricity remained unknown to the people who most needed them. Every inventor ached to understand what happened to current flowing through wires of different lengths and thickness, and they continued to struggle for more than a decade after Georg Ohm, in Germany, worked out a precise mathematical theory for current, voltage, and resistance. Such news traveled slowly.

It was in this context that Samuel Morse and Alfred Vail, in the United

States, and, in England, William Cooke and Charles Wheatstone made the electric telegraph a reality and a business. In one way or another, all of them later claimed to have "invented" the telegraph, though none of them had done so—certainly not Morse. Their partnerships were destined to end in brutal, turbulent, and bitter patent disputes embroiling most of the leading electrical scientists on two continents. The trail of invention, leading through so many countries, had been poorly recorded and even more poorly communicated.

In England, Cooke was a young entrepreneur—he saw a prototype needle telegraph while traveling in Heidelberg—and Wheatstone a King's College, London, physicist with whom Cooke formed a partnership in 1837. Wheatstone had performed experiments on the velocity of sound and of electricity, and once again the real problem lay in connecting the physics with language. They consulted England's authority on electricity, Michael Faraday, and Peter Roget, author of a *Treatise on Electro-Magnetism* as well as the system of verbal classification he called the *Thesaurus*. The Cooke-Wheatstone telegraph went through a series of prototypes. One used six wires to form three circuits, each controlling a magnetic needle. "I worked out every possible permutation and practical combination of the signals given by the three needles, and I thus obtained an alphabet of twenty-six signals," noted Cooke, somewhat obscurely. There was also an alarm, in case the operator's attention wandered from the apparatus; Cooke said he had been inspired by the only mechanical device he knew well: a musical snuffbox. In the next version, a synchronized pair of rotating clockwork disks displayed the letters of the alphabet through a slot. More ingenious still, and just as awkward, was a five-needle design: twenty letters were arranged on a diamond-shaped grid and an operator, by depressing numbered buttons, would cause two of five needles to point, uniquely, to the desired letter. This Cooke-Wheatstone telegraph managed to do without *C, J, Q, U, X,* and *Z*. Their American competitor, Vail, later described the operation as follows:

Suppose the message to be sent from the Paddington station to the Slough station, is this, "We have met the enemy and they are ours." The operator at Paddington presses down the buttons, 11 and 18, for signalizing upon the dial of the Slough station, the letter W. The operator there, who is supposed to be constantly on watch, observes the two needles pointing at W. He writes it down, or calls it aloud, to another, who records it, taking, according to a calculation given in a recent account, two seconds at least for each signal.

Vail considered this inefficient. He was in a position to be smug.

As for Samuel Finley Breese Morse, his later recollections came in the context of controversy—what his son called "the wordy battles waged in the scientific world over the questions of priority, exclusive discovery or invention, indebtedness to others, and conscious or unconscious plagiarism." All these thrived on failures of communication and record-keeping. Educated at Yale College, the son of a Massachusetts preacher, Morse was an artist, not a scientist. In the 1820s and 1830s he spent much of his time traveling in England, France, Switzerland, and Italy to study painting. It was on one of these trips that he first heard about electric telegraphy or, in the terms of his memoirs, had his sudden insight: "like a flash of the subtle fluid which afterwards became his servant," as his son put it. Morse told a friend who was rooming with him in Paris: "The mails in our country are too slow; this French telegraph is better, and would do even better in our clear atmosphere than here, where half the time fogs obscure the skies. But this will not be fast enough—the lightning would serve us better." As he described his epiphany, it was an insight not about lightning but about signs: "It would not be difficult to construct *a system of signs* by which intelligence could be instantaneously transmitted."

TELEGRAPHIC WRITING BY MORSE'S FIRST INSTRUMENT

Morse had a great insight from which all the rest flowed. Knowing nothing about pith balls, bubbles, or litmus paper, he saw that a sign could be made from something simpler, more funda-mental, and less tangible—the most minimal event, the closing and opening

ALFRED VAIL'S TELEGRAPH "KEY"

of a circuit. Never mind needles. The electric current flowed and was interrupted, and the interruptions could be organized to create meaning. The idea was simple, but Morse's first devices were convoluted, involving clockwork, wooden pendulums, pencils, ribbons of paper, rollers, and cranks. Vail, an experienced machinist, cut all this back. For the sending end, Vail devised what became an iconic piece of user interface: a simple spring-loaded lever, with which an operator could control the circuit by the touch of a finger. First he called this lever a "correspondent"; then just a "key." Its simplicity made it at least an order of magnitude faster than the buttons and cranks employed by Wheatstone and Cooke. With the telegraph key, an operator could send signals—which were, after all, mere interruptions of the current—at a rate of hundreds per minute.

So at one end they had a lever, for closing and opening the circuit, and at the other end the current controlled an electromagnet. One of them, probably Vail, thought of putting the two together. The magnet could operate the lever. This combination (invented more or less simul-taneously by Joseph Henry at Princeton and Edward Davy in England) was named the "relay," from the word for a fresh horse that replaced an exhausted one. It removed the greatest obstacle standing in the way of long-distance electrical telegraphy: the weakening of currents as they passed through lengths of wire. A weakened current could still operate a relay, enabling a new circuit, powered by a new battery. The relay had greater potential than its inventors realized. Besides letting a signal prop-agate itself, a relay might reverse the signal. And relays might combine signals from more than one source. But that was for later.

The turning point came in 1844, both in England and the United States. Cooke and Wheatstone had their first line up and running along the railway from the Paddington station. Morse and Vail had theirs from Washington to the Pratt Street railway station in Baltimore, on wires wrapped in yarn and tar, suspended from twenty-foot wooden posts. The communications traffic was light at first, but Morse was able to report proudly to Congress that an instrument could transmit thirty characters per minute and that the lines had "remained undisturbed from the wantonness or evil disposition of any one." From the outset the communications content diverged sharply—comically—from the martial and official dispatches familiar to French telegraphists. In England the first messages recorded in the telegraph book at Paddington concerned lost luggage and retail transactions. "Send a messenger to Mr Harris, Duke-street, Manchester-square, and request him to send 6 lbs of white bait and 4 lbs of sausages by the 5.30 train to Mr Finch of Windsor; they must be sent by 5.30 down train, or not at all." At the stroke of the new year, the superintendent at Paddington sent salutations to his counterpart in Slough and received a reply that the wish was a half-minute early; midnight had not yet arrived there. That morning, a druggist in Slough named John Tawell poisoned his mistress, Sarah Hart, and ran for the train to Paddington. A telegraph message outraced him with his description ("in the garb of a kwaker, with a brown great coat on"—no Q's in the English system); he was captured in London and hanged in March. The drama filled the newspapers for months. It was later said of the telegraph wires, "Them's the cords that hung John Tawell." In April, a Captain Kennedy, at the South-Western Railway terminus, played a game of chess with a Mr. Staunton, at Gosport; it was reported that "in conveying the moves, the electricity travelled backward and forward during the game upwards of 10,000 miles." The newspapers loved that story, too—and, more and more, they valued any story revealing the marvels of the electric telegraph.

When the English and the American enterprises opened their doors to the general public, it was far from clear who, besides the police and the occasional chess player, would line up to pay the tariff. In Washington, where pricing began in 1845 at one-quarter cent per letter, total revenues for the first three months amounted to less than two hundred dollars. The next year, when a Morse line opened between New York and Phila-delphia, the traffic grew a little faster. "When you consider that business is extremely dull [and] we have not yet the confidence of the public," a company official wrote, "you will see we are all well satisfied with results so far." He predicted that revenues would soon rise to fifty dollars a day. Newspaper reporters caught on. In the fall of 1846 Alexander Jones sent his first story by wire from New York City to the Washington Union: an account of the launch of the USS *Albany* at the Brooklyn Navy Yard. In England a writer for *The Morning Chronicle* described the thrill of receiv-ing his first report across the Cooke-Wheatstone telegraph line,

> the first instalment of the intelligence by a sudden stir of the station-ary needle, and the shrill ring of the alarum. We looked delightedly into the taciturn face of our friend, the mystic dial, and pencilled down with rapidity in our note-book, what were his utterances some ninety miles off.

This was contagious. Some worried that the telegraph would be the death of newspapers, heretofore "the rapid and indispensable carrier of commercial, political and other intelligence," as an American journal-ist put it.

> For this purpose the newspapers will become emphatically useless. Antici-pated at every point by the lightning wings of the Telegraph, they can only deal in local "items" or abstract speculations. Their power to cre-ate sensations, even in election campaigns, will be greatly lessened—as the infallible Telegraph will contradict their falsehoods as fast as they can publish them.

Undaunted, newspapers could not wait to put the technology to work. Editors found that any dispatch seemed more urgent and thrilling with the label "Communicated by Electric Telegraph." Despite the expense—at first, typically, fifty cents for ten words—the newspapers became the telegraph services' most enthusiastic patrons. Within a few years, 120 provincial newspapers were getting reports from Parliament nightly. News bulletins from the Crimean War radiated from London to Liverpool, York, Manchester, Leeds, Bristol, Birmingham, and Hull. "Swifter than a rocket could fly the distance, like a rocket it bursts and is again carried by the diverging wires into a dozen neighbouring towns," one journalist noted. He saw dangers, though: "Intelligence, thus hastily gathered and transmitted, has also its drawbacks, and is not so trustworthy as the news which starts later and travels slower." The relationship between the telegraph and the newspaper was symbiotic. Positive feedback loops amplified the effect. Because the telegraph was an information technology, it served as an agent of its own ascendency.

The global expansion of the telegraph continued to surprise even its backers. When the first telegraph office opened in New York City on Wall Street, its biggest problem was the Hudson River. The Morse system ran a line sixty miles up the eastern side until it reached a point narrow enough to stretch a wire across. Within a few years, though, an insulated cable was laid under the harbor. Across the English Channel, a submarine cable twenty-five miles long made the connection between Dover and Calais in 1851. Soon after, a knowledgeable authority warned: "All idea of connecting Europe with America, by lines extending directly across the Atlantic, is utterly impracticable and absurd." That was in 1852; the impossible was accomplished by 1858, at which point Queen Victoria and President Buchanan exchanged pleasantries and *The New York Times* announced "a result so practical, yet so inconceivable . . . so full of hopeful prognostics for the future of mankind . . . one of the grand way-marks in the onward and upward march of the human intellect." What was the essence of the achievement? "The transmission of thought, the vital impulse of matter." The excitement was global but the effects were local. Fire brigades and police stations linked their communications. Proud shopkeepers advertised their ability to take telegraph orders.

Information that just two years earlier had taken days to arrive at its destination could now be there—anywhere—in seconds. This was not a doubling or tripling of transmission speed; it was a leap of many orders of magnitude. It was like the bursting of a dam whose presence had not even been known. The social consequences could not have been predicted, but some were observed and appreciated almost immediately. People's sense of the weather began to change—weather, that is, as a generalization, an abstraction. Simple weather reports began crossing the wires on behalf of corn speculators: *Derby, very dull; York, fine; Leeds, fine; Nottingham, no rain but dull and cold.* The very idea of a "weather report" was new. It required some approximation of instant knowledge of a distant place. The telegraph enabled people to think of weather as a widespread and interconnected affair, rather than an assortment of local surprises. "The phenomena of the atmosphere, the mysteries of meteors, the cause and effect of skiey combinations, are no longer matters of superstition or of panic to the husbandman, the sailor or the shepherd," noted an enthusiastic commentator in 1848:

> The telegraph comes in to tell him, for his every-day uses and observances, not only that "fair weather cometh out of the north," but the electric wire can tell him in a moment the character of the weather simultaneously in all quarters of our island. . . . In this manner, the telegraph may be made a vast national barometer, electricity becoming the handmaid of the mercury.

This was a transformative idea. In 1854 the government established a Meteorological Office in the Board of Trade. The department's chief, Admiral Robert FitzRoy, formerly a captain of HMS *Beagle*, moved into an office on King Street, furnished it with barometers, aneroids, and stormglasses, and dispatched observers equipped with the same instruments to ports all around the coastline. They telegraphed their cloud and wind reports twice daily. FitzRoy began issuing weather predictions, which he dubbed "forecasts," and in 1860 *The Times* began publishing these daily. Meteorologists began to understand that all great winds, when seen in the large, were circular, or at least "highly curved."

The most fundamental concepts were now in play as a consequence of instantaneous communication between widely separated points. Cultural observers began to say that the telegraph was "annihilating" time and space. It "enables us to send communications, by means of the mysterious fluid, with the quickness of thought, and to annihilate time as well as space," announced an American telegraph official in 1860. This was an exaggeration that soon became a cliché. The telegraph did seem to vitiate or curtail time in one specific sense: time as an obstacle or encumbrance to human intercourse. "For all practical purposes," one newspaper announced, "time, in the transit, may be regarded as entirely eliminated." It was the same with space. "Distance and time have been so changed in our imaginations," said Josiah Latimer Clark, an English telegraph engineer, "that the globe has been practically reduced in magnitude, and there can be no doubt that our conception of its dimensions is entirely different to that held by our forefathers."

Formerly all time was local: when the sun was highest, that was noon. Only a visionary (or an astronomer) would know that people in a different place lived by a different clock. Now time could be either local or standard, and the distinction baffled most people. The railroads required standard time, and the telegraph made it feasible. For standard time to prevail took decades; the process could only begin in the 1840s, when the Astronomer Royal arranged wires from the Observatory in Greenwich to the Electric Telegraph Company in Lothbury, intending to synchronize the clocks of the nation. Previously, the state of the art in time-signaling technology was a ball dropped from a mast atop the observatory dome. When faraway places were coordinated in time, they could finally measure their longitude precisely. The key to measuring longitude was knowing the time someplace else and the distance to that place. Ships therefore carried clocks, preserving time in imperfect mechanical capsules. Lieutenant Charles Wilkes of the U.S. Exploring Expedition used the first Morse line in 1844 to locate the Battle Monument in Baltimore at 1 minute, 34.868 seconds east of the Capitol in Washington.

Far from annihilating time, synchrony extended its dominion. The very idea of synchrony, and the awareness that the idea was new, made heads spin. *The New York Herald* declared:

> Professor Morse's telegraph is not only an era in the transmission of intelligence, but it has originated in the mind an entirely new class of ideas, a new species of consciousness. Never before was any one conscious that he knew with certainty what events were at that moment passing in a distant city—40, 100, or 500 miles off.

Imagine, continued this exhilarated writer, that it is *now* 11 o'clock. The telegraph relays what a legislator is *now* saying in Washington.

> It requires no small intellectual effort to realize that this is a fact that *now* is, and not one that *has been*.

This is a fact that *now* is.

History (and history making) changed, too. The telegraph caused the preservation of quantities of minutiae concerning everyday life. For a while, until it became impractical, the telegraph companies tried to maintain a record of every message. This was information storage without precedent. "Fancy some future Macaulay rummaging among such a store, and painting therefrom the salient features of the social and commercial life of England in the nineteenth century," mused one essayist. "What might not be gathered some day in the twenty-first century from a record of the correspondence of an entire people?" In 1845, after a year's experience with the line between Washington and Baltimore,

Alfred Vail attempted a catalogue of all the telegraph had conveyed thus far. "Much important information," he wrote,

> consisting of messages to and from merchants, members of Congress, officers of the government, banks, brokers, police officers; parties, who by agreement had met each other at the two stations, or had been sent for by one of the parties; items of news, election returns, announcement of deaths, inquiries respecting the health of families and individuals, the daily proceedings of the Senate and House of Representatives, orders for goods, inquiries respecting the sailing of vessels, proceedings of cases in the various courts, summoning of witnesses, messages in relation to special and express trains, invitations, the receipt of money at one station and its payment at the other, for persons requesting the transmission of funds from debtors, consultations of physicians . . .

These diverse items had never before been aggregated under one heading. The telegraph gave them their commonality. In patent applications and legal agreements, too, the inventors had reason to think about their topic in the broadest possible terms: e.g., the giving, printing, stamping, or otherwise transmitting of signals, or the sounding of alarms, or the communication of intelligence.

In this time of conceptual change, mental readjustments were needed to understand the telegraph itself. Confusion inspired anecdotes, which often turned on awkward new meanings of familiar terms: innocent words like *send*, and heavily laden ones, like *message*. There was the woman who brought a dish of sauerkraut into the telegraph office in Karlsruhe to be "sent" to her son in Rastatt. She had heard of soldiers being "sent" to the front by telegraph. There was the man who brought a "message" into the telegraph office in Bangor, Maine. The operator manipulated the telegraph key and then placed the paper on the hook. The customer complained that the message had not been sent, because he could still see it hanging on the hook. To *Harper's New Monthly Magazine*, which recounted this story in 1873, the point was that even the "intelligent and well-informed" continued to find these matters inscrutable:

> The difficulty of forming a clear conception of the subject is increased by the fact that while we have to deal with novel and strange facts, we have also to use old words in novel and inconsistent senses.

A message had seemed to be a physical object. That was always an illusion; now people needed consciously to divorce their conception of the message from the paper on which it was written. Scientists, *Harper's* explained, will say that the electric current "*carries* a message," but one must not imagine that anything—any *thing*—is transported. There is only "the action and reaction of an imponderable force, and the making of intelligible signals by its means at a distance." No wonder people were misled. "Such language the world must, perhaps for a long time to come, continue to employ."

The physical landscape changed, too. Wires everywhere made for strange ornamentation, on city streets and country roads. "Telegraphic companies are running a race to take possession of the air over our heads," wrote an English journalist, Andrew Wynter. "Look where we will aloft, we cannot avoid seeing either thick cables suspended by gossamer threads, or parallel lines of wire in immense numbers sweeping from post to post, fixed on the house-tops and suspended over long distances." They did not for some time fade into the background. People looked at the wires and thought of their great invisible cargo. "They string an instrument against the sky," said Robert Frost, "Wherein words whether beaten out or spoken / Will run as hushed as when they were a thought."

The wires resembled nothing in architecture and not much in nature. Writers seeking similes thought of spiders and their webs. They thought of labyrinths and mazes. And one more word seemed appropriate: the earth was being covered, people said, with an iron *net-work*. "A net-work of nerves of iron wire, strung with lightning, will ramify from the brain, New York, to the distant limbs and members," said the *New York Tribune*. "The whole net-work of wires," wrote *Harper's*, "all quivering from end to end with signals of human intelligence."

Wynter offered a prediction. "The time is not distant," he wrote, "when everybody will be able to talk with everybody without going out of the house." He meant "talk" metaphorically.

In more ways than one, using the telegraph meant writing in code.

The Morse system of dots and dashes was not called a code at first. It was just called an alphabet: "the Morse Telegraphic Alphabet," typically. But it was not an alphabet. It did not represent sounds by signs. The Morse scheme took the alphabet as a starting point and leveraged it, by substitution, replacing signs with new signs. It was a meta-alphabet, an alphabet once removed. This process—the transferring of meaning from one symbolic level to another—already had a place in mathematics. In a way it was the very essence of mathematics. Now it became a familiar part of the human toolkit. Entirely because of the telegraph, by the late nineteenth century people grew comfortable, or at least familiar, with the idea of codes: signs used for other signs, words used for other words. Movement from one symbolic level to another could be called *encoding*.

Two motivations went hand in glove: secrecy and brevity. Short messages saved money—that was simple. So powerful was that impulse that English prose style soon seemed to be feeling the effects. *Telegraphic* and *telegraphese* described the new way of writing. Flowers of rhetoric cost too much, and some regretted it. "The telegraphic style banishes all the forms of politeness," wrote Andrew Wynter:

> "May I ask you to do me the favour" is 6*d.* for a distance of 50 miles. How many of those fond adjectives therefore must our poor fellow relentlessly strike out to bring his billet down to a reasonable charge?

Almost immediately, newspaper reporters began to contrive methods for transmitting more information with fewer billable words. "We early invented a short-hand system, or cipher," boasted one, "so arranged, that the receipts of produce and the sales and prices of all leading articles of breadstuffs, provisions, &c., could be sent from Buffalo and Albany daily, in twenty words, for both cities, which, when written out, would make one hundred or more words." The telegraph companies tried to push back, on the grounds that private codes were gaming the system, but ciphers flourished. One typical system assigned dictionary words to whole phrases, organizing them semantically and alphabetically. For example, all words starting with B referred to the flour market: baal = "The transactions are smaller than yesterday"; babble = "There is a good business doing"; baby = "Western is firm, with moderate demand for home trade and export"; button = "market quiet and prices easier." It was necessary, of course, for sender and recipient to work from identical word lists. To the telegraph operators themselves, the encoded messages looked like nonsense, and that, in itself, proved an extra virtue.

As soon as people conceived of sending messages by telegraph, they worried that their communication was exposed to the world—at the very least, to the telegraph operators, unreliable strangers who could not help but read the words they fed through their devices. Compared to handwritten letters, folded and sealed with wax, the whole affair seemed public and insecure—the messages passing along those mysterious conduits, the electric wires. Vail himself wrote in 1847, "The great advantage which this telegraph possesses in transmitting messages with the rapidity of lightning, annihilating time and space, would perhaps be much lessened in its usefulness, could it not avail itself of the application of a secret alphabet." There were, he said, "systems"—

> by which a message may pass between two correspondents, through the medium of the telegraph, and yet the contents of that message remain a profound secret to all others, not excepting the operators of the telegraphic stations, through whose hands it must pass.

This was all very difficult. The telegraph served not just as a device but as a *medium*—a middle, intermediary state. The message passes through this medium. Distinct from the message, one must also consider the contents of that message. Even when the message must be exposed, the contents could be concealed. Vail explained what he meant by *secret alphabet:* an alphabet whose characters have been "transposed and interchanged."

Then the representative of *a*, in the *permanent* alphabet, may be represented by *y*, or *c*, or *x*, in the *secret* alphabet; and so of every other letter.

Thus, "The firm of G. Barlow & Co. have failed" becomes "Ejn stwz ys & qhwkyf p iy jhan shtknr." For less sensitive occasions, Vail proposed using abbreviated versions of common phrases. Instead of "give my love to," he suggested sending "gmlt." He offered a few more suggestions:

mhii	My health is improving
shf	Stocks have fallen
ymir	Your message is received
wmietg	When may I expect the goods?
wyegfef	Will you exchange gold for eastern funds?

All these systems required prearrangement between sender and recipient: the message was to be supplemented, or altered, by preexisting knowledge shared at both ends. A convenient repository for this knowledge was a code book, and when the first Morse line opened for business, one of its key investors and promoters, the Maine congressman Francis O. J. Smith, known as Fog, produced one: *The Secret Corresponding Vocabulary; adapted for use to Morse's Electro-Magnetic Telegraph: and also in conducting written correspondence, transmitted by the mails, or otherwise.* It was nothing but a numbered, alphabetical list of 56,000 English words, *Aaronic* to *zygodactylous,* plus instructions. "We will suppose the person writing, and the person written to, are each in possession of a copy of this work," Smith explained. "Instead of sending their communications in words, they send numbers only, or partly in numbers, and partly in words." For greater security, they might agree in advance to add or subtract a private number of their own choosing, or different numbers for alternate words. "A few such conventional substitutes," he promised, "will render the whole language a perfectly dead letter to all persons not conusant to the concerted arrangement."

Cryptographers had a mysterious history, their secrets handed along in clandestine manuscripts, like the alchemists'. Now code making emerged into the light, exposed in the hardware of commerce, inspiring the popular imagination. In the succeeding decades, many other schemes were contrived and published. They ranged from penny pamphlets to volumes of hundreds of pages of densely packed type. From London came E. Erskine Scott's *Three Letter Code for Condensed Telegraphic and Inscrutably Secret Messages and Correspondence.* Scott was an actuary and accountant and, like so many in the code business, a man evidently driven by an obsession with data. The telegraph opened up a world of possibilities for such people—cataloguers and taxonomists, wordsmiths and numerologists, completists of all kinds. Scott's chapters included not only a vocabulary of common words and two-word combinations, but also geographic names, Christian names, names of all shares quoted on the London Stock Exchange, all the days in the year, all regiments belonging to the British army, registries of shipping, and the names of all the peers of the realm. Organizing and numbering all this data made possible a form of compression, too. Shortening messages meant saving money. Customers found that the mere substitution of numbers for words helped little if at all: it cost just as much to send "3747" as "azotite." So code books became phrase books. Their object was a sort of packing of messages into capsules, impenetrable to prying eyes and suitable for efficient transmission. And of course, at the recipient's end, for unpacking.

An especially successful volume in the 1870s and '80s was *The A B C Universal Commercial Electric Telegraphic Code,* devised by William Clauson-Thue. He advertised his code to "financiers, merchants, shipowners, brokers, agents, &c." His motto: "Simplicity and

Economy Palpable, Secrecy Absolute." Clauson-Thue, another information obsessive, tried to arrange the entire language—or at least the language of commerce—into phrases, and to organize the phrases by keyword. The result is a peculiar lexicographic achievement, a window into a nation's economic life, and a trove of odd nuance and unwitting lyricism. For the keyword *panic* (assigned numbers 10054–10065), the inventory includes:

A great panic prevails in ————
The panic is settling down
The panic still continues
The worst of the panic is over
The panic may be considered over

For *rain* (11310–11330):

Cannot work on account of rain
The rain has done much good
The rain has done a great amount of damage
The rain is now pouring down in good earnest
Every prospect of the rain continuing
Rain much needed
Rain at times
Rainfall general

For *wreck* (15388–15403):

Parted from her anchors and became a wreck
I think it best to sell the wreck as it lies
Every attention will be made to save wreck
Must become a total wreck
Customs authorities have sold the wreck
Consul has engaged men to salve wreck

The world being full of things as well as words, he endeavored, too, to assign numbers to as many proper names as he could list: names of railways, banks, mines, commodities, vessels, ports, and stocks (British, colonial, and foreign).

As the telegraph networks spread under the oceans and across the globe, and international tariffs ran to many dollars per word, the code books thrived. Economy mattered even more than secrecy. The original trans-Atlantic rate was about one hundred dollars for a message—a "cable," as it was metonymically called—of ten words. For not much less, messages could travel between England and India, by way of Turkey or Persia and Russia. To save on the tariff, clever middlemen devised a practice called "packing." A packer would collect, say, four messages of five words each and bundle them into a fixed-price telegram of twenty words. The code books got bigger and they got smaller. In 1885 W. H. Beer & Company in Covent Garden published a popular *Pocket Telegraphic Code*, price one penny, containing "more than 300 one-word telegrams," neatly organized by subject matter. Essential subjects were Betting ("To what amount shall I back for you at present odds?"), Bootmaker ("These boots don't fit, send for them directly"), Washerwoman ("Call for the washing to-day"), and Weather—In Connexion with Voyages ("It is far too rough for you to cross to-day"). And a blank page was provided for "Secret Code. (Fill up by arrangement with friends.)" There were specialized codes for railways and yachts and trades from pharmacist to carpetmaker. The grandest and most expensive code books borrowed freely from one another. "It has been brought to the Author's knowledge that some persons have purchased a single copy of the 'A B C Telegraphic Code' for service in compiling Codes of their own," complained Clauson-Thue. "The Author would intimate that such an operation is a breach of the Copyright Act, and liable to become a matter of legal and unpleasant procedure." This was just bluster. By the turn of the century, the world's telegraphers, through the medium of International Telegraphic Conferences held in Berne and in London, had systematized codes with words in English, Dutch, French, German, Italian, Latin, Portuguese, and Spanish. The code books prospered and expanded through the first decades of the twentieth century and then vanished into obscurity.

Those who used the telegraph codes slowly discovered an unanticipated side effect of their efficiency and brevity. They were perilously vulnerable to the smallest errors. Because they lacked the natural redundancy of English prose—even the foreshortened prose of telegraphese—these cleverly encoded messages could be disrupted by a mistake in a single character. By a single dot, for that matter. For example, on June 16, 1887, a Philadelphia wool dealer named Frank Primrose telegraphed his agent in Kansas to say that he had *bought*—abbreviated in their agreed code as BAY—500,000 pounds of wool. When the message arrived, the key word had become BUY. The agent began buying wool, and before long the error cost Primrose $20,000, according to the lawsuit he filed against the Western Union Telegraph Company. The legal battle dragged on for six years, until finally the Supreme Court upheld the fine print on the back of the telegraph blank, which spelled out a procedure for protecting against errors:

> To guard against mistakes or delays, the sender of a message should order it REPEATED; that is telegraphed back to the originating office for comparison. . . . Said company shall not be liable for mistakes in . . . any UNREPEATED message . . . nor in any case for errors in cipher or obscure messages.

The telegraph company had to tolerate ciphers but did not have to like them. The court found in favor of Primrose in the amount of $1.15, the price of sending the telegram.

———

Secret writing was as old as writing. When writing began, it was in itself secret to all but the few. As the mystery dissolved, people found new ways to keep their words privileged and recondite. They rearranged words into anagrams. They reversed their script in the mirror. They invented ciphers.

In 1641, just as the English Civil War began, an anonymous little book catalogued the many known methods of what it called "cryptographia." These included special paper and ink: the juice of lemons or onions, raw egg, or "the distilled Juice of Gloworms," which might or might not be visible in the dark. Alternatively, writing could be obscured by substituting letters for other letters, or inventing new symbols, or writing from right to left, or "transposing each Letter, according to some unusual Order, as, suppose the first Letter should be at the latter End of the Line, the second at the Beginning, or the like." Or a message could be written across two lines:

    T e o l i r a e l m s f m s e s p l u o w e u t e l
    h s u d e s r a l o t a i h d, u p y s r e m s y i d

The Souldiers are allmost famished, supply us or wee must yeild.

Through transposition and substitution of letters, the Romans and the Jews had devised other methods, more intricate and thus more obscure.

This little book was titled *Mercury: or the Secret and Swift Messenger. Shewing, How a Man may with Privacy and Speed communicate his Thoughts to a Friend at any Distance*. The author eventually revealed himself as John Wilkins, a vicar and mathematician, later to become master of Trinity College, Cambridge, and a founder of the Royal Society. "He was a very ingenious man and had a very mechanical head," one contemporary said. "One of much and deep thinking, . . . lusty, strong grown, well set, broad shouldered." He was also thorough. If he could not mention every cipher tried since ancient times, he nonetheless included all that could have been known to a scholar in seventeenth-century England. He surveyed secret writing both as a primer and a compendium.

For Wilkins the issues of cryptography stood near the fundamental problem of communication. He considered writing and secret writing as essentially the same. Leaving secrecy aside, he expressed the problem this way: "How a Man may with the greatest Swiftness and Speed, discover his Intentions to one that is far distant from him." By *swiftness* and *speed* he meant, in 1641, something philosophical; the birth of Isaac Newton was a year away. "There is nothing (we say) so swift as Thought," he noted. Next to thought, the swiftest action seemed to be that of sight. As a clergyman, he observed that the swiftest motion of all must belong to angels and spirits. If only a man could send an angel on an errand, he could dispatch business at any distance. The rest of us, stuck with Organical Bodies, "cannot communicate their Thoughts so easie and immediate a way." No wonder, Wilkins wrote, that angels are called messengers.

As a mathematician, he considered the problem from another side. He set out to determine how a restricted set of symbols—perhaps just two, three, or five—might be made to stand for a whole alphabet. They would have to be used in combination. For example, a set of five symbols—a, b, c, d, e—used in pairs could replace an alphabet of twenty-five letters:

A	B	C	D	E	F	G	H	I	K	L	M	N	O	P	Q	R	S	T	V	W	X	Y	Z	&
aa	ab	ac	ad	ae	ba	bb	bc	bd	be	ca	cb	cc	cd	ce	da	db	dc	dd	de	ea	eb	ec	ed	ee

"According to which," wrote Wilkins, "these words, *I am betrayed*, may be thus described: *Bd aacb abaedddbaaecaead.*" So even a small symbol set could be arranged to express any message at all. However, with a small symbol set, a given message requires a longer string of characters—"more Labour and Time," he wrote. Wilkins did not explain that $25 = 5^2$, nor that three symbols taken in threes (aaa, aab, aac, . . .) produce twenty-seven possibilities because $3^3 = 27$. But he clearly understood the underlying mathematics. His last example was a binary code, awkward though this was to express in words:

Two Letters of the Alphabet being transposed through five Places, will yield thirty two Differences, and so will more than serve for the Four and twenty Letters; unto which they may be thus applied.

A	B	C	D	E	F	G
aaaaa	*aaaab*	*aaaba*	*aaabb*	*aabaa*	*aabab*	*aabba*
H	I	K	L	M	N	O
aabbb	*abaaa*	*abaab*	*ababa*	*ababb*	*abbaa*	*abbab*
P	Q	R	S	T	V	W
abbba	*abbbb*	*baaaa*	*baaab*	*baaba*	*baabb*	*babaa*
X	Y	Z				
babab	*babba*	*babbb*				

Two symbols. In groups of five. "Yield thirty two Differences."

That word, *differences*, must have struck Wilkins's readers (few though they were) as an odd choice. But it was deliberate and pregnant with meaning. Wilkins was reaching for a conception of information in its purest, most general form. Writing was only a special case: "For in the general we must note, That *whatever is capable of a competent Difference, perceptible to any Sense, may be a sufficient Means whereby to express the Cogitations.*" A difference could be "two Bells of different Notes"; or "any Object of Sight, whether Flame, Smoak, &c."; or trumpets, cannons, or drums. Any difference meant a binary choice. Any binary choice began the expressing of cogitations. Here, in this arcane and anonymous treatise of 1641, the essential idea of information theory poked to the surface of human thought, saw its shadow, and disappeared again for three hundred years.

The contribution of the dilettantes is what the historian of cryptography David Kahn calls the excited era triggered by the advent of the telegraph. A new public interest in ciphers arose just as the subject bloomed in certain intellectual circles. Ancient methods of secret writing appealed to an odd assortment of people, puzzle makers and game players, mathematically or poetically inclined. They analyzed ancient methods of secret writing and invented new ones. Theorists debated who should prevail, the best code maker or the best code breaker. The great American popularizer of cryptography was Edgar Allan Poe. In his fantastic tales and magazine essays he publicized the ancient art and boasted of his own skill as a practitioner. "We can scarcely imagine a time when there did not exist a necessity, or at least a desire," he wrote in *Graham's Magazine* in 1841, "of transmitting information from one individual to another, in such manner as to elude general comprehension." For Poe, code making was more than just a historical or technical enthusiasm; it was an obsession. It reflected his sense of how we communicate our selves to the world. Code makers and writers are trafficking in the same goods. "The soul is a cypher, in the sense of a cryptograph; and the shorter a cryptograph is, the more difficulty there is in comprehension," he wrote. Secrecy was in Poe's nature; he preferred mystery to transparency.

"Secret intercommunication must have existed almost contemporaneously with the invention of letters," he declared. This was for Poe a bridge between science and the occult, between the rational mind and the savant. To analyze cryptography—"a serious thing, as the means of imparting information"—required a special form of mental power, a penetrating mind, and might well be taught in academies. He said again and again that "a peculiar mental action is called into play." He published as challenges to his readers a series of substitution ciphers.

Along with Poe, Jules Verne and Honoré de Balzac also introduced ciphers into their fiction. In 1868, Lewis Carroll had a card printed on two sides with what he called "The Telegraph-Cipher," which employed a "key-alphabet" and a "message-alphabet," to be transposed according to a secret word agreed on by the correspondents and carried in their memories. But the most advanced cryptanalyst in Victorian England was Charles Babbage. The process of substituting symbols, crossing levels of meaning, lay near the heart of so many issues. And he enjoyed the challenge. "One of the most singular characteristics of the art of deciphering," he asserted, "is the strong conviction possessed by every person, even moderately acquainted with it, that he is able to construct a cipher which nobody else can decipher. I have also observed that the cleverer the person, the more intimate is his conviction." He believed that himself, at first, but later switched to the side of the code break-ers. He planned an authoritative work to be known as *The Philosophy of Decyphering* but never managed to complete it. He did solve, among others, a polyalphabetic cipher known as the Vigenère, *le chiffre indéchif-frable*, thought to be the most secure in Europe. As in his other work, he applied algebraic methods, expressing cryptanalysis in the form of equa-tions. Even so, he remained a dilettante and knew it.

When Babbage attacked cryptography with a calculus, he was employ-ing the same tools he had explored more conventionally in their home, mathematics, and less conventionally in the realm of machinery, where he created a symbolism for the moving parts of gears and levers and switches. Dionysius Lardner had said of the mechanical notation, "The various parts of the machinery being once expressed on paper by proper symbols, the enquirer dismisses altogether from his thoughts the mecha-nism itself and attends only to the symbols . . . an almost metaphysical system of abstract signs, by which the motion of the hand performs the office of the mind." Two younger Englishmen, Augustus De Morgan and George Boole, turned the same methodology to work on an even more abstract material: the propositions of logic. De Morgan was Bab-bage's friend and Ada Byron's tutor and a professor at University College, London. Boole was the son of a Lincolnshire cobbler and a lady's maid and became, by the 1840s, a professor at Queen's College, Cork. In 1847 they published separately and simultaneously books that amounted to the greatest milestone in the development of logic since Aristotle: Boole's *Mathematical Analysis of Logic, Being an Essay Towards a Calculus of Deductive Reasoning*, and De Morgan's *Formal Logic: or, the Calculus of Inference, Necessary and Probable*. The subject, esoteric as it was, had stagnated for centuries.

De Morgan knew more about the scholastic traditions of the subject, but Boole was the more original and free-thinking mathematician. By post, for years, they exchanged ideas about converting language, or truth, into algebraic symbols. *X* could mean "cow" and *Y* "horse." That might be one cow, or a member of the set of all cows. (The same?) In the algebraic fashion the symbols were to be manipulated. *XY* could be "name of everything which is *both X* and *Y* " while *X,Y* stood in for "name of everything which is either *X* or *Y*." Simple enough—but language is not simple and complications reared up. "Now some *Z*s are not *X*s, the *ZY*s," wrote De Morgan at one point. "But they are *nonexistent*. You may say that *nonexistents* are not *X*s. A nonexistent horse is not even a horse; and (*a fortiori?*) not a cow."

He added wistfully, "I do not despair of seeing you give meaning to this new kind of negative quantity." He did not post this and he did not throw it away.

Boole thought of his system as a mathematics without numbers. "It is simply a fact," he wrote, "that the ultimate laws of logic—those alone on which it is possible to construct a science of logic—are mathematical in their form and expression, although not belonging to the mathematics of quantity." The only numbers allowed, he proposed, were zero and one. It was all or nothing: "The respective interpretation of the symbols 0 and 1 in the system of logic are *Nothing* and *Universe*." Until now logic had belonged to philosophy. Boole was claiming possession on behalf of mathematics. In doing so, he devised a new form of encoding. Its code book paired two types of symbolism, each abstracted far from the world of things. On one side was a set of characters drawn from the formalism of mathematics: *p*'s and *q*'s, +'s and –'s, braces and brackets. On the other were operations, propositions, relations ordinarily expressed in the fuzzy and mutable speech of everyday life: words about truth and falsity, membership in classes, premises and conclusions. There were "particles": *if, either, or*. These were the elements of Boole's credo:

That Language is an instrument of human reason, and not merely a medium for the expression of thought.

The elements of which all language consists are signs or symbols.

Words are signs. Sometimes they are said to represent things; sometimes the operations by which the mind combines together the simple notions of things into complex conceptions.

Words . . . are not the only signs which we are capable of employing. Arbitrary marks, which speak only to the eye, and arbitrary sounds or actions . . . are equally of the nature of signs.

The encoding, the conversion from one modality to the other, served a purpose. In the case of Morse code, the purpose was to turn everyday language into a form suitable for near-instantaneous transmission across miles of copper wire. In the case of symbolic logic, the new form was suitable for manipulation by a calculus. The symbols were like little capsules, protecting their delicate cargo from the wind and fog of everyday communication. How much safer to write:

$$1 - x = y(1 - z) + z(1 - y) + (1 - y)(1 - z)$$

than the real-language proposition for which, in a typical Boolean example, it stood:

Unclean beasts are all which divide the hoof without chewing the cud, all which chew the cud without dividing the hoof, and all which neither divide the hoof nor chew the cud.

The safety came in no small part from draining the words of meaning. Signs and symbols were not just placeholders; they were operators, like the gears and levers in a machine. Language, after all, is an instrument.

It was seen distinctly now as an instrument with two separate functions: expression and thought. Thinking came first, or so people assumed. To Boole, logic *was* thought—polished and purified. He chose *The Laws of Thought* as the title for his 1854 masterwork. Not coincidentally, the telegraphists also felt they were generating insight into messaging within the brain. "A word is a tool for thinking, before the thinker uses it as a signal for communicating his thought," asserted an essayist in *Harper's New Monthly Magazine* in 1873.

> Perhaps the most extended and important influence which the telegraph is destined to exert upon the human mind is that which it will ultimately work out through its influence on language. . . . By the principle which Darwin describes as natural selection short words are gaining the advantage over long words, direct forms of expression are gaining the advantage over indirect, words of precise meaning the advantage of the ambiguous, and local idioms are everywhere at a disadvantage.

Boole's influence was subtle and slow. He corresponded only briefly with Babbage; they never met. One of his champions was Lewis Carroll, who, at the very end of his life, a quarter century after *Alice in Wonderland*, wrote two volumes of instruction, puzzles, diagrams, and exercises in symbolic logic. Although his symbolism was impeccable, his syllogisms ran toward whimsy:

(1) Babies are illogical;
(2) Nobody is despised who can manage a crocodile;
(3) Illogical persons are despised.
(Concl.) Babies cannot manage crocodiles.

The symbolic version—$b_1 d_0$ † ac_0 † $d'_1 c'_0$; $b\underline{d}$ † $\underline{d}' \underline{c}'$ † $a\underline{c}$ ¶ ba_0 † b_1, i.e. ¶ $b_1 a_0$—having been suitably drained of meaning, allowed the user to reach the desired conclusion without tripping over awkward intermediate propositions along the lines of "babies are despised."

As the century turned, Bertrand Russell paid George Boole an extraordinary compliment: "Pure mathematics was discovered by Boole, in a work which he called the *Laws of Thought*." It has been quoted often.

What makes the compliment extraordinary is the seldom quoted disparagement that follows on its heels:

> He was also mistaken in supposing that he was dealing with the laws of thought: the question how people actually think was quite irrelevant to him, and if his book had really contained the laws of thought, it was curious that no one should ever have thought in such a way before.

One might almost think Russell enjoyed paradoxes.

6 | NEW WIRES, NEW LOGIC

(No Other Thing Is More Enswathed in the Unknown)

> *The perfect symmetry of the whole apparatus—the wire in the middle, the two telephones at the ends of the wire, and the two gossips at the ends of the telephones—may be very fascinating to a mere mathematician.*
> —James Clerk Maxwell (1878)

A CURIOUS CHILD IN A COUNTRY TOWN in the 1920s might naturally form an interest in the sending of messages along wires, as Claude Shannon did in Gaylord, Michigan. He saw wires every day, fencing the pastures—double strands of steel, twisted and barbed, stretched from post to post. He scrounged what parts he could and jerry-rigged his own barbed-wire telegraph, tapping messages to another boy a half mile away. He used the code devised by Samuel F. B. Morse. That suited him. He liked the very idea of codes—not just secret codes, but codes in the more general sense, words or symbols standing in for other words or symbols. He was an inventive and playful spirit. The child stayed with the man. All his life, he played games and invented games. He was a gadgeteer. The grown-up Shannon juggled and devised theories about juggling. When researchers at the Massachusetts Institute of Technology or Bell Laboratories had to leap aside to let a unicycle pass, that was Shannon. He had more than his share of playfulness, and as a child he had a large portion of loneliness, too, which along with his tinkerer's ingenuity helped motivate his barbed-wire telegraph.

Gaylord amounted to little more than a few streets and stores interrupting the broad northern farmland of the Michigan peninsula. Here and onward across the plains and prairie to the Rocky Mountains barbed wire had spread like a vine, begetting industrial fortunes though it was not a particularly glamorous technology amid the excitement of what was already called the Age of Electricity. Beginning in 1874, when an Illinois farmer received U. S. Patent No. 157,124 for "a new and valuable Improvement in Wire-Fences," battles for ownership raged, ultimately reaching the Supreme Court, while the wire defined territory and closed the open range. At the peak, American farmers, ranchers, and railroads laid more than a million miles a year. Taken collectively the nation's fence wire formed no web or network, just a broken lattice. Its purpose had been to separate, not to connect. For electricity it made a poor conductor even in dry weather. But wire was wire, and Claude Shannon was not the first to see this wide-ranging lattice as a potential communications grid. Thousands of farmers in remote places had the same idea. Unwilling to wait for the telephone companies to venture out from the cities, rural folk formed barbed-wire telephone cooperatives. They replaced metal staples with insulated fasteners. They attached dry batteries and speaking tubes and added spare wire to bridge the gaps. In the summer of 1895 *The New York Times* reported: "There can be no doubt that many rough-and-ready utilizations of the telephone are now being made. For instance, a number of South Dakota farmers have helped themselves to a telephone system covering eight miles of wire by supplying themselves with transmitters and making connections with the barb wire which constitutes the fence in that part of the country." The reporter observed: "The idea is gaining ground that the day of cheap telephones for the million is at hand. Whether this impression is soundly based is an open question." Clearly people wanted the connections. Cattlemen who despised fences for making parcels of the free range now hooked up their speaking tubes to hear market quotations, weather reports, or just, crackling along the wires, the attenuated simulacrum of the human voice, a thrill in itself.

Three great waves of electrical communication crested in sequence: telegraphy, telephony, and radio. People began to feel that it was natural to possess machines dedicated to the sending and receiving of messages. These devices changed the topology—ripped the social fabric and reconnected it, added gateways and junctions where there had only been blank distance. Already at the turn of the twentieth century there was worry about unanticipated effects on social behavior. The superintendent of the line in Wisconsin fretted about young men and women "constantly sparking over the wire" between Eau Claire and Chippewa Falls. "This free use of the line for flirtation purposes has grown to an alarming extent," he wrote, "and if it is to go on somebody must pay for it." The Bell companies tried to discourage frivolous telephony, particularly by women and servants. A freer spirit prevailed at the farmer cooperatives, which avoided paying the telephone companies well into the 1920s. The Montana East Line Telephone Association—eight members—sent "up to the minute" news reports around its network, because the men also owned a radio. Children wanted to play this game, too.

Claude Elwood Shannon, born in 1916, was given the full name of his father, a self-made businessman—furniture, undertaking, and real estate—and probate judge, already well into middle age. Claude's grandfather, a farmer, had invented a machine for washing clothes: a waterproof tub, a wooden arm, and a plunger. Claude's mother, Mabel Catherine Wolf, daughter of German immigrants, worked as a language teacher and sometime principal of the high school. His older sister, Catherine Wolf Shannon (the parents doled out names parsimoniously), studied mathematics and regularly entertained Claude with puzzles. They lived on Center Street a few blocks north of Main Street. The town of Gaylord boasted barely three thousand souls, but this was enough to support a band with Teutonic uniforms and shiny instruments, and in grade school Claude played an E-flat alto horn broader than his chest. He had Erector Sets and books. He made model planes and earned money delivering telegrams for the local Western Union office. He solved cryptograms.

Left on his own, he read and reread books; a story he loved was Edgar Allan Poe's "The Gold-Bug," set on a remote southern island, featuring a peculiar William Legrand, a man with an "excitable brain" and "unusual powers of mind" but "subject to perverse moods of alternate enthusiasm and melancholy"—in other words, a version of his creator. Such ingenious protagonists were required by the times and duly conjured by Poe and other prescient writers, like Arthur Conan Doyle and H. G. Wells. The hero of "The Gold-Bug" finds buried treasure by deciphering a cryptograph written on parchment. Poe spells out the string of numerals and symbols ("rudely traced, in a red tint, between the death's-head and the goat")—53‡‡†305))6* ;4826)4‡.)4‡) ;806* ;48†8¶60))85;1‡(;:‡*8†83(88) 5*‡ ;46(;88*96*?;8) *‡(;485) ;5*†2:*‡(;4956*2(5*–4) 8§8* ;4069285) ;)6†8)4‡‡;1 (‡9;48081 ;8:8‡1 ;48†85;4)485†528806*81 (‡9:48;(88;4 (†?34;48)4‡;161;:188; ‡?;—and walks the reader through every twist of its construction and deconstruction. "Circumstances, and a certain bias of mind, have led me to take interest in such riddles," his dark hero proclaims, thrilling a reader who might have the same bias of mind. The solution leads to the gold, but no one cares about the gold, really. The thrill is in the code: mystery and transmutation.

Claude finished Gaylord High School in three years instead of four and went on in 1932 to the University of Michigan, where he studied electrical engineering and mathematics. Just before graduating, in 1936, he saw a postcard on a bulletin board advertising a graduate-student job at the Massachusetts Institute of Technology. Vannevar Bush, then the dean of engineering, was looking for a research assistant to run a new machine with a peculiar name: the Differential Analyzer. This was a 100-ton iron platform of rotating shafts and gears. In the newspapers it was being called a "mechanical brain" or "thinking machine"; a typical headline declared:

> "Thinking Machine" Does Higher Mathematics;
> Solves Equations That Take Humans Months

Charles Babbage's Difference Engine and Analytical Engine loomed as ancestral ghosts, but despite the echoes of nomenclature and the similarity of purpose, the Differential Analyzer owed virtually nothing to Babbage. Bush had barely heard of him. Bush, like Babbage, hated the numbing, wasteful labor of mere calculation. "A mathematician is not a man who can readily manipulate figures; often he cannot," Bush wrote. "He is primarily an individual who is skilled in the use of symbolic logic on a high plane, and especially he is a man of intuitive judgment."

MIT in the years after World War I was one of the nation's three focal points for the burgeoning practical science of electrical engineering, along with the Bell Telephone Laboratories and General Electric. It was also a place with a voracious need for the solving of equations—especially differential equations, and particularly differential equations of the second order. Differential equations express rates of change, as in ballistic projectiles and oscillating electric currents. Second-order differential equations concern rates of change in rates of change: from position to velocity to acceleration. They are hard to solve analytically, and they pop up everywhere. Bush designed his machine to handle this entire class of problems and thus the whole range of physical systems that generated them. Like Babbage's machines, it was essentially mechanical, though it used electric motors to drive the weighty apparatus and, as it evolved, more and more electromechanical switches to control it.

Unlike Babbage's machine, it did not manipulate numbers. It worked on quantities—generating curves, as Bush liked to say, to represent the future of a dynamical system. We would say now that it was analog rather than digital. Its wheels and disks were arranged to produce a physical analog of the differential equations. In a way it was a monstrous descendant of the planimeter, a little measuring contraption that translated the integration of curves into the motion of a wheel. Professors and students came to the Differential Analyzer as supplicants, and when it could solve their equations with 2 percent accuracy, the operator, Claude Shannon, was happy. In any case he was utterly captivated by this "computer," and

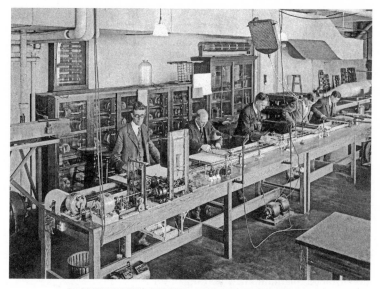

THE DIFFERENTIAL ANALYZER OF VANNEVAR BUSH AT MIT

not just by the grinding, rasping, room-filling analog part, but by the nearly silent (save for the occasional click and tap) electrical controls.

These were of two kinds: ordinary switches and the special switches called relays—the telegraph's progeny. The relay was an electrical switch controlled by electricity (a looping idea). For the telegraph, the point was to reach across long distances by making a chain. For Shannon, the point was not distance but control. A hundred relays, intricately inter-connected, switching on and off in particular sequence, coordinated the Differential Analyzer. The best experts on complex relay circuits were telephone engineers; relays controlled the routing of calls through telephone exchanges, as well as machinery on factory assembly lines. Relay circuitry was designed for each particular case. No one had thought to study the idea systematically, but Shannon was looking for a topic for his master's thesis, and he saw a possibility. In his last year of college he had taken a course in symbolic logic, and, when he tried to make an orderly list of the possible arrangements of switching circuits, he had a sudden feeling of déjà vu. In a deeply abstract way, these problems lined up. The peculiar artificial notation of symbolic logic, Boole's "algebra," could be used to describe circuits.

This was an odd connection to make. The worlds of electricity and logic seemed incongruous. Yet, as Shannon realized, what a relay passes onward from one circuit to the next is not really electricity but rather a fact: the fact of whether the circuit is open or closed. If a circuit is open, then a relay may cause the next circuit to open. But the reverse arrangement is also possible, the negative arrangement: when a circuit is open, a relay may cause the next circuit to close. It was clumsy to describe the possibilities with words; simpler to reduce them to symbols, and natural, for a mathematician, to manipulate the symbols in equations. (Charles Babbage had taken steps down the same path with his mechanical notation, though Shannon knew nothing of this.)

"A calculus is developed for manipulating these equations by simple mathematical processes"—with this clarion call, Shannon began his thesis in 1937. So far the equations just represented combinations of circuits. Then, "the calculus is shown to be exactly analogous to the calculus of propositions used in the symbolic study of logic." Like Boole, Shannon showed that he needed only two numbers for his equations: zero and one. Zero represented a closed circuit; one represented an open circuit. On or off. Yes or no. True or false. Shannon pursued the consequences. He began with simple cases: two-switch circuits, in series or in parallel. Circuits in series, he noted, corresponded to the logical connective *and;* whereas circuits in parallel had the effect of *or.* An operation of logic that could be matched electrically was negation, converting a value into its opposite. As in logic, he saw that circuitry could make "if . . . then" choices. Before he was done, he had analyzed "star" and "mesh" networks of increasing complexity, by setting down postulates and theorems to handle systems of simultaneous equations. He followed this tower of abstraction with practical examples—inventions, on paper, some practical and some just quirky.

He diagrammed the design of an electric combination lock, to be made from five push-button switches. He laid out a circuit that would "automatically add two numbers, using only relays and switches"; for convenience, he suggested arithmetic using base two. "It is possible to perform complex mathematical operations by means of relay circuits," he wrote. "In fact, any operation that can be completely described in a finite number of steps using the words *if, or, and,* etc. can be done automatically with relays." As a topic for a student in electrical engineering this was unheard of: a typical thesis concerned refinements to electric motors or transmission lines. There was no practical call for a machine that could solve puzzles of logic, but it pointed to the future. Logic circuits. Binary arithmetic. Here in a master's thesis by a research assistant was the essence of the computer revolution yet to come.

Shannon spent a summer working at the Bell Telephone Laboratories in New York City and then, at Vannevar Bush's suggestion, switched from electrical engineering to mathematics at MIT. Bush also suggested that he look into the possibility of applying an algebra of symbols—his "queer algebra"—to the nascent science of genetics, whose basic elements, genes and chromosomes, were just dimly understood. So Shannon began work on an ambitious doctoral dissertation to be called "An Algebra for Theoretical Genetics." Genes, as he noted, were a theoretical construct. They were thought to be carried in the rodlike bodies known as chromosomes, which could be seen under a microscope, but no one knew exactly how genes were structured or even if they were real. "Still," as Shannon noted, "it is possible for our purposes to act as though they were. . . . We shall speak therefore as though the genes actually exist and as though our simple representation of hereditary phenomena were really true, since so far as we are concerned, this might just as well be so." He devised an arrangement of letters and numbers to represent "genetic formulas" for an individual; for example, two chromosome pairs and four gene positions could be represented thus:

$$A_1\ B_2\ C_3\ D_5 \qquad E_4\ F_1\ G_6\ H_1$$
$$A_3\ B_1\ C_4\ D_3 \qquad E_4\ F_2\ G_6\ H_2$$

Then, the processes of genetic combination and cross-breeding could be predicted by a calculus of additions and multiplications. It was a sort of road map, far abstracted from the messy biological reality. He explained: "To non-mathematicians we point out that it is a commonplace of modern algebra for symbols to represent concepts other than numbers." The result was complex, original, and quite detached from anything people in the field were doing.* He never bothered to publish it.

Meanwhile, late in the winter of 1939, he wrote Bush a long letter about an idea closer to his heart:

> Off and on I have been working on an analysis of some of the fundamental properties of general systems for the transmission of intellegence, including telephony, radio, television, telegraphy, etc. Practically all systems of communication may be thrown into the following general form:

$$f_1(t) \quad \rightarrow \quad \boxed{T} \quad \rightarrow \quad F(t) \quad \rightarrow \quad \boxed{R} \quad \rightarrow \quad f_2(t)$$

T and R were a transmitter and a receiver. They mediated three "functions of time," $f(t)$: the "intelligence to be transmitted," the signal, and the final output, which, of course, was meant to be as nearly identical to the input as possible. ("In an ideal system it would be an exact replica.") The problem, as Shannon saw it, was that real systems always suffer *distortion*—a term for which he proposed to give a rigorous definition in mathematical form. There was also *noise* ("e.g., static"). Shannon told Bush he was trying to prove some theorems. Also, and not incidentally, he was working on a machine for performing symbolic mathematical

---

* In an evaluation forty years later the geneticist James F. Crow wrote: "It seems to have been written in complete isolation from the population genetics community. . . . [Shannon] discovered principles that were rediscovered later. . . . My regret is that [it] did not become widely known in 1940. It would have changed the history of the subject substantially, I think."

operations, to do the work of the Differential Analyzer and more, entirely by means of electric circuits. He had far to go. "Although I have made some progress in various outskirts of the problem I am still pretty much in the woods, as far as actual results are concerned," he said.

I have a set of circuits drawn up which actually will perform symbolic differentiation and integration on most functions, but the method is not quite general or natural enough to be perfectly satisfactory. Some of the general philosophy underlying the machine seems to evade me completely.

He was painfully thin, almost gaunt. His ears stuck out a little from his close-trimmed wavy hair. In the fall of 1939, at a party in the Garden Street apartment he shared with two roommates, he was standing shyly in his own doorway, a jazz record playing on the phonograph, when a young woman started throwing popcorn at him. She was Norma Levor, an adventurous nineteen-year-old Radcliffe student from New York. She had left school to live in Paris that summer but returned when Nazi Germany invaded Poland; even at home, the looming war had begun to unsettle people's lives. Claude struck her as dark in temperament and sparkling in intellect. They began to see each other every day; he wrote sonnets for her, uncapitalized in the style of E. E. Cummings. She loved the way he loved words, the way he said *Boooooooolean* algebra. By January they were married (Boston judge, no ceremony), and she followed him to Princeton, where he had received a postdoctoral fellowship.

The invention of writing had catalyzed logic, by making it possible to reason about reasoning—to hold a train of thought up before the eyes for examination—and now, all these centuries later, logic was reanimated with the invention of machinery that could work upon symbols. In logic and mathematics, the highest forms of reasoning, everything seemed to be coming together.

By melding logic and mathematics in a system of axioms, signs, formulas, and proofs, philosophers seemed within reach of a kind of perfection—a rigorous, formal certainty. This was the goal of Bertrand Russell and Alfred North Whitehead, the giants of English rationalism, who published their great work in three volumes from 1910 to 1913. Their title, *Principia Mathematica*, grandly echoed Isaac Newton; their ambition was nothing less than the perfection of all mathematics. This was finally possible, they claimed, through the instrument of symbolic logic, with its obsidian signs and implacable rules. Their mission was to prove every mathematical fact. The process of proof, when carried out properly, should be mechanical. In contrast to words, *symbolism* (they declared) enables "perfectly precise expression." This elusive quarry had been pursued by Boole, and before him, Babbage, and long before either of them, Leibniz, all believing that the perfection of reasoning could come with the perfect encoding of thought. Leibniz could only imagine it: "a certain script of language," he wrote in 1678, "that perfectly represents the relationships between our thoughts." With such encoding, logical falsehoods would be instantly exposed.

> The characters would be quite different from what has been imagined up to now. . . . The characters of this script should serve invention and judgment as in algebra and arithmetic. . . . It will be impossible to write, using these characters, chimerical notions [*chimères*].

Russell and Whitehead explained that symbolism suits the "highly abstract processes and ideas" used in logic, with its trains of reasoning. Ordinary language works better for the muck and mire of the ordinary world. A statement like *a whale is big* uses simple words to express "a complicated fact," they observed, whereas *one is a number* "leads, in language, to an intolerable prolixity." Understanding whales, and bigness, requires knowledge and experience of real things, but to manage *1*, and *number*, and all their associated arithmetical operations, when properly expressed in desiccated symbols, should be automatic.

They had noticed some bumps along the way, though—some of the *chimères* that should have been impossible. "A very large part of the labour," they said in their preface, "has been expended on the contradictions and paradoxes which have infected logic." "Infected" was a strong word but barely adequate to express the agony of the paradoxes. They were a cancer.

Some had been known since ancient times:

Epimenides the Cretan said that all Cretans were liars, and all other statements made by Cretans were certainly lies. Was this a lie?

A cleaner formulation of Epimenides' paradox—cleaner because one need not worry about Cretans and their attributes—is the liar's paradox: *This statement is false.* The statement cannot be true, because then it is false. It cannot be false, because then it becomes true. It is neither true nor false, or it is both at once. But the discovery of this twisting, backfiring, mind-bending circularity does not bring life or language crashing to a halt—one grasps the idea and moves on—because life and language lack the perfection, the absolutes, that give them force. In real life, all Cretans cannot be liars. Even liars often tell the truth. The pain begins only with the attempt to build an airtight vessel. Russell and Whitehead aimed for perfection—for proof—otherwise the enterprise had little point. The more rigorously they built, the more paradoxes they found. "It was in the air," Douglas Hofstadter has written, "that truly peculiar things could happen when modern cousins of various ancient paradoxes cropped up inside the rigorously logical world of numbers, . . . a pristine paradise in which no one had dreamt paradox might arise."

One was Berry's paradox, first suggested to Russell by G. G. Berry, a librarian at the Bodleian. It has to do with counting the syllables needed to specify each integer. Generally, of course, the larger the number the more syllables are required. In English, the smallest integer requiring two syllables is seven. The smallest requiring three syllables is eleven. The number 121 seems to require six syllables ("one hundred twenty-one"), but actually four will do the job, with some cleverness: "eleven squared." Still, even with cleverness, there are only a finite number of possible syllables and therefore a finite number of names, and, as Russell put it, "Hence the names of some integers must consist of at least nineteen syllables, and among these there must be a least. Hence *the least integer not nameable in fewer than nineteen syllables* must denote a definite integer."* Now comes the paradox. This phrase, *the least integer not nameable in fewer than nineteen syllables*, contains only eighteen syllables. So the least integer not nameable in fewer than nineteen syllables has just been named in fewer than nineteen syllables.

Another paradox of Russell's is the Barber paradox. The barber is the man (let us say) who shaves all the men, and only those, who do not shave themselves. Does the barber shave himself? If he does he does not, and if he does not he does. Few people are troubled by such puzzles, because in real life the barber does as he likes and the world goes on. We tend to feel, as Russell put it, that "the whole form of words is just a noise without meaning." But the paradox cannot be dismissed so easily when a mathematician examines the subject known as set theory, or the theory of classes. Sets are groups of things—for example, integers. The set 0, 2, 4 has integers as its members. A set can also be a member of other sets. For example, the set 0, 2, 4 belongs to the set of *sets of integers* and the set of *sets with three members* but not the set of *sets of prime numbers*. So Russell defined a certain set this way:

S is the set of all sets that are not members of themselves.

This version is known as Russell's paradox. It cannot be dismissed as noise.

To eliminate Russell's paradox Russell took drastic measures. The enabling factor seemed to be the peculiar recursion within the offending

* In standard English, as Russell noted, it is one hundred and eleven thousand seven hundred and seventy-seven.

statement: the idea of sets belonging to sets. Recursion was the oxygen feeding the flame. In the same way, the liar paradox relies on statements about statements. "This statement is false" is meta-language: language about language. Russell's paradoxical set relies on a meta-set: a set of sets. So the problem was a crossing of levels, or, as Russell termed it, a mixing of types. His solution: declare it illegal, taboo, out of bounds. No mixing different levels of abstraction. No self-reference; no self-containment. The rules of symbolism in *Principia Mathematica* would not allow the reaching-back-around, snake-eating-its-tail feedback loop that seemed to turn on the possibility of self-contradiction. This was his firewall.

Enter Kurt Gödel.

He was born in 1906 in Brno, at the center of the Czech province of Moravia. He studied physics at the University of Vienna, seventy-five miles south, and as a twenty-year-old became part of the Vienna Circle, a group of philosophers and mathematicians who met regularly in smoky coffeehouses like the Café Josephinum and the Café Reichsrat to propound logic and realism as a bulwark against metaphysics—by which they meant spiritualism, phenomenology, irrationality. Gödel talked to them about the New Logic (this term was in the air) and before long about metamathematics—*der Metamathematik.* Metamathematics was not to mathematics what metaphysics was to physics. It was mathematics once removed—mathematics about mathematics—a formal system "looked at from the outside" (*"äußerlich betrachtet"*). He was about to make the most important statement, prove the most important theorem about knowledge in the twentieth century. He was going to kill Russell's dream of a perfect logical system. He was going to show that the paradoxes were not excrescences; they were fundamental.

Gödel praised the Russell and Whitehead project before he buried it: mathematical logic was, he wrote, "a science prior to all others, which contains the ideas and principles underlying all sciences." *Principia Mathematica*, the great opus, embodied a formal system that had become, in its brief lifetime, so comprehensive and so dominant that

Gödel referred to it in shorthand: PM. By PM he meant the system, as opposed to the book. In PM, mathematics had been contained—a ship in a bottle, no longer buffeted and turned by the vast unruly seas. By 1930, when mathematicians proved something, they did it according to PM. In PM, as Gödel said, "one can prove any theorem using nothing but a few mechanical rules."

Any theorem: for the system was, or claimed to be, complete. *Mechanical* rules: for the logic operated inexorably, with no room for varying human interpretation. Its symbols were drained of meaning. Anyone could verify a proof step by step, by following the rules, without understanding it. Calling this quality mechanical invoked the dreams of Charles Babbage and Ada Lovelace, machines grinding through numbers, and numbers standing for anything at all.

Amid the doomed culture of 1930 Vienna, listening to his new friends debate the New Logic, his manner reticent, his eyes magnified by black-framed round spectacles, the twenty-four-year-old Gödel believed in the perfection of the bottle that was PM but doubted whether mathematics could truly be contained. This slight young man turned his doubt into a great and horrifying discovery. He found that lurking within PM— and within any consistent system of logic—there must be monsters of a kind hitherto unconceived: statements that can never be proved, and yet can never be disproved. There must be *truths*, that is, that cannot be proved—and Gödel could prove it.

He accomplished this with iron rigor disguised as sleight of hand. He employed the formal rules of PM and, as he employed them, also approached them metamathematically—viewed them, that is, from the outside. As he explained, all the symbols of PM—numbers, operations of arithmetic, logical connectors, and punctuation—constituted a limited alphabet. Every statement or formula of PM was written in this alphabet. Likewise every proof comprised a finite sequence of formulas—just a longer passage written in the same alphabet. This is where metamathematics came in. Metamathematically, Gödel pointed out, one sign is as good as another; the choice of a particular alphabet is arbitrary. One could use the traditional assortment of numerals and glyphs (from arithmetic: +, −, =, ×; from logic: ¬, ∨, ⊃, ∃), or one could use letters, or one could use dots and dashes. It was a matter of encoding, slipping from one symbol set to another.

Gödel proposed to use numbers for all his signs. Numbers were his alphabet. And because numbers can be combined using arithmetic, any sequence of numbers amounts to one (possibly very large) number. So every statement, every formula of PM can be expressed as a single number, and so can every proof. Gödel outlined a rigorous scheme for doing the encoding—an algorithm, mechanical, just rules to follow, no intelligence necessary. It works forward and backward: given any formula, following the rules generates one number, and given any number, following the rules produces the corresponding formula.

Not every number translates into a correct formula, however. Some numbers decode back into gibberish, or formulas that are false within the rules of the system. The string of symbols "0 0 0 = = =" does not make a formula at all, though it translates to some number. The statement "0 = 1" is a recognizable formula, but it is false. The formula "$0 + x = x + 0$" is true, and it is provable.

This last quality—the property of *being provable according to PM*—was not meant to be expressible in the language of PM. It seems to be a statement from outside the system, a metamathematical statement. But Gödel's encoding reeled it in. In the framework he constructed, the natural numbers led a double life, as numbers and also as statements. A statement could assert that a given number is *even*, or *prime*, or *a perfect square*, and a statement could also assert that a given number is *a provable formula*. Given the number 1,044,045,317,700, for example, one could make various statements and test their truth or falsity: this number is even, it is not a prime, it is not a perfect square, it is greater than 5, it is divisible by 121, and (when decoded according to the official rules) it is a provable formula.

Gödel laid all this out in a little paper in 1931. Making his proof watertight required complex logic, but the basic argument was simple and elegant. Gödel showed how to construct a formula that said *A certain number, x, is not provable*. That was easy: there were infinitely many such formulas. He then demonstrated that, in at least some cases, the number *x* would happen to represent that very formula. This was just the looping self-reference that Russell had tried to forbid in the rules of PM—

This statement is not provable

—and now Gödel showed that such statements must exist anyway. The Liar returned, and it could not be locked out by changing the rules. As Gödel explained (in one of history's most pregnant footnotes),

> Contrary to appearances, such a proposition involves no faulty circularity, for it only asserts that a certain well-defined formula . . . is unprovable. Only subsequently (and so to speak by chance) does it turn out that this formula is precisely the one by which the proposition itself was expressed.

Within PM, and within any consistent logical system capable of elementary arithmetic, there must always be such accursed statements, true but unprovable. Thus Gödel showed that a consistent formal system must be incomplete; no complete and consistent system can exist.

The paradoxes were back, nor were they mere quirks. Now they struck at the core of the enterprise. It was, as Gödel said afterward, an "amazing fact"—"that our logical intuitions (i.e., intuitions concerning such notions as: truth, concept, being, class, etc.) are self-contradictory." It was, as Douglas Hofstadter says, "a sudden thunderbolt from the bluest of skies," its power arising not from the edifice it struck down but the lesson it contained about numbers, about symbolism, about encoding:

> Gödel's conclusion sprang not from a weakness in PM but from a strength. That strength is the fact that numbers are so flexible or "chameleonic"

that their patterns can mimic patterns of reasoning. . . . PM's *expressive power* is what gives rise to its incompleteness.

The long-sought universal language, the *characteristica universalis* Leibniz had pretended to invent, had been there all along, in the numbers. Numbers could encode all of reasoning. They could represent any form of knowledge.

Gödel's first public mention of his discovery, on the third and last day of a philosophical conference in Königsberg in 1930, drew no response; only one person seems to have heard him at all, a Hungarian named Neumann János. This young mathematician was in the process of moving to the United States, where he would soon and for the rest of his life be called John von Neumann. He understood Gödel's import at once; it stunned him, but he studied it and was persuaded. No sooner did Gödel's paper appear than von Neumann was presenting it to the mathematics colloquium at Princeton. Incompleteness was real. It meant that mathematics could never be proved free of self-contradiction. And "the important point," von Neumann said, "is that this is not a philosophical principle or a plausible intellectual attitude, but the result of a rigorous mathematical proof of an extremely sophisticated kind." Either you believed in mathematics or you did not.

Bertrand Russell (who, of course, *did*) had moved on to more gentle sorts of philosophy. Much later, as an old man, he admitted that Gödel had troubled him: "It made me glad that I was no longer working at mathematical logic. If a given set of axioms leads to a contradiction, it is clear that at least one of the axioms must be false." On the other hand, Vienna's most famous philosopher, Ludwig Wittgenstein (who, fundamentally, *did not*), dismissed the incompleteness theorem as trickery ("*Kunststücken*") and boasted that rather than try to refute it, he would simply pass it by:

Mathematics cannot be incomplete; any more than a *sense* can be incomplete. Whatever I can understand, I must completely understand.

Gödel's retort took care of them both. "Russell evidently misinterprets my result; however, he does so in a very interesting manner," he wrote. "In contradistinction Wittgenstein . . . advances a completely trivial and uninteresting misinterpretation."

In 1933 the newly formed Institute for Advanced Study, with John von Neumann and Albert Einstein among its first faculty members, invited Gödel to Princeton for the year. He crossed the Atlantic several more times that decade, as fascism rose and the brief glory of Vienna began to fade. Gödel, ignorant of politics and naïve about history, suffered depressive breakdowns and bouts of hypochondria that forced him into sanatoria. Princeton beckoned but Gödel vacillated. He stayed in Vienna in 1938, through the *Anschluss*, as the Vienna Circle ceased to be, its members murdered or exiled, and even in 1939, when Hitler's army occupied his native Czechoslovakia. He was not a Jew, but mathematics was *verjudet* enough. He finally managed to leave in January 1940 by way of the Trans-Siberian Railway, Japan, and a ship to San Francisco. His name was recoded by the telephone company as "K. Goedel" when he arrived in Princeton, this time to stay.

Claude Shannon had also arrived at the Institute for Advanced Study, to spend a postdoctoral year. He found it a lonely place, occupying a new red-brick building with clocktower and cupola framed by elms on a former farm a mile from Princeton University. The first of its fifteen or so professors was Einstein, whose office was at the back of the first floor; Shannon seldom laid eyes on him. Gödel, who had arrived in March, hardly spoke to anyone but Einstein. Shannon's nominal supervisor was Hermann Weyl, another German exile, the most formidable mathematical theorist of the new quantum mechanics. Weyl was only mildly interested in Shannon's thesis on genetics—"your biomathematical problems"—but thought Shannon might find common ground with the institute's other great young mathematician, von Neumann. Mostly Shannon stayed moodily in his room in Palmer Square. His twenty-year-old wife, having left Radcliffe to be with him, found it increasingly grim, staying home while Claude played clarinet accompaniment to his Bix Beiderbecke record on the phonograph. Norma thought he was depressed and wanted him to see a psychiatrist. Meeting Einstein was nice, but the thrill wore off. Their marriage was over; she was gone by the end of the year.

Nor could Shannon stay in Princeton. He wanted to pursue the transmission of intelligence, a notion poorly defined and yet more pragmatic than the heady theoretical physics that dominated the institute's agenda. Furthermore, war approached. Research agendas were changing everywhere. Vannevar Bush was now heading the National Defense Research Committee, which assigned Shannon "Project 7": the mathematics of fire-control mechanisms for antiaircraft guns—"the job," as the NDRC reported dryly, "of applying corrections to the gun control so that the shell and the target will arrive at the same position at the same time." Airplanes had suddenly rendered obsolete almost all the mathematics used in ballistics: for the first time, the targets were moving at speeds not much less than the missiles themselves. The problem was complex and critical, on ships and on land. London was organizing batteries of heavy guns firing 3.7-inch shells. Aiming projectiles at fast-moving aircraft needed either intuition and luck or a vast amount of implicit computation by gears and linkages and servos. Shannon analyzed physical problems as well as computational problems: the machinery had to track rapid paths in three dimensions, with shafts and gears controlled by rate finders and integrators. An antiaircraft gun in itself behaved as a dynamical system, subject to "backlash" and oscillations that might or might not be predictable. (Where the differential equations were nonlinear, Shannon made little headway and knew it.)

He had spent two of his summers working for Bell Telephone Laboratories in New York; its mathematics department was also taking on the fire-control project and asked Shannon to join. This was work for which the Differential Analyzer had prepared him well. An automated antiaircraft gun was already an analog computer: it had to convert what were, in effect, second-order differential equations into mechanical motion; it had to accept input from rangefinder sightings or new, experimental radar; and it had to smooth and filter this data, to compensate for errors.

At Bell Labs, the last part of this problem looked familiar. It resembled an issue that plagued communication by telephone. The noisy data looked like static on the line. "There is an obvious analogy," Shannon and his colleagues reported, "between the problem of smoothing the data to eliminate or reduce the effect of tracking errors and the problem of separating a signal from interfering noise in communications systems." The data constituted a signal; the whole problem was "a special case of the transmission, manipulation, and utilization of intelligence." Their specialty, at Bell Labs.

Transformative as the telegraph had been, miraculous as the wireless radio now seemed, electrical communication now meant the telephone. The "electrical speaking telephone" first appeared in the United States with the establishment of a few experimental circuits in the 1870s. By the turn of the century, the telephone industry surpassed the telegraph by every measure—number of messages, miles of wire, capital invested—and telephone usage was doubling every few years. There was no mystery about why: anyone could use a telephone. The only skills required were talking and listening: no writing, no codes, no keypads. Everyone responded to the sound of the human voice; it conveyed not just words but feeling.

The advantages were obvious—but not to everyone. Elisha Gray, a telegraph man who came close to trumping Alexander Graham Bell as inventor of the telephone, told his own patent lawyer in 1875 that the work was hardly worthwhile: "Bell seems to be spending all his energies in [the] talking telegraph. While this is very interesting scientifically it has no commercial value at present, for they can do much more business over a line by methods already in use." Three years later, when Theodore

N. Vail quit the Post Office Department to become the first general manager (and only salaried officer) of the new Bell Telephone Company, the assistant postmaster general wrote angrily, "I can scarce believe that a man of your sound judgment . . . should throw it up for a d——d old Yankee notion (a piece of wire with two Texan steer horns attached to the ends, with an arrangement to make the concern blate like a calf) called a telephone!" The next year, in England, the chief engineer of the General Post Office, William Preece, reported to Parliament: "I fancy the descriptions we get of its use in America are a little exaggerated, though there are conditions in America which necessitate the use of such instruments more than here. Here we have a superabundance of messengers, errand boys and things of that kind. . . . I have one in my office, but more for show. If I want to send a message—I use a sounder or employ a boy to take it."

One reason for these misguesses was just the usual failure of imagination in the face of a radically new technology. The telegraph lay in plain view, but its lessons did not extrapolate well to this new device. The telegraph demanded literacy; the telephone embraced orality. A message sent by telegraph had first to be written, encoded, and tapped out by a trained intermediary. To employ the telephone, one just talked. A child could use it. For that very reason it seemed like a toy. In fact, it seemed like a familiar toy, made from tin cylinders and string. The telephone left no permanent record. *The Telephone* had no future as a newspaper name. Business people thought it unserious. Where the telegraph dealt in facts and numbers, the telephone appealed to emotions.

The new Bell company had little trouble turning this into a selling point. Its promoters liked to quote Pliny, "The living voice is that which sways the soul," and Thomas Middleton, "How sweetly sounds the voice of a good woman." On the other hand, there was anxiety about the notion of capturing and reifying voices—the phonograph, too, had just arrived. As one commentator said, "No matter to what extent a man may close his doors and windows, and hermetically seal his key-holes and furnace-registers with towels and blankets, whatever he may say, either to himself or a companion, will be overheard." Voices, hitherto, had remained mostly private.

The new contraption had to be explained, and generally this began by comparison to telegraphy. There were a transmitter and receiver, and wires connected them, and *something* was carried along the wire in the form of electricity. In the case of the telephone, that thing was sound, simply converted from waves of pressure in the air to waves of electric current. One advantage was apparent: the telephone would surely be useful to musicians. Bell himself, traveling around the country as impresario for the new technology, encouraged this way of thinking, giving demonstrations in concert halls, where full orchestras and choruses played "America" and "Auld Lang Syne" into his gadgetry. He encouraged people to think of the telephone as a broadcasting device, to send music and sermons across long distances, bringing the concert hall and the church into the living room. Newspapers and commentators mostly went along. That is what comes of analyzing a technology in the abstract. As soon as people laid their hands on telephones, they worked out what to do. They talked.

In a lecture at Cambridge, the physicist James Clerk Maxwell offered a scientific description of the telephone conversation: "The speaker talks to the transmitter at one end of the line, and at the other end of the line the listener puts his ear to the receiver, and hears what the speaker said. The process in its two extreme states is so exactly similar to the old-fashioned method of speaking and hearing that no preparatory practice is required on the part of either operator." He, too, had noticed its ease of use.

So by 1880, four years after Bell conveyed the words "Mr. Watson, come here, I want to see you," and three years after the first pair of telephones rented for twenty dollars, more than sixty thousand telephones were in use in the United States. The first customers bought pairs of telephones for communication point to point: between a factory and its business office, for example. Queen Victoria installed one at Windsor Castle and one at Buckingham Palace (fabricated in ivory; a gift from the savvy Bell). The topology changed when the number of sets reachable by other sets passed a critical threshold, and that happened surprisingly soon. Then community networks arose, their multiple connections managed through a new apparatus called a switch-board.

The initial phase of ignorance and skepticism passed in an eyeblink. The second phase of amusement and entertainment did not last much longer. Businesses quickly forgot their qualms about the device's seriousness. Anyone could be a telephone prophet now—some of the same predictions had already been heard in regard to the telegraph—but the most prescient comments came from those who focused on the exponential power of interconnection. *Scientific American* assessed "The Future of the Telephone" as early as 1880 and emphasized the forming of "little clusters of telephonic communicants." The larger the network and the more diverse its interests, the greater its potential would be.

> What the telegraph accomplished in years the telephone has done in months. One year it was a scientific toy, with infinite possibilities of practical use; the next it was the basis of a system of communication the most rapidly expanding, intricate, and convenient that the world has known. . . . Soon it will be the rule and not the exception for business houses, indeed for the dwellings of well-to-do people as well, to be interlocked by means of telephone exchange, not merely in our cities, but in all outlying regions. The result can be nothing less than a new organization of society—a state of things in which every individual, however secluded, will have at call every other individual in the community, to the saving of no end of social and business complications, of needless goings to and fro, of disappointments, delays, and a countless host of those great and little evils and annoyances.
>
> The time is close at hand when the scattered members of civilized communities will be as closely united, so far as instant telephonic communication is concerned, as the various members of the body now are by the nervous system.

The scattered members using telephones numbered half a million by 1890; by 1914, 10 million. The telephone was already thought, correctly, to be responsible for rapid industrial progress. The case could hardly be overstated. The areas depending on "instantaneous communication across space" were listed by the United States Commerce Department in 1907: "agriculture, mining, commerce, manufacturing, transportation, and, in fact, all the various branches of production and distribution of natural and artificial resources." Not to mention "cobblers, cleaners of clothing, and even laundresses." In other words, every cog in the engine of the economy. "Existence of telephone traffic is essentially an indication that time is being saved," the department commented. It observed changes in the structure of life and society that would still seem new a century later: "The last few years have seen such an extension of telephone lines through the various summer-resort districts of the country that it has become practicable for business men to leave their offices for several days at a time, and yet keep in close touch with their offices." In 1908 John J. Carty, who became the first head of the Bell Laboratories, offered an information-based analysis to show how the telephone had shaped the New York skyline—arguing that the telephone, as much as the elevator, had made skyscrapers possible.

> It may sound ridiculous to say that Bell and his successors were the fathers of modern commercial architecture—of the skyscraper. But wait a minute. Take the Singer Building, the Flatiron Building, the Broad Exchange, the Trinity, or any of the giant office buildings. How many messages do you suppose go in and out of those buildings every day? Suppose there was no telephone and every message had to be carried by a personal messenger? How much room do you think the necessary elevators would leave for offices? Such structures would be an economic impossibility.

To enable the fast expansion of this extraordinary network, the telephone demanded new technologies and new science. They were broadly of two kinds. One had to do with electricity itself: measuring electrical quantities; controlling the electromagnetic wave, as it was now understood—its modulation in amplitude and in frequency. Maxwell had established in the 1860s that electrical pulses and magnetism and light itself were all manifestations of a single force: "affectations of the same substance," light being one more case of "an electromagnetic disturbance propagated through the field according to electromagnetic laws." These were the laws that electrical engineers now had to apply, unifying telephone and radio among other technologies. Even the telegraph employed a simple kind of amplitude modulation, in which only two values mattered, a maximum for "on" and a minimum for "off." To convey sound required far stronger current, far more delicately controlled. The engineers had to understand feedback: a coupling of the output of a power amplifier, such as a telephone mouthpiece, with its input. They had to design vacuum-tube repeaters to carry the electric current over long distance, making possible the first transcontinental line in 1914, between New York and San Francisco, 3,400 miles of wire suspended from 130,000 poles. The engineers also discovered how to modulate individual currents so as to combine them in a single channel—multiplexing—without losing their identity. By 1918 they could get four conversations into a single pair of wires. But it was not *currents* that preserved identity. Before the engineers quite realized it, they were thinking in terms of the transmission of a *signal*, an abstract entity, quite distinct from the electrical waves in which it was embodied.

A second, less well defined sort of science concerned the organizing of connections—switching, numbering, and logic. This branch descended from Bell's original realization, dating from 1877, that telephones need not be sold in pairs; that each individual telephone could be connected to many other telephones, not by direct wires but through a central "exchange." George W. Coy, a telegraph man in New Haven, Connecticut, built the first "switch-board" there, complete with "switch-pins" and "switch-plugs" made from carriage bolts and wire from discarded bustles. He patented it and served as the world's first telephone "operator." With all the making and breaking of connections, switch-pins wore out quickly. An early improvement was a hinged two-inch plate resembling a jackknife: the "jack-knife switch," or as it was soon called, the "jack." In January 1878, Coy's switchboard could manage two simultaneous conversations between any of the exchange's twenty-one customers. In

February, Coy published a list of subscribers: himself and some friends; several physicians and dentists; the post office, police station, and mercantile club; and some meat and fish markets. This has been called the world's first telephone directory, but it was hardly that: one page, not alphabetized, and no numbers associated with the names. The telephone number had yet to be invented.

That innovation came the next year in Lowell, Massachusetts, where by the end of 1879 four operators managed the connections among two hundred subscribers by shouting to one another across the switchboard room. An epidemic of measles broke out, and Dr. Moses Greeley Parker worried that if the operators succumbed, they would be hard to replace. He suggested identifying each telephone by number. He also suggested listing the numbers in an alphabetical directory of subscribers. These ideas could not be patented and arose again in telephone exchanges across the country, where the burgeoning networks were creating clusters of data in need of organization. Telephone books soon represented the most comprehensive listings of, and directories to, human populations ever attempted. (They became the thickest and densest of the world's books—four volumes for London; a 2,600-page tome for Chicago—and seemed a permanent, indispensable part of the world's information ecology until, suddenly, they were not. They went obsolete, effectively, at the turn of the twenty-first century. American telephone companies were officially phasing them out by 2010; in New York, the end of automatic delivery of telephone directories was estimated to save 5,000 tons of paper.)

At first, customers resented the impersonality of telephone numbers, and engineers doubted whether people could remember a number of more than four or five digits. The Bell Company finally had to insist. The first telephone operators were teenage boys, cheaply hired from the ranks of telegraph messengers, but exchanges everywhere discovered that boys were wild, given to clowning and practical jokes, and more likely to be found wrestling on the floor than sitting on stools to perform the exacting, repetitive work of a switchboard operator. A new source of cheap labor was available, and by 1881 virtually every telephone operator was a woman. In Cincinnati, for example, W. H. Eckert reported hiring sixty-six "young ladies" who were "very much superior" to boys: "They are steadier, do not drink beer, and are always on hand." He hardly needed to add that the company could pay a woman as little as or less than a teenage boy. It was challenging work that soon required training. Operators had to be quick in distinguishing many different voices and accents, had to maintain a polite equilibrium in the face of impatience and rudeness, as they engaged in long hours of athletic upper-body exercise, wearing headsets like harnesses. Some men thought this was good for them. "The action of stretching her arms up above her head, and to the right and left of her, develops her chest and arms," said *Every Woman's Encyclopedia*, "and turns thin and weedy girls into strong ones. There are no anaemic, unhealthy looking girls in the operating rooms." Along with another new technology, the typewriter, the telephone switchboard catalyzed the introduction of women into the white-collar workforce, but battalions of human operators could not sustain a network on the scale now arising. Switching would have to be performed automatically.

This meant a mechanical linkage to take from callers not just the sound of their voice but also a number—identifying a person, or at least another telephone. The challenge of converting a number into electrical form still required ingenuity: first push buttons were tried, then an awkward-seeming rotary dial, with ten finger positions for the decimal digits, sending pulses down the line. Then the coded pulses served as an agent of control at the central exchange, where another mechanism selected from an array of circuits and set up a connection. Altogether this made for an unprecedented degree of complexity in the translations between human and machine, number and circuitry. The point was not lost on the company, which liked to promote its automatic switches as "electrical brains." Having borrowed from telegraphy the electro-mechanical relay—using one circuit to control another—the telephone companies had reduced it in size and weight to less than four ounces and now manufactured several million each year.

"The telephone remains the acme of electrical marvels," wrote a historian in 1910—a historian of the telephone, already. "No other thing does so much with so little energy. No other thing is more enswathed in the unknown." New York City had several hundred thousand listed telephone customers, and *Scribner's Magazine* highlighted this astounding fact: "Any two of that large number can, in five seconds, be placed in communication with each other, so well has engineering science kept pace with public needs." To make the connections, the switchboard had grown to a monster of 2 million soldered parts, 4,000 miles of wire, and 15,000 signal lamps. By 1925, when an assortment of telephone research groups were formally organized into the Bell Telephone Laboratories, a mechanical "line finder" with a capacity of 400 lines was replacing 22-point electromechanical rotary switches. The American Telephone & Telegraph Company was consolidating its monopoly. Engineers struggled to minimize the hunt time. At first, long-distance calling required reaching a second, "toll" operator and waiting for a call back; soon the interconnection of local exchanges would have to allow for automatic dialing. The complexities multiplied. Bell Labs needed mathematicians.

What began as the Mathematics Consulting Department grew into a center of practical mathematics like none other. It was not like the prestigious citadels, Harvard and Princeton. To the academic world it was barely visible. Its first director, Thornton C. Fry, enjoyed the tension between theory and practice—the clashing cultures. "For the mathematician, an argument is either perfect in every detail or else it is wrong," he wrote in 1941. "He calls this 'rigorous thinking.' The typical engineer calls it 'hair-splitting.'"

> The mathematician also tends to idealize any situation with which he is confronted. His gases are "ideal," his conductors "perfect," his surfaces "smooth." He calls this "getting down to essentials." The engineer is likely to dub it "ignoring the facts."

In other words, the mathematicians and engineers could not do without each other. Every electrical engineer could now handle the basic analysis of waves treated as sinusoidal signals. But new difficulties arose in understanding the action of networks; network theorems were devised to handle these mathematically. Mathematicians applied queuing theory to usage conflicts; developed graphs and trees to manage issues of intercity trunks and lines; and used combinatorial analysis to break down telephone probability problems.

Then there was noise. This did not at first (to Alexander Graham Bell, for example) seem like a problem for theorists. It was just there, always crowding the line—pops, hisses, crackles interfering with, or degrading, the voice that had entered the mouthpiece. It plagued radio, too. At best it stayed in the background and people hardly noticed; at worst the weedy profusion spurred the customers' imaginations:

> There was sputtering and bubbling, jerking and rasping, whistling and screaming. There was the rustling of leaves, the croaking of frogs, the hissing of steam, and the flapping of birds' wings. There were clicks from telegraph wires, scraps of talk from other telephones, curious little squeals that were unlike any known sound. . . . The night was noisier than the day, and at the ghostly hour of midnight, for what strange reasons no one knows, the babel was at its height.

But engineers could now *see* the noise on their oscilloscopes, interfering with and degrading their clean waveforms, and naturally they wanted to measure it, even if there was something quixotic about measuring a nuisance so random and ghostly. There was a way, in fact, and Albert Einstein had shown what it was.

In 1905, his finest year, Einstein published a paper on Brownian motion, the random, jittery motion of tiny particles suspended in a fluid. Antony van Leeuwenhoek had discovered it with his early microscope, and the phenomenon was named after Robert Brown, the Scottish botanist who studied it carefully in 1827: first pollen in water, then soot and powdered rock. Brown convinced himself that these particles were not alive—they were not animalcules—yet they would not sit still. In a mathematical tour de force, Einstein explained this as a consequence of the heat energy of molecules, whose existence he thereby proved. Microscopically visible particles, like pollen, are bombarded by molecular collisions and are light enough to be jolted randomly this way and that. The fluctuations of the particles, individually unpredictable, collectively express the laws of statistical mechanics. Although the fluid may be at rest and the system in thermodynamic equilibrium, the irregular motion perseveres, as long as the temperature is above absolute zero. By the same token, he showed that random thermal agitation would also affect free electrons in any electrical conductor—making noise.

Physicists paid little attention to the electrical aspects of Einstein's work, and it was not until 1927 that thermal noise in circuits was put on a rigorous mathematical footing, by two Swedes working at Bell Labs. John B. Johnson was the first to measure what he realized was noise intrinsic to the circuit, as opposed to evidence of flawed design. Then Harry Nyquist explained it, deriving formulas for the fluctuations in current and in voltage in an idealized network. Nyquist was the son of a farmer and shoemaker who was originally called Lars Jonsson but had to find a new name because his mail was getting mixed up with another Lars Jonsson's. The Nyquists immigrated to the United States when Harry was a teenager; he made his way from North Dakota to Bell Labs by way of Yale, where he got a doctorate in physics. He always seemed to have his eye on the big picture—which did not mean telephony per se. As early as 1918, he began working on a method for transmitting pictures by wire: "telephotography." His idea was to mount a photograph on a spinning drum, scan it, and generate currents proportional to the lightness or darkness of the image. By 1924 the company had a working prototype that could send a five-by-seven-inch picture in seven minutes.

But Nyquist meanwhile was looking backward, too, and that same year, at an electrical engineers' convention in Philadelphia, gave a talk with the modest title "Certain Factors Affecting Telegraph Speed."

It had been known since the dawn of telegraphy that the fundamental units of messaging were discrete: dots and dashes. It became equally obvious in the telephone era that, on the contrary, useful information was continuous: sounds and colors, shading into one another, blending seamlessly along a spectrum of frequencies. So which was it? Physicists like Nyquist were dealing with electric currents as waveforms, even when they were conveying discrete telegraph signals. Nowadays most of the current in a telegraph line was being wasted. In Nyquist's way of thinking, if those continuous signals could represent anything as complex as voices, then the simple stuff of telegraphy was just a special case. Specifically, it was a special case of amplitude modulation, in which the only interesting amplitudes were *on* and *off*. By treating the telegraph signals as pulses in the shape of waveforms, engineers could speed their transmission and could combine them in a single circuit—could combine them, too, with voice channels. Nyquist wanted to know *how much*—how much telegraph data, how fast. To answer that question he found an ingenious approach to converting continuous waves into data that was discrete, or "digital." Nyquist's method was to sample the waves at intervals, in effect converting them into countable pieces.

A circuit carried waves of many different frequencies: a "band" of waves, engineers would say. The range of frequencies—the width of that band, or "band width"—served as a measure of the capacity of the circuit. A telephone line could handle frequencies from about 400 to 3,400 hertz, or waves per second, for a bandwidth of 3,000 hertz. (That would cover most of the sound from an orchestra, but the high notes of the piccolo would be cut off.) Nyquist wanted to put this as generally as he could. He calculated a formula for the "speed of transmission of intelligence." To transmit intelligence at a certain speed, he showed, a channel needs a certain, measurable bandwidth. If the bandwidth was too small, it would be necessary to slow down the transmission. (But with time and ingenuity, it was realized later, even complex messages could be sent across a channel of very small bandwidth: a drum, for example, beaten by hand, sounding notes of only two pitches.)

Nyquist's colleague Ralph Hartley, who had begun his career as an expert on radio receivers, extended these results in a presentation in the summer of 1927, at an international congress on the shore of Lake Como, Italy. Hartley used a different word, "information." It was a good occasion for grand ideas. Scientists had gathered from around the world for the centennial of Alessandro Volta's death. Niels Bohr spoke on the new quantum theory and introduced for the first time his concept of complementarity. Hartley offered his listeners both a fundamental theorem and a new set of definitions.

The theorem was an extension of Nyquist's formula, and it could be expressed in words: the most information that can be transmitted in any given time is proportional to the available frequency range (he did not yet use the term *bandwidth*). Hartley was bringing into the open a set of ideas and assumptions that were becoming part of the unconscious culture of electrical engineering, and the culture of Bell Labs especially. First was the idea of information itself. He needed to pin a butterfly to the board. "As commonly used," he said, "information is a very elastic term." It is the stuff of communication—which, in turn, can be direct speech, writing, or anything else. Communication takes place by means of symbols—Hartley cited for example "words" and "dots and dashes." The symbols, by common agreement, convey "meaning." So far, this was one slippery concept after another. If the goal was to "eliminate the psychological factors involved" and to establish a measure "in terms of purely physical quantities," Hartley needed something definite and countable. He began by counting symbols—never mind what they meant. Any transmission contained a countable number of symbols. Each symbol represented a choice; each was selected from a certain set of possible symbols—an alphabet, for example—and the number of possibilities, too, was countable. The number of possible words is not so easy to count, but even in ordinary language, each word represents a selection from a set of possibilities:

> For example, in the sentence, "Apples are red," the first word eliminated other kinds of fruit and all other objects in general. The second directs attention to some property or condition of apples, and the third eliminates other possible colors. . . .
>
> The number of symbols available at any one selection obviously varies widely with the type of symbols used, with the particular communicators and with the degree of previous understanding existing between them.

Hartley had to admit that some symbols might convey more information, as the word was *commonly* understood, than others. "For example, the single word 'yes' or 'no,' when coming at the end of a protracted discussion, may have an extraordinarily great significance." His listeners could think of their own examples. But the point was to subtract human knowledge from the equation. Telegraphs and telephones are, after all, stupid.

It seemed intuitively clear that the amount of information should be proportional to the number of symbols: twice as many symbols, twice as much information. But a dot or dash—a symbol in a set with just two members—carries less information than a letter of the alphabet and much less information than a word chosen from a thousand-word dictionary. The more possible symbols, the more information each selection carries. How much more? The equation, as Hartley wrote it, was this:

$$H = n \log s$$

where H is the amount of information, n is the number of symbols transmitted, and s is the size of the alphabet. In a dot-dash system, s is just 2. A single Chinese character carries so much more weight than a Morse dot or dash; it is so much more valuable. In a system with a symbol for every word in a thousand-word dictionary, s would be 1,000.

The amount of information is not proportional to the alphabet size, however. That relationship is logarithmic: to double the amount of information, it is necessary to square the alphabet size. Hartley illustrated this in terms of a printing telegraph—one of the hodgepodge of devices, from obsolete to newfangled, being hooked up to electrical circuits. Such telegraphs used keypads arranged according to a system devised in France by Émile Baudot. The human operators used keypads, that is—the device translated these key presses, as usual, into the opening and closing of telegraph contacts. The Baudot code used five units to transmit each character, so the number of possible characters was 2^5 or 32. In terms of information content, each such character was five times as valuable—not thirty-two times—as its basic binary units.

Telephones, meanwhile, were sending their human voices across the network in happy, curvaceous analog waves. Where were the symbols in those? How could they be counted?

Hartley followed Nyquist in arguing that the continuous curve should be thought of as the limit approached by a succession of discrete steps, and that the steps could be recovered, in effect, by sampling the waveform at intervals. That way telephony could be made subject to the same mathematical treatment as telegraphy. By a crude but convincing analysis, he showed that in both cases the total amount of information would depend on two factors: the time available for transmission and the bandwidth of the channel. Phonograph records and motion pictures could be analyzed the same way.

These odd papers by Nyquist and Hartley attracted little immediate attention. They were hardly suitable for any prestigious journal of mathematics or physics, but Bell Labs had its own, *The Bell System Technical Journal*, and Claude Shannon read them there. He absorbed the mathematical insights, sketchy though they were—first awkward steps toward a shadowy goal. He noted also the difficulties both men had in defining their terms. "By the speed of transmission of intelligence is meant the number of characters, representing different letters, figures, etc., which

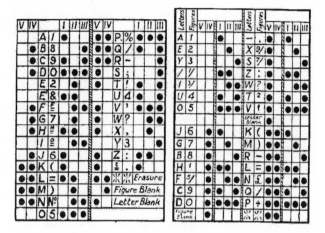

THE BAUDOT CODE

can be transmitted in a given length of time." Characters, letters, figures: hard to count. There were concepts, too, for which terms had yet to be invented: "the capacity of a system to transmit a particular sequence of symbols . . ."

Shannon felt the promise of unification. The communications engineers were talking not just about wires but also the air, the "ether," and even punched tape. They were contemplating not just words but also sounds and images. They were representing the whole world as symbols, in electricity.

7 | INFORMATION THEORY

(All I'm After Is Just a Mundane Brain)

> *Perhaps coming up with a theory of information and its processing is a bit like building a transcontinental railway. You can start in the east, trying to understand how agents can* process *anything, and head west. Or you can start in the west, with trying to understand what* information *is, and then head east. One hopes that these tracks will meet.*
>
> —Jon Barwise (1986)

AT THE HEIGHT OF THE WAR, in early 1943, two like-minded thinkers, Claude Shannon and Alan Turing, met daily at teatime in the Bell Labs cafeteria and said nothing to each other about their work, because it was secret. Both men had become cryptanalysts. Even Turing's presence at Bell Labs was a sort of secret. He had come over on the *Queen Elizabeth*, zigzagging to elude U-boats, after a clandestine triumph at Bletchley Park in deciphering Enigma, the code used by the German military for its critical communication (including signals to the U-boats). Shannon was working on the X System, used for encrypting voice conversations between Franklin D. Roosevelt at the Pentagon and Winston Churchill in his War Rooms. It operated by sampling the analog voice signal fifty times a second—"quantizing" or "digitizing" it—and masking it by applying a random key, which happened to bear a strong resemblance to the circuit noise with which the engineers were so familiar. Shannon did not design the system; he was assigned to analyze it theoretically and—it was hoped—prove it to be unbreakable. He accomplished this. It was clear later that these men, on their respective sides of the Atlantic, had done more than anyone else to turn cryptography from an art into a science, but for now the code makers and code breakers were not talking to each other.

With that subject off the table, Turing showed Shannon a paper he had written seven years earlier, called "On Computable Numbers," about the powers and limitations of an idealized computing machine. They talked about another topic that turned out to be close to their hearts, the possibility of machines learning to think. Shannon proposed feeding "cultural things," such as music, to an electronic brain, and they outdid each other in brashness, Turing exclaiming once, "No, I'm not interested in developing a *powerful* brain. All I'm after is just a *mundane* brain, something like the president of the American Telephone & Telegraph Company." It bordered on impudence to talk about thinking machines in 1943, when both the transistor and the electronic computer had yet to be born. The vision Shannon and Turing shared had nothing to do with electronics; it was about logic.

Can machines think? was a question with a relatively brief and slightly odd tradition—odd because machines were so adamantly physical in themselves. Charles Babbage and Ada Lovelace lay near the beginning of this tradition, though they were all but forgotten, and now the trail led to Alan Turing, who did something really outlandish: thought up a machine with ideal powers in the mental realm and showed what it could *not* do. His machine never existed (except that now it exists everywhere). It was only a thought experiment.

Running alongside the issue of what a machine could do was a parallel issue: what tasks were *mechanical* (an old word with new significance). Now that machines could play music, capture images, aim antiaircraft guns, connect telephone calls, control assembly lines, and perform mathematical calculations, the word did not seem quite so pejorative. But only the fearful and superstitious imagined that machines could be creative or original or spontaneous; those qualities were opposite to *mechanical*, which meant automatic, determined, and routine. This concept now came in handy for philosophers. An example of an intellectual object that could be called mechanical was the algorithm: another new term for something that had always existed (a recipe, a set of instructions, a step-by-step procedure) but now demanded formal recognition. Babbage and Lovelace trafficked in algorithms without naming them. The twentieth century gave algorithms a central role—beginning here.

Turing was a fellow and a recent graduate at King's College, Cambridge, when he presented his computable-numbers paper to his professor in 1936. The full title finished with a flourish in fancy German: it was "On Computable Numbers, with an Application to the *Entscheidungsproblem*." The "decision problem" was a challenge that had been posed by David Hilbert at the 1928 International Congress of Mathematicians. As perhaps the most influential mathematician of his time, Hilbert, like Russell and Whitehead, believed fervently in the mission of rooting all mathematics in a solid logical foundation—*"In der Mathematik gibt es kein Ignorabimus,"* he declared. ("In mathematics there is no *we will not know*.") Of course mathematics had many unsolved problems, some quite famous, such as Fermat's Last Theorem and the Goldbach conjecture—statements that seemed true but had not been proved. Had not *yet* been proved, most people thought. There was an assumption, even a faith, that any mathematical truth would be provable, someday.

The *Entscheidungsproblem* was to find a rigorous step-by-step procedure by which, given a formal language of deductive reasoning, one could perform a proof automatically. This was Leibniz's dream revived once again: the expression of all valid reasoning in mechanical rules. Hilbert posed it in the form of a question, but he was an optimist. He thought or hoped that he knew the answer. It was just then, at this watershed moment for mathematics and logic, that Gödel threw his incompleteness theorem into the works. In flavor, at least, Gödel's result seemed a perfect antidote to Hilbert's optimism, as it was to Russell's. But Gödel actually left the *Entscheidungsproblem* unanswered. Hilbert had distinguished among three questions:

Is mathematics complete?
Is mathematics consistent?
Is mathematics decidable?

Gödel showed that mathematics could not be both complete and consistent but had not definitively answered the third question, at least not for all mathematics. Even though a particular closed system of formal logic must contain statements that could neither be proved nor disproved from within the system, it might conceivably be decided, as it were, by an outside referee—by external logic or rules.*

Alan Turing, just twenty-two years old, unfamiliar with much of the relevant literature, so alone in his work habits that his professor worried about his becoming "a confirmed solitary," posed an entirely different question (it seemed): Are all numbers computable? This was an unexpected question to begin with, because hardly anyone had considered the idea of an *un*computable number. Most numbers that people work with, or think about, are computable by definition. The rational numbers are computable because they can be expressed as the quotient of two integers, a/b. The algebraic numbers are computable because they are solutions of polynomial equations. Famous numbers like π and e are computable; people compute them all the time. Nonetheless Turing made the seemingly mild statement that numbers might exist that are somehow nameable, definable, and *not* computable.

What did this mean? He defined a computable number as one whose decimal expression can be calculated by finite means. "The justification," he said, "lies in the fact that the human memory is necessarily

---

* Toward the end of his life Gödel wrote, "It was only by Turing's work that it became completely clear, that my proof is applicable to *every* formal system containing arithmetic."

limited." He also defined *calculation* as a mechanical procedure, an algorithm. Humans solve problems with intuition, imagination, flashes of insight—arguably nonmechanical calculation, or then again perhaps just computation whose steps are hidden. Turing needed to eliminate the ineffable. He asked, quite literally, what would a machine do? "According to my definition, a number is computable if its decimal can be written down by a machine."

No actual machine offered a relevant model. "Computers" were, as ever, people. Nearly all the world's computation was still performed through the act of writing marks on paper. Turing did have one information machine for a starting point: the typewriter. As an eleven-year-old sent to boarding school he had imagined inventing one. "You see," he wrote to his parents, "the funny little rounds are letters cut out on one side slide along to the round Ⓐ along an ink pad and stamp down and make the letter, thats not nearly all though." Of course, a typewriter is not automatic; it is more a tool than a machine. It does not flow a stream of language onto the page; rather, the page shifts its position space by space under the hammer, where one character is laid down after another. With this model in mind, Turing imagined another kind of machine, of the utmost purity and simplicity. Being imaginary, it was unencumbered by the real-world details one would need for a blueprint, an engineering specification, or a patent application. Turing, like Babbage, meant his machine to compute numbers, but he had no need to worry about the limitations of iron and brass. Turing did not plan ever to build his machine.

He listed the very few items his machine must possess: tape, symbols, and states. Each of these required definition.

Tape is to the Turing machine what paper is to a typewriter. But where a typewriter uses two dimensions of its paper, the machine uses only one—thus, a tape, a long strip, divided into squares. "In elementary arithmetic the two-dimensional character of the paper is sometimes used," he wrote. "But such a use is always avoidable, and I think that it will be agreed that the two-dimensional character of paper is no essential of computation." The tape is to be thought of as infinite: there is always more when needed. But just one square is "in the machine" at any given time. The tape (or the machine) can move left or right, to the next square.

Symbols can be written onto the tape, one per square. How many symbols could be used? This required some thought, especially to make sure the number was finite. Turing observed that words—in European languages, at least—behaved as individual symbols. Chinese, he said, "attempts to have an enumerable infinity of symbols." Arabic numerals might also be considered infinite, if 17 and 999,999,999,999,999 were treated as single symbols, but he preferred to treat them as compound: "It is always possible to use sequences of symbols in the place of single symbols." In fact, in keeping with the machine's minimalist spirit, he favored the absolute minimum of two symbols: binary notation, zeroes and ones. Symbols were not only to be written but also read from the tape—"scanned" was the word Turing used. In reality, of course, no technology could yet scan symbols written on paper back into a machine, but there were equivalents: for example, punched cards, now used in tabulating machines. Turing specified one more limitation: the machine is "aware" (only the anthropomorphic word would do) of one symbol at a time—the one on the square that is in the machine.

States required more explaining. Turing used the word "configurations" and pointed out that these resembled "states of mind." The machine has a few of these—some finite number. In any given state, the machine takes one or more actions depending on the current symbol. For example, in state *a*, the machine might move one square to the right if the current symbol is 1, or move one square to the left if the current symbol is 0, or print 1 if the current symbol is blank. In state *b*, the machine might erase the current symbol. In state *c*, if the symbol is 0 or 1, the machine might move to the right, and otherwise stop. After each action, the machine finishes in a new state, which might be the same or different. The various states used for a given calculation were stored in a table—how this was to be managed physically did not matter. The state table was, in effect, the machine's set of instructions.

And this was all.

Turing was *programming* his machine, though he did not yet use that word. From the primitive actions—moving, printing, erasing, changing state, and stopping—larger processes were built up, and these were used again and again: "copying down sequences of symbols, comparing sequences, erasing all symbols of a given form, etc." The machine can see just one symbol at a time, but can in effect use parts of the tape to store information temporarily. As Turing put it, "Some of the symbols written down . . . are just rough notes 'to assist the memory.'" The tape, unfurling to the horizon and beyond, provides an unbounded record. In this way all arithmetic lies within the machine's grasp. Turing showed how to add a pair of numbers—that is, he wrote out the necessary table of states. He showed how to make the machine print out (endlessly) the binary representation of π. He spent considerable time working out what the machine could do and how it would accomplish particular tasks. He demonstrated that this short list covers everything a person does in computing a number. No other knowledge or intuition is necessary. Anything computable can be computed by this machine.

Then came the final flourish. Turing's machines, stripped down to a finite table of states and a finite set of input, could themselves be represented as numbers. Every possible state table, combined with its initial tape, represents a different machine. Each machine itself, then, can be described by a particular number—a certain state table combined with its initial tape. Turing was encoding his machines just as Gödel had encoded the language of symbolic logic. This obliterated the distinction between data and instructions: in the end they were all numbers. For every computable number, there must be a corresponding machine number.

Turing produced (still in his mind's eye) a version of the machine that could simulate every other possible machine—every digital computer. He called this machine *U*, for "universal," and mathematicians fondly use the name *U* to this day. It takes machine numbers as input. That is, it reads the descriptions of other machines from its tape—their algorithms and their own input. No matter how complex a digital computer may grow, its description can still be encoded on tape to be read by *U*. If a problem can be solved by any digital computer—encoded in symbols and solved algorithmically—the universal machine can solve it as well.

Now the microscope is turned onto itself. The Turing machine sets about examining every number to see whether it corresponds to a computable algorithm. Some will prove computable. Some might prove uncomputable. And there is a third possibility, the one that most interested Turing. Some algorithms might defy the inspector, causing the machine to march along, performing its inscrutable business, never coming to a halt, never obviously repeating itself, and leaving the logical observer forever in the dark about whether it *would* halt.

By now Turing's argument, as published in 1936, has become a knotty masterpiece of recursive definitions, symbols invented to represent other symbols, numbers standing in for numbers, for state tables, for algorithms, for machines. In print it looked like this:

> By combining the machines *D* and *U* we could construct a machine *M* to compute the sequence β'. The machine *D* may require a tape. We may suppose that it uses the *E*-squares beyond all symbols on *F*-squares, and that when it has reached its verdict all the rough work done by *D* is erased. . . .
>
> We can show further that *there can be no machine E which, when applied with the S.D of an arbitrary machine M*, will determine whether *M* ever prints a given symbol (0 say).

Few could follow it. It seems paradoxical—it *is* paradoxical—but Turing proved that some numbers are uncomputable. (In fact, most are.)

Also, because every number corresponds to an encoded proposition of mathematics and logic, Turing had resolved Hilbert's question about whether every proposition is decidable. He had proved that the *Entscheidungsproblem* has an answer, and the answer is no. An uncomputable number is, in effect, an undecidable proposition.

So Turing's computer—a fanciful, abstract, wholly imaginary machine—led him to a proof parallel to Gödel's. Turing went further than Gödel by defining the general concept of a formal system. Any mechanical procedure for generating formulas is essentially a Turing machine. *Any* formal system, therefore, must have undecidable propositions. Mathematics is not decidable. Incompleteness follows from uncomputability.

Once again, the paradoxes come to life when numbers gain the power to encode the machine's own behavior. That is the necessary recursive twist. The entity being reckoned is fatally entwined with the entity doing the reckoning. As Douglas Hofstadter put it much later, "The thing hinges on getting this halting inspector to try to predict its own behavior when looking at itself trying to predict its own behavior when looking at itself trying to predict its own behavior when . . ." A conundrum that at least smelled similar had lately appeared in physics, too: Werner Heisenberg's new uncertainty principle. When Turing learned about that, he expressed it in terms of self-reference: "It used to be supposed in Science that if everything was known about the Universe at any particular moment then we can predict what it will be through all the future. . . . More modern science however has come to the conclusion that when we are dealing with atoms and electrons we are quite unable to know the exact state of them; our instruments being made of atoms and electrons themselves."

A century had passed between Babbage's Analytical Engine and Turing's Universal Machine—a grand and unwieldy contraption and an elegant unreal abstraction. Turing never even tried to be a machinist. "One can picture an industrious and diligent clerk, well supplied with scratch paper, tirelessly following his instructions," as the mathematician and logician Herbert Enderton remarked years later. Like Ada Lovelace, Turing was a programmer, looking inward to the step-by-step logic of his own mind. He imagined himself as a computer. He distilled mental procedures into their smallest constituent parts, the atoms of information processing.

———

Alan Turing and Claude Shannon had codes in common. Turing encoded instructions as numbers. He encoded decimal numbers as zeroes and ones. Shannon made codes for genes and chromosomes and relays and switches. Both men applied their ingenuity to mapping one set of objects onto another: logical operators and electric circuits; algebraic functions and machine instructions. The play of symbols and the idea of *mapping*, in the sense of finding a rigorous correspondence between two sets, had a prominent place in their mental arsenals. This kind of coding was not meant to obscure but to illuminate: to discover that apples and oranges were after all equivalent, or if not equivalent then fungible. The war brought both men to cryptography in its most riddling forms.

Turing's mother often asked him what use his mathematics had, and he told her as early as 1936 that he had discovered a possible application: "a lot of particular and interesting codes." He added, "I expect I could sell them to H. M. Government for quite a substantial sum, but am rather doubtful about the morality of such things." Indeed, a Turing machine could *make* ciphers. But His Majesty's Government turned out to have a different problem. As war loomed, the task of reading messages intercepted from German cable and wireless traffic fell to the Government Code and Cypher School, originally part of the Admiralty, with a staff at first composed of linguists, clerks, and typists, but no mathematicians. Turing was recruited in the summer of 1938. When the Code and Cypher School evacuated from London to Bletchley Park, a country mansion in Buckinghamshire, he went along with a team that also included some champions at chess and crossword-puzzle solving. It was clear now that classical language scholarship had little to contribute to cryptanalysis.

The German system, named Enigma, employed a polyalphabetic cipher implemented by a rotor machine the size of a suitcase, with a typewriter keyboard and signal lamps. The cipher had evolved from a famous ancestor, the Vigenère cipher, thought to be unbreakable until Charles Babbage cracked it in 1854, and Babbage's mathematical insight gave Bletchley early help, as did work by Polish cryptographers who had the first hard years of experience with the Wehrmacht's signal traffic. Working from a warren known as Hut 8, Turing took the theoretical lead and solved the problem, not just mathematically but physically.

This meant building a machine to invert the enciphering of any number of Enigmas. Where his first machine was a phantasm of hypothetical tape, this one, dubbed the Bombe, filled ninety cubic feet with a ton of wire and metal leaking oil and effectively mapping the rotors of the German device onto electric circuitry. The scientific triumph at Bletchley—secret for the duration of the war and for thirty years after—had a greater effect on the outcome than even the Manhattan Project, the real bomb. By the war's end, the Turing Bombes were deciphering thousands of military intercepts every day: processing information, that is, on a scale never before seen.

Although nothing of this passed between Turing and Shannon when they met for meals at Bell Labs, they did talk indirectly about a notion of Turing's about how to measure all this *stuff*. He had watched analysts weigh the messages passing through Bletchley, some uncertain and some contradictory, as they tried to assess the probability of some fact—a particular Enigma code setting, for example, or the location of a submarine.

A CAPTURED ENIGMA MACHINE

He felt that something here needed measuring, mathematically. It was not the probability, which would traditionally be expressed as an odds ratio (such as three to two) or a number from zero to one (such as 0.6, or 60 percent). Rather, Turing cared about the data that *changed* the probability: a probability factor, something like the weight of evidence. He invented a unit he named a "ban." He found it convenient to use a logarithmic scale, so that bans would be added rather than multiplied. With a base of ten, a ban was the weight of evidence needed to make a fact ten times as likely. For more fine-grained measurement there were "decibans" and "centibans."

Shannon had a notion along similar lines.

Working in the old West Village headquarters, he developed theoretical ideas about cryptography that helped him focus the dream he had intimated to Vannevar Bush: his "analysis of some of the fundamental properties of general systems for the transmission of intelligence." He followed parallel tracks all during the war, showing his supervisors the cryptography work and concealing the rest. Concealment was the order of the day. In the realm of pure mathematics, Shannon treated some of the same ciphering systems that Turing was attacking with real intercepts and brute hardware—for example, the specific question of the safety of Vigenère cryptograms when "the enemy knows the system being used." (The Germans were using just such cryptograms, and the British were the enemy who knew the system.) Shannon was looking at the most general cases, all involving, as he put it, "discrete information." That meant sequences of symbols, chosen from a finite set, mainly letters of the alphabet but also words of a language and even "quantized speech," voice signals broken into packets with different amplitude levels. To conceal these meant substituting wrong symbols for the right ones, according to some systematic procedure in which a *key* is known to the receiver of the message, who can use it to reverse the substitutions. A secure system works even when the enemy knows the procedure, as long as the key remains secret.

The code breakers see a stream of data that looks like junk. They want to find the real signal. "From the point of view of the cryptanalyst," Shannon noted, "a secrecy system is almost identical with a noisy communication system." (He completed his report, "A Mathematical Theory of Cryptography," in 1945; it was immediately classified.) The data stream is meant to look stochastic, or random, but of course it is not: if it were truly random the signal would be lost. The cipher must transform a patterned thing, ordinary language, into something apparently without pattern. But pattern is surprisingly persistent. To analyze and categorize the transformations of ciphering, Shannon had to understand the patterns of language in a way that scholars—linguists, for example—had never done before. Linguists had, however, begun to focus their discipline on structure in language—system to be found amid the vague billowing shapes and sounds. The linguist Edward Sapir wrote of "symbolic atoms" formed by a language's underlying phonetic patterns. "The mere sounds of speech," he wrote in 1921, "are not the essential fact of language, which lies rather in the classification, in the formal patterning. . . . Language, as a structure, is on its inner face the mold of thought." *Mold of thought* was exquisite. Shannon, however, needed to view language in terms more tangible and countable.

Pattern, as he saw it, equals redundancy. In ordinary language, redundancy serves as an aid to understanding. In cryptanalysis, that same redundancy is the Achilles' heel. Where is this redundancy? As a simple example in English, wherever the letter *q* appears, the *u* that follows is redundant. (Or almost—it would be entirely redundant were it not for rare borrowed items like *qin* and *Qatar*.) After *q*, a *u* is expected. There is no surprise. It contributes no information. After the letter *t*, an *h* has a certain amount of redundancy, because it is the likeliest letter to appear. Every language has a certain statistical structure, Shannon argued, and with it a certain redundancy. Let us call this (he suggested) *D*. "*D* measures, in a sense, how much a text in the language can be reduced in length without losing any information."

Shannon estimated that English has redundancy of about 50 percent.*
Without computers to process masses of text, he could not be sure, but
his estimate proved correct. Typical passages can be shortened by half
without loss of information. (*If u cn rd ths . . .*) With the simplest early
substitution ciphers, this redundancy provided the point of first weakness. Edgar Allan Poe knew that when a cryptogram contained more
z's than any other letter, then z was probably the substitute for e, since
e is the most frequent letter in English. As soon as q was solved, so was
u. A code breaker looked for recurring patterns that might match common words or letter combinations: *the, and, -tion.* To perfect this kind
of frequency analysis, code breakers needed better information about
letter frequencies than Alfred Vail or Samuel Morse had been able to
get by examining printers' type trays, and anyway, more clever ciphers
overcame this weakness, by constantly varying the substitution alphabet,
so that every letter had many possible substitutes. The obvious, recognizable patterns vanished. But as long as a cryptogram retained any trace of
patterning—any form or sequence or statistical regularity—a mathematician could, in theory, find a way in.

What all secrecy systems had in common was the use of a key: a code
word, or phrase, or an entire book, or something even more complex, but
in any case a source of characters known to both the sender and receiver—
knowledge shared apart from the message itself. In the German Enigma
system, the key was internalized in hardware and changed daily; Bletchley
Park had to rediscover it anew each time, its experts sussing out the patterns
of language freshly transformed. Shannon, meanwhile, removed himself to
the most distant, most general, most theoretical vantage point. A secrecy
system comprised a finite (though possibly very large) number of possible
messages, a finite number of possible cryptograms, and in between, transforming one to the other, a finite number of keys, each with an associated
probability. This was his schematic diagram:

---

* "not considering statistical structure over greater distances than about eight letters."

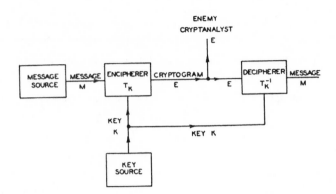

The enemy and the recipient are trying to arrive at the same target: the message. By framing it this way, in terms of mathematics and probabilities, Shannon had utterly abstracted the idea of the message from its physical details. Sounds, waveforms, all the customary worries of a Bell Labs engineer—none of these mattered. The message was seen as a choice: one alternative selected from a set. At Old North Church the night of Paul Revere's ride, the number of possible messages was two. Nowadays the numbers were almost uncountable—but still susceptible to statistical analysis.

Still in the dark about the very real and utterly relevant experience at Bletchley Park, Shannon built an edifice of algebraic methods, theorems, and proofs that gave cryptologists what they had never before possessed: a rigorous way of assessing the security of any secrecy system. He established the scientific principles of cryptography. Among other things, he proved that perfect ciphers were possible—"perfect" meaning that even an infinitely long captured message would not help a code breaker ("the enemy is no better off after intercepting any amount of material than before"). But as he gave, so he took away, because he also proved that the requirements were so severe as to make them practically useless. In a perfect cipher, all keys must be equally likely, in effect, a random stream of characters; each key can be used only once; and, worst of all, each key must be as long as the entire message.

Also in this secret paper, almost in passing, Shannon used a phrase he had never used before: "information theory."

First Shannon had to eradicate "meaning." The germicidal quotation marks were his. "The 'meaning' of a message is generally irrelevant," he proposed cheerfully.

He offered this provocation in order to make his purpose utterly clear. Shannon needed, if he were to create a theory, to hijack the word *information*. " 'Information' here," he wrote, "although related to the everyday meaning of the word, should not be confused with it." Like Nyquist and Hartley before him, he wished to leave aside "the psychological factors" and focus only on "the physical." But if information was divorced from semantic content, what was left? A few things could be said, and at first blush they all sounded paradoxical. Information is uncertainty, surprise, difficulty, and entropy:

- "Information is closely associated with uncertainty." Uncertainty, in turn, can be measured by counting the number of possible messages. If only one message is possible, there is no uncertainty and thus no information.
- Some messages may be likelier than others, and information implies surprise. Surprise is a way of talking about probabilities. If the letter following *t* (in English) is *h*, not so much information is conveyed, because the probability of *h* was relatively high.
- "What is significant is the difficulty in transmitting the message from one point to another." Perhaps this seemed backward, or tautological, like defining mass in terms of the force needed to move an object. But then, mass *can* be defined that way.
- Information is entropy. This was the strangest and most powerful notion of all. Entropy—already a difficult and poorly understood concept—is a measure of disorder in thermodynamics, the science of heat and energy.

Fire control and cryptography aside, Shannon had been pursuing this haze of ideas all through the war. Living alone in a Greenwich Village apartment, he seldom socialized with his colleagues, who mainly worked now in the New Jersey headquarters, while Shannon preferred the old West Street hulk. He did not have to explain himself. His war work got him deferred from military service and the deferment continued after the war ended. Bell Labs was a rigorously male enterprise, but in wartime the computing group, especially, badly needed competent staff and began hiring women, among them Betty Moore, who had grown up on Staten Island. It was like a typing pool for math majors, she thought. After a

THE WEST STREET HEADQUARTERS OF BELL LABORATORIES,
WITH TRAINS OF THE HIGH LINE RUNNING THROUGH

year she was promoted to the microwave research group, in the former Nabisco building—the "cracker factory"—across West Street from the main building. The group designed tubes on the second floor and built them on the first floor and every so often Claude wandered over to visit. He and Betty began dating in 1948 and married early in 1949. Just then he was the scientist everyone was talking about.

Few libraries carried *The Bell System Technical Journal*, so researchers heard about "A Mathematical Theory of Communication" the traditional way, by word of mouth, and obtained copies the traditional way, by writing directly to the author for an offprint. Many scientists used preprinted postcards for such requests, and these arrived in growing volume over the next year. Not everyone understood the paper. The mathematics was difficult for many engineers, and mathematicians meanwhile lacked the engineering context. But Warren Weaver, the director of natural sciences for the Rockefeller Foundation uptown, was already telling his president that Shannon had done for communication theory "what Gibbs did for physical chemistry." Weaver had headed the government's applied mathematics research during the war, supervising the fire-control project as well as nascent work in electronic calculating machines. In 1949 he wrote up an appreciative and not too technical essay about Shannon's theory for *Scientific American*, and late that year the two pieces—Weaver's essay and Shannon's monograph—were published together as a book, now titled with a grander first word *The Mathematical Theory of Communication*. To John Robinson Pierce, the Bell Labs engineer who had been watching the simultaneous gestation of the transistor and Shannon's paper, it was the latter that "came as a bomb, and something of a delayed action bomb."

Where a layman might have said that the fundamental problem of communication is to make oneself understood—to convey meaning—Shannon set the stage differently:

> The fundamental problem of communication is that of reproducing at one point either exactly or approximately a message selected at another point.

"Point" was a carefully chosen word: the origin and destination of a message could be separated in space or in time; information storage, as in a phonograph record, counts as a communication. Meanwhile, the message is not created; it is selected. It is a choice. It might be a card dealt from a deck, or three decimal digits chosen from the thousand possibilities, or a combination of words from a fixed code book. He could hardly overlook meaning altogether, so he dressed it with a scientist's definition and then showed it the door:

> Frequently the messages have *meaning;* that is they refer to or are correlated according to some system with certain physical or conceptual entities. These semantic aspects of communication are irrelevant to the engineering problem.

Nonetheless, as Weaver took pains to explain, this was not a narrow view of communication. On the contrary, it was all-encompassing: "not only written and oral speech, but also music, the pictorial arts, the theatre, the ballet, and in fact all human behavior." Nonhuman as well: why should machines not have messages to send?

Shannon's model for communication fit a simple diagram—essentially the same diagram, by no coincidence, as in his secret cryptography paper.

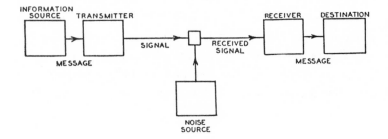

A communication system must contain the following elements:

- The information source is the person or machine generating the message, which may be simply a sequence of characters, as in a telegraph or teletype, or may be expressed mathematically as functions—$f(x, y, t)$—of time and other variables. In a complex example like color television, the components are three functions in a three-dimensional continuum, Shannon noted.
- The transmitter "operates on the message in some way"—that is, *encodes* the message—to produce a suitable signal. A telephone converts sound pressure into analog electric current. A telegraph encodes characters in dots, dashes, and spaces. More complex messages may be sampled, compressed, quantized, and interleaved.
- The channel: "merely the medium used to transmit the signal."
- The receiver inverts the operation of the transmitter. It decodes the message, or reconstructs it from the signal.
- The destination "is the person (or thing)" at the other end.

In the case of ordinary speech, these elements are the speaker's brain, the speaker's vocal cords, the air, the listener's ear, and the listener's brain.

As prominent as the other elements in Shannon's diagram—because for an engineer it is inescapable—is a box labeled "Noise Source." This covers everything that corrupts the signal, predictably or unpredictably: unwanted additions, plain errors, random disturbances, static, "atmospherics," interference, and distortion. An unruly family under any circumstances, and Shannon had two different types of systems to deal with, continuous and discrete. In a discrete system, message and signal take the form of individual detached symbols, such as characters or digits or dots and dashes. Telegraphy notwithstanding, continuous systems of waves and functions were the ones facing electrical engineers every day. Every engineer, when asked to push more information through a channel, knew what to do: boost the power. Over long distances, however, this approach was failing, because amplifying a signal again and again leads to a crippling buildup of noise.

Shannon sidestepped this problem by treating the signal as a string of discrete symbols. Now, instead of boosting the power, a sender can overcome noise by using extra symbols for error correction—just as an African drummer makes himself understood across long distances, not by banging the drums harder, but by expanding the verbosity of his discourse. Shannon considered the discrete case to be more fundamental in a mathematical sense as well. And he was considering another point: that treating messages as discrete had application not just for traditional communication but for a new and rather esoteric subfield, the theory of computing machines.

So back he went to the telegraph. Analyzed precisely, the telegraph did not use a language with just two symbols, dot and dash. In the real world telegraphers used dot (one unit of "line closed" and one unit of "line open"), dash (three units, say, of line closed and one unit of line open), and also two distinct spaces: a letter space (typically three units of line open) and a longer space separating words (six units of line open). These four symbols have unequal status and probability. For example, a space can never follow another space, whereas a dot or dash can follow anything. Shannon expressed this in terms of *states*. The system has two states: in one, a space was the previous symbol and only a dot or dash is allowed, and the state then changes; in the other, any symbol is allowed, and the state changes only if a space is transmitted. He illustrated this as a graph:

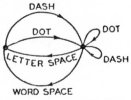

This was far from a simple, binary system of encoding. Nonetheless Shannon showed how to derive the correct equations for information content and channel capacity. More important, he focused on the effect of the statistical structure of the language of the message. The very existence of this structure—the greater frequency of *e* than *q*, of *th* than *xp*, and so forth—allows for a saving of time or channel capacity.

> This is already done to a limited extent in telegraphy by using the shortest channel sequence, a dot, for the most common English letter E; while the infrequent letters, Q, X, Z are represented by longer sequences of dots and dashes. This idea is carried still further in certain commercial codes where common words and phrases are represented by four- or five-letter code groups with a considerable saving in average time. The standardized greeting and anniversary telegrams now in use extend this to the point of encoding a sentence or two into a relatively short sequence of numbers.

To illuminate the structure of the message Shannon turned to some methodology and language from the physics of stochastic processes, from Brownian motion to stellar dynamics. (He cited a landmark 1943 paper by the astrophysicist Subrahmanyan Chandrasekhar in *Reviews of Modern Physics*.) A stochastic process is neither deterministic (the next event can be calculated with certainty) nor random (the next event is totally free). It is governed by a set of probabilities. Each event has a probability that depends on the state of the system and perhaps also on its previous history. If for *event* we substitute *symbol*, then a natural written language like English or Chinese is a stochastic process. So is digitized speech; so is a television signal.

Looking more deeply, Shannon examined statistical structure in terms of how much of a message influences the probability of the next symbol. The answer could be none: each symbol has its own probability but does not depend on what came before. This is the first-order case. In the second-order case, the probability of each symbol depends on the symbol immediately before, but not on any others. Then each two-character combination, or digram, has its own probability: *th* greater than *xp*, in English. In the third-order case, one looks at trigrams, and so forth. Beyond that, in ordinary text, it makes sense to look at the level of words rather than individual characters, and many types of statistical facts come into play. Immediately after the word *yellow*, some words have a higher probability than usual and others virtually zero. After the word *an*, words beginning with consonants become exceedingly rare. If the letter *u* ends a word, the word is probably *you*. If two consecutive letters are the same, they are probably *ll, ee, ss*, or *oo*. And structure can extend over long distances: in a message containing the word *cow*, even after many other characters intervene, the word *cow* is relatively likely to occur again. As is the word *horse*. A message, as Shannon saw, can behave like a dynamical system whose future course is conditioned by its past history.

To illustrate the differences between these different orders of structure, he wrote down—computed, really—a series of "approximations" of English text. He used an alphabet of twenty-seven characters, the letters plus a space between words, and generated strings of characters with the help of a table of random numbers. (These he drew from a book newly published for such purposes by Cambridge University Press: 100,000 digits for three shillings nine pence, and the authors "have furnished a guarantee of the random arrangement.") Even with random numbers presupplied, working out the sequences was painstaking. The sample texts looked like this:

- "Zero-order approximation"—that is, random characters, no structure or correlations.

    XFOML RXKHRJFFJUJ ZLPWCFWKCYJ
    FFJEYVKCQSGHYD QPAAMKBZAACIBZLHJQD.

- First order—each character is independent of the rest, but the frequencies are those expected in English: more *e*'s and *t*'s, fewer *z*'s and *j*'s, and the word lengths look realistic.

    OCRO HLI RGWR NMIELWIS EU LL NBNESEBYA
    TH EEI ALHENHTTPA OOBTTVA NAH BRL.

- Second order—the frequencies of each character match English and so also do the frequencies of each digram, or letter pair. (Shannon found the necessary statistics in tables constructed for use by code breakers. The most common digram in English is *th*, with a frequency of 168 per thousand words, followed by *he, an, re,* and *er*. Quite a few digrams have zero frequency.)

> ON IE ANTSOUTINYS ARE T INCTORE ST BE S DEAMY ACHIN
> D ILONASIVE TUCOOWE AT TEASONARE FUSO TIZIN ANDY
> TOBESEACE CTISBE.

• Third order—trigram structure.

> IN NO IST LAT WHEY CRATICT FROURE BIRS GROCID
> PONDENOME OF DEMONSTURES OF THE REPTAGIN IS
> REGOACTIONA OF CRE.

• First-order word approximation.

> REPRESENTING AND SPEEDILY IS AN GOOD APT OR COME CAN
> DIFFERENT NATURAL HERE HE THE A IN CAME THE TO OF TO
> EXPERT GRAY COME TO FURNISHES THE LINE MESSAGE HAD
> BE THESE.

• Second-order word approximation—now pairs of words appear in the expected frequency, so we do not see "a in" or "to of."

> THE HEAD AND IN FRONTAL ATTACK ON AN ENGLISH
> WRITER THAT THE CHARACTER OF THIS POINT IS
> THEREFORE ANOTHER METHOD FOR THE LETTERS THAT
> THE TIME OF WHO EVER TOLD THE PROBLEM FOR AN
> UNEXPECTED.

These sequences increasingly "look" like English. Less subjectively, it turns out that touch typists can handle them with increasing speed—another indication of the ways people unconsciously internalize a language's statistical structure.

Shannon could have produced further approximations, given enough time, but the labor involved was becoming enormous. The point was to represent a message as the outcome of a process that generated events with discrete probabilities. Then what could be said about the amount of information, or the rate at which information is generated? For each event, the possible choices each have a known probability (represented as p_1, p_2, p_3, and so on). Shannon wanted to define the measure of information (represented as H) as the measure of uncertainty: "of how much 'choice' is involved in the selection of the event or of how uncertain we are of the outcome." The probabilities might be the same or different, but generally more choices meant more uncertainty—more information. Choices might be broken down into successive choices, with their own probabilities, and the probabilities had to be additive; for example, the probability of a particular digram should be a weighted sum of the probabilities of the individual symbols. When those probabilities were equal, the amount of information conveyed by each symbol was simply the logarithm of the number of possible symbols—Nyquist and Hartley's formula:

$$H = n \log s$$

For the more realistic case, Shannon reached an elegant solution to the problem of how to measure information as a function of probabilities—an equation that summed the probabilities with a logarithmic weighting (base 2 was most convenient). It is the average logarithm of the improbability of the message; in effect, a measure of unexpectedness:

$$H = -\sum p_i \log_2 p_i$$

where p_i is the probability of each message. He declared that we would be seeing this again and again: that quantities of this form "play a central role in information theory as measures of information, choice, and uncertainty." Indeed, H is ubiquitous, conventionally called the entropy of a message, or the Shannon entropy, or, simply, the information.

A new unit of measure was needed. Shannon said: "The resulting units may be called binary digits, or more briefly, *bits*." As the smallest possible quantity of information, a bit represents the amount of uncertainty that exists in the flipping of a coin. The coin toss makes a choice between two possibilities of equal likelihood: in this case p_1 and p_2 each equal ½; the base 2 logarithm of ½ is −1; so H = 1 bit. A single character chosen randomly from an alphabet of 32 conveys more information: 5 bits, to be exact, because there are 32 possible messages and the logarithm of 32 is 5. A string of 1,000 such characters carries 5,000 bits—not just by simple multiplication, but because the amount of information represents the amount of uncertainty: the number of possible choices. With 1,000 characters in a 32-character alphabet, there are 32^{1000} possible messages, and the logarithm of that number is 5,000.

This is where the statistical structure of natural languages reenters the picture. If the thousand-character message is known to be English text, the number of possible messages is smaller—*much* smaller. Looking at correlations extending over eight letters, Shannon estimated that English has a built-in redundancy of about 50 percent: that each new character of a message conveys not 5 bits but only about 2.3. Considering longer-range statistical effects, at the level of sentences and paragraphs, he raised that estimate to 75 percent—warning, however, that such estimates become "more erratic and uncertain, and they depend more critically on the type of text involved." One way to measure redundancy was crudely empirical: carry out a psychology test with a human subject. This method "exploits the fact that anyone speaking a language possesses, implicitly, an enormous knowledge of the statistics of the language."

> Familiarity with the words, idioms, clichés and grammar enables him to fill in missing or incorrect letters in proof-reading, or to complete an unfinished phrase in conversation.

He might have said "her," because in point of fact his test subject was his wife, Betty. He pulled a book from the shelf (it was a Raymond Chandler detective novel, *Pickup on Noon Street*), put his finger on a short passage at random, and asked Betty to start guessing the letter, then the next letter, then the next. The more text she saw, of course, the better her chances of guessing right. After "A SMALL OBLONG READING LAMP ON THE" she got the next letter wrong. But once she knew it was *D,* she had no trouble guessing the next three letters. Shannon observed, "The errors, as would be expected, occur most frequently at the beginning of words and syllables where the line of thought has more possibility of branching out."

Quantifying predictability and redundancy in this way is a backward way of measuring information content. If a letter can be guessed from what comes before, it is redundant; to the extent that it is redundant, it provides no new information. If English is 75 percent redundant, then a thousand-letter message in English carries only 25 percent as much information as one thousand letters chosen at random. Paradoxical though it sounded, random messages carry *more* information. The implication was that natural-language text could be encoded more efficiently for transmission or storage.

Shannon demonstrated one way to do this, an algorithm that exploits differing probabilities of different symbols. And he delivered a stunning package of fundamental results. One was a formula for channel capacity, the absolute speed limit of any communication channel (now known simply as the Shannon limit). Another was the discovery that, within that limit, it must always be possible to devise schemes of error correction that will overcome any level of noise. The sender may have to devote more and more bits to correcting errors, making transmission slower and slower, but the message will ultimately get through. Shannon did not show how to design such schemes; he only proved that it was possible, thereby inspiring a future branch of computer science. "To make the chance of error as small as you wish? Nobody had thought of that," his colleague Robert Fano recalled years later. "How he got that insight, how he came to believe such a thing, I don't know. But almost all modern communication theory is based on that work." Whether removing redundancy to increase efficiency or adding redundancy to enable error correction, the encoding depends on knowledge of the language's statistical structure to do the encoding. Information cannot be separated from probabilities. A bit, fundamentally, is always a coin toss.

If the two sides of a coin were one way of representing a bit, Shannon offered a more practical hardware example as well:

> A device with two stable positions, such as a relay or a flip-flop circuit, can store one bit of information. N such devices can store N bits, since the total number of possible states is 2^N and $\log_2 2^N = N$.

Shannon had seen devices—arrays of relays, for example—that could store hundreds, even thousands of bits. That seemed like a great many. As he was finishing his write-up, he wandered one day into the office of a Bell Labs colleague, William Shockley, a physicist in his thirties. Shockley belonged to a group of solid-state physicists working on alternatives to vacuum tubes for electronics, and sitting on his desk was a tiny prototype, a piece of semiconducting crystal. "It's a solid-state amplifier," Shockley told Shannon. At that point it still needed a name.

One day in the summer of 1949, before the book version of *The Mathematical Theory of Communication* appeared, Shannon took a pencil and a piece of notebook paper, drew a line from top to bottom, and wrote the powers of ten from 10^0 to 10^{13}. He labeled this axis "bits storage capacity." He began listing some items that might be said to "store" information. A digit wheel, of the kind used in a desktop adding machine—ten decimal digits—represents just over 3 bits. At just under 10^3 bits, he wrote "punched card (all config. allowed)." At 10^4 he put "page single spaced typing (32 possible symbols)." Near 10^5 he wrote something offbeat: "genetic constitution of man." There was no real precedent for this in current scientific thinking. James D. Watson was a twenty-one-year-old student of zoology in Indiana; the discovery of the structure of DNA lay several years in the future. This was the first time anyone suggested the genome was an information store measurable in bits. Shannon's guess was conservative, by at least four orders of magnitude. He thought a "phono record (128 levels)" held more information: about 300,000 bits. To the 10 million level he assigned a thick professional journal (*Proceedings of the Institute of Radio Engineers*) and to 1 billion the *Encyclopaedia Britannica*. He estimated one hour of broadcast television at 10^{11} bits and one hour of "technicolor movie" at more than a trillion. Finally, just under his pencil mark for 10^{14}, 100 trillion bits, he put the largest information stockpile he could think of: the Library of Congress.

8 | THE INFORMATIONAL TURN

(The Basic Ingredient in Building a Mind)

> *It is probably dangerous to use this theory of information in fields for which it was not designed, but I think the danger will not keep people from using it.*
>
> —J. C. R. Licklider (1950)

MOST MATHEMATICAL THEORIES take shape slowly; Shannon's information theory sprang forth like Athena, fully formed. Yet the little book of Shannon and Weaver drew scant public attention when it appeared in 1949. The first review came from a mathematician, Joseph L. Doob, who complained that it was more "suggestive" than mathematical—"and it is not always clear that the author's mathematical intentions are honorable." A biology journal said, "At first glance, it might appear that this is primarily an engineering monograph with little or no application to human problems. Actually, the theory has some rather exciting implications." *The Philosophical Review* said it would be a mistake for philosophers to overlook this book: "Shannon develops a concept of *information* which, surprisingly enough, turns out to be an extension of the thermodynamic concept of *entropy*." The strangest review was barely a review at all: five paragraphs in *Physics Today*, September 1950, signed by Norbert Wiener, Massachusetts Institute of Technology.

Wiener began with a faintly patronizing anecdote:

Some fifteen years ago, a very bright young student came to the authorities at MIT with an idea for a theory of electric switching dependent on the algebra of logic. The student was Claude E. Shannon.

In the present book (Wiener continued), Shannon, along with Warren Weaver, "has summed up his views on communication engineering."

The fundamental idea developed by Shannon, said Wiener, "is that of the amount of information as negative entropy." He added that he himself—"the author of the present review"—had developed the same idea at about the same time.

Wiener declared the book to be work "whose origins were independent of my own work, but which has been bound from the beginning to my investigations by cross influences spreading in both directions." He mentioned "those of us who have tried to pursue this analogy into the study of Maxwell's demon" and added that much work remained to be done.

Then he suggested that the treatment of language was incomplete without greater emphasis on the human nervous system: "nervous reception and the transmission of language into the brain. I say these things not as a hostile criticism."

Finally, Wiener concluded with a paragraph devoted to another new book: "my own *Cybernetics*." Both books, he said, represent opening salvos in a field that promises to grow rapidly.

> In my book, I have taken the privilege of an author to be more speculative, and to cover a wider range than Drs. Shannon and Weaver have chosen to do. . . . There is not only room, but a definite need for different books.

He saluted his colleagues for their well-worked and independent approach—to cybernetics.

Shannon, meanwhile, had already contributed a short review of Wiener's book to the *Proceedings of the Institute of Radio Engineers*, offering praise that could be described as faint. It is "an excellent introduction," he said. There was a little tension between these men. It could be felt weighing down the long footnote that anchored the opening page of Weaver's portion of *The Mathematical Theory of Communication*:

> Dr. Shannon has himself emphasized that communication theory owes a great debt to Professor Norbert Wiener for most of its basic philosophy. Professor Wiener, on the other hand, points out that much of Shannon's early work on switching and mathematical logic antedated his own interest in this field; and generously adds that Shannon certainly deserves credit for independent development of such fundamental aspects of the theory as the introduction of entropic ideas.

Shannon's colleague John Pierce wrote later: "Wiener's head was full of his own work. . . . Competent people have told me that Wiener, under the misapprehension that he already knew what Shannon had done, never actually found out."

Cybernetics was a coinage, future buzzword, proposed field of study, would-be philosophical movement entirely conceived by this brilliant and prickly thinker. The word he took from the Greek for *steersman*: κυβερνήτησ, *kubernites*, from which comes also (not coincidentally) the word *governor*. He meant cybernetics to be a field that would synthesize the study of communication and control, also the study of human and machine. Norbert Wiener had first become known to the world as a curiosity: a sport, a prodigy, driven and promoted by his father, a professor at Harvard. "A lad who has been proudly termed by his friends the brightest boy in the world," *The New York Times* reported on page 1 when he was fourteen years old, "will graduate next month from Tufts College. . . . Aside from the fact that Norbert Wiener's capacity for learning is phenomenal, he is as other boys. . . . His intense black eyes are his most striking feature." When he wrote his memoirs, he always used the word *prodigy* in the titles: *Ex-Prodigy: My Childhood and Youth* and *I Am a Mathematician: The Later Life of a Prodigy*.

After Tufts (mathematics), Harvard graduate school (zoology), Cornell (philosophy), and Harvard again, Wiener left for Cambridge, England, where he studied symbolic logic and *Principia Mathematica* with Bertrand Russell himself. Russell was not entirely charmed. "An infant prodigy named Wiener, Ph.D. (Harvard), aged 18, turned up," he wrote a friend. "The youth has been flattered, and thinks himself God Almighty—there is a perpetual contest between him and me as to which is to do the teaching." For his part, Wiener detested Russell: "He is an iceberg. His mind impresses one as a keen, cold, narrow logical machine, that cuts the universe into neat little packets, that measure, as it were, just three inches each way." On his return to the United States, Wiener joined the faculty of MIT in 1919, the same year as Vannevar Bush. When Shannon got there in 1936, he took one of Wiener's mathematics courses. When war loomed, Wiener was one of the first to join the hidden, scattered teams of mathematicians working on antiaircraft fire control.

He was short and rotund, with heavy glasses and a Mephistophelian goatee. Where Shannon's fire-control work drilled down to the signal amid the noise, Wiener stayed with the noise: swarming fluctuations

NORBERT WIENER (1956)

in the radar receiver, unpredictable deviations in flight paths. The noise behaved statistically, he understood, like Brownian motion, the "extremely lively and wholly haphazard movement" that van Leeuwenhoek had observed through his microscope in the seventeenth century. Wiener had undertaken a thoroughgoing mathematical treatment of Brownian motion in the 1920s. The very discontinuity appealed to him—not just the particle trajectories but the mathematical functions, too, seemed to misbehave. This was, as he wrote, discrete chaos, a term that would not be well understood for several generations. On the fire-control project, where Shannon made a modest contribution to the Bell Labs team, Wiener and his colleague Julian Bigelow produced a legendary 120-page monograph, classified and known to the several dozen people allowed to see it as the Yellow Peril because of the color of its binder and the difficulty of its treatment. The formal title was *Extrapolation, Interpolation, and Smoothing of Stationary Time Series*. In it Wiener developed a statistical method for predicting the future from noisy, uncertain, and corrupted data about the past. It was too ambitious for the existing gun machinery, but he tested it on Vannevar Bush's Differential Analyzer. Both the antiaircraft gun, with its operator, and the target airplane, with its pilot, were hybrids of machine and human. One had to predict the behavior of the other.

Wiener was as worldly as Shannon was reticent. He was well traveled and polyglot, ambitious and socially aware; he took science personally and passionately. His expression of the second law of thermodynamics, for example, was a cry of the heart:

> We are swimming upstream against a great torrent of disorganization, which tends to reduce everything to the heat death of equilibrium and sameness. . . . This heat death in physics has a counterpart in the ethics of Kierkegaard, who pointed out that we live in a chaotic moral universe. In this, our main obligation is to establish arbitrary enclaves of order and system. . . . Like the Red Queen, we cannot stay where we are without running as fast as we can.

He was concerned for his place in intellectual history, and he aimed high. Cybernetics, he wrote in his memoirs, amounted to "a new interpretation of man, of man's knowledge of the universe, and of society." Where Shannon saw himself as a mathematician and an engineer, Wiener considered himself foremost a philosopher, and from his fire-control work he drew philosophical lessons about purpose and behavior. If one defines behavior cleverly—"any change of an entity with respect to its surroundings"—then the word can apply to machines as well as animals. Behavior directed toward a goal is purposeful, and the purpose can sometimes be imputed to the machine rather than a human operator: for example, in the case of a target-seeking mechanism. "The term servomechanisms has been coined precisely to designate machines with an intrinsic purposeful behavior." The key was control, or self-regulation.

To analyze it properly he borrowed an obscure term from electrical engineering: "feed-back," the return of energy from a circuit's output back to its input. When feedback is positive, as when the sound from loudspeakers is re-amplified through a microphone, it grows wildly out of control. But when feedback is negative—as in the original mechanical governor of steam engines, first analyzed by James Clerk Maxwell—it can guide a system toward equilibrium; it serves as an agent of stability. Feedback can be mechanical: the faster Maxwell's governor spins, the wider its arms extend, and the wider its arms extend, the slower it must spin. Or it can be electrical. Either way, the key to the process is information. What governs the antiaircraft gun, for example, is information about the plane's coordinates and about the previous position of the gun itself. Wiener's friend Bigelow emphasized this: "that it was not some particular physical thing such as energy or length or voltage, but only information (conveyed by any means)."

Negative feedback must be ubiquitous, Wiener felt. He could see it at work in the coordination of eye and hand, guiding the nervous system of a person performing an action as ordinary as picking up a pencil. He focused especially on neurological disorders, maladies that disrupted physical coordination or language. He saw them quite specifically as cases of information feedback gone awry: varieties of ataxia, for example, where sense messages are either interrupted in the spinal cord or misinterpreted in the cerebellum. His analysis was detailed and mathematical, with equations—almost unheard of in neurology. Meanwhile, feedback control systems were creeping into factory assembly lines, because a mechanical system, too, can modify its own behavior. Feedback is the governor, the steersman.

So *Cybernetics* became the title of Wiener's first book, published in the fall of 1948 in both the United States and France. Subtitle: *Control and Communication in the Animal and the Machine.* The book is a hodgepodge of notions and analysis, and, to the astonishment of its publishers, it became the year's unexpected bestseller. The popular American news magazines, *Time* and *Newsweek*, both featured it. Wiener and cybernetics were identified with a phenomenon that was bursting into public consciousness just at that moment: computing machines. With the end of the war, a veil had been lifted from the first urgent projects in electronic calculation, particularly the ENIAC, a thirty-ton monster of vacuum tubes, relays, and hand-soldered wires stretching across eighty feet at the University of Pennsylvania's electrical engineering school. It could store and multiply up to twenty numbers of ten decimal digits; the army used it to calculate artillery firing tables. The International Business Machines company, IBM, which provided punched card machines for the army projects, also built a giant calculating machine at Harvard, the Mark I. In Britain, still secret, the code breakers at Bletchley Park had gone on to build a vacuum-tube computing machine called the Colossus. Alan Turing was beginning work on another, at the University of Manchester. When the public learned about these machines, they were naturally thought of as "brains." Everyone asked the same question: Can machines think?

"They are growing with fearful speed," declared *Time* in its year-end issue. "They started by solving mathematical equations with flash-of-lightning rapidity. Now they are beginning to act like genuine mechanical brains." Wiener encouraged the speculation, if not the wild imagery:

Dr. Wiener sees no reason why they can't learn from experience, like monstrous and precocious children racing through grammar school. One such mechanical brain, ripe with stored experience, might run a whole industry, replacing not only mechanics and clerks but many of the executives too. . . .

As men construct better calculating machines, explains Wiener, and as they explore their own brains, the two seem more & more alike. Man, he thinks, is recreating himself, monstrously magnified, in his own image.

Much of the success of his book, abstruse and ungainly as it was, lay in Wiener's always returning his focus to the human, not the machine. He was not as interested in shedding light on the rise of computing—to which, in any case, his connections were peripheral—as in how computing might shed light on humanity. He cared profoundly, it turned out, about understanding mental disorders; about mechanical prostheses; and about the social dislocations that might follow the rise of smart machinery. He worried that it would devalue the human brain as factory machinery had devalued the human hand.

He developed the human-machine parallels in a chapter titled "Computing Machines and the Nervous System." First he laid out a distinction between two types of computing machines: analog and digital, though he did not yet use those words. The first type, like the Bush Differential Analyzer, represented numbers as measurements on a continuous scale; they were analogy machines. The other kind, which he called numerical machines, represented numbers directly and exactly, as desk calculators did. Ideally, these devices would use the binary number system for simplicity. For advanced calculations they would need to employ a form of logic. What form? Shannon had answered that question in his master's thesis of 1937, and Wiener offered the same answer:

the algebra of logic *par excellence*, or the Boolean algebra. This algorithm, like the binary arithmetic, is based on the dichotomy, the choice between *yes* and *no*, the choice between being in a class and outside.

The brain, too, he argued, is at least partly a logical machine. Where computers employ relays—mechanical, or electromechanical, or purely electrical—the brain has neurons. These cells tend to be in one of two states at any given moment: active (firing) or at rest (in repose). So they may be considered relays with two states. They are connected to one another in vast arrays, at points of contact known as synapses. They transmit messages. To store the messages, brains have memory; computing machines, too, need physical storage that can be called memory. (He knew well that this was a simplified picture of a complex system, that other sorts of messages, more analog than digital, seemed to be carried chemically by hormones.) Wiener suggested, too, that functional disorders such as "nervous breakdowns" might have cousins in electronics. Designers of computing machines might need to plan for untimely floods of data—perhaps the equivalent of "traffic problems and overloading in the nervous system."

Brains and electronic computers both use quantities of energy in performing their work of logic—"all of which is wasted and dissipated in heat," to be carried away by the blood or by ventilating and cooling apparatus. But this is really beside the point, Wiener said. "Information is information, not matter or energy. No materialism which does not admit this can survive at the present day."

Now came a time of excitement.

"We are again in one of those prodigious periods of scientific progress—in its own way like the pre-Socratic period," declared the gnomic, white-bearded neurophysiologist Warren McCulloch to a meeting of British philosophers. He told them that listening to Wiener and von Neumann put him in mind of the debates of the ancients. A new physics of communication had been born, he said, and metaphysics would never be the same: "For the first time in the history of science we know how we know and hence are able to state it clearly." He offered them heresy: that the knower was a computing machine, the brain composed of relays, perhaps ten billion of them, each receiving signals from other relays and sending them onward. The signals are quantized: they either happen or do not happen. So once again the stuff of the world, he said, turns out to be the atoms of Democritus—"indivisibles—leasts—which go batting about in the void."

> It is a world for Heraclitus, always "on the move." I do not mean merely that every relay is itself being momentarily destroyed and re-created like a flame, but I mean that its business is with information which pours into it over many channels, passes through it, eddies within it and emerges again to the world.

That these ideas were spilling across disciplinary borders was due in large part to McCulloch, a dynamo of eclecticism and cross-fertilization. Soon after the war he began organizing a series of conferences at the Beekman Hotel on Park Avenue in New York City, with money from the Josiah Macy Jr. Foundation, endowed in the nineteenth century by heirs of Nantucket whalers. A host of sciences were coming of age all at once—so-called social sciences, like anthropology and psychology, looking for new mathematical footing; medical offshoots with hybrid names, like neurophysiology; not-quite-sciences like psychoanalysis—and McCulloch invited experts in all these fields, as well as mathematics and electrical engineering. He instituted a Noah's Ark rule, inviting two of each species so that speakers would always have someone present who could see through their jargon. Among the core group were the already famous anthropologist Margaret Mead and her then-husband Gregory Bateson, the psychologists Lawrence K. Frank and Heinrich Klüver, and that formidable, sometimes rivalrous pair of mathematicians, Wiener and von Neumann.

Mead, recording the proceedings in a shorthand no one else could read, said she broke a tooth in the excitement of the first meeting and did not realize it till afterward. Wiener told them that all these sciences, the social sciences especially, were fundamentally the study of communication, and that their unifying idea was the *message*. The meetings began with the unwieldy name of Conferences for Circular Causal and Feedback Mechanisms in Biological and Social Systems and then, in deference to Wiener, whose new fame they enjoyed, changed that to Conference on Cybernetics. Throughout the conferences, it became habitual to use the new, awkward, and slightly suspect term *information theory*. Some of the disciplines were more comfortable than others. It was far from clear where information belonged in their respective worldviews.

The meeting in 1950, on March 22 and 23, began self-consciously. "The subject and the group have provoked a tremendous amount of external interest," said Ralph Gerard, a neuroscientist from the University of Chicago's medical school, "almost to the extent of a national fad. They have prompted extensive articles in such well known scientific magazines as *Time, News-Week*, and *Life*." He was referring, among others, to *Time*'s cover story earlier that winter titled "The Thinking Machine" and featuring Wiener:

> Professor Wiener is a stormy petrel (he looks more like a stormy puffin) of mathematics and adjacent territory. . . . The great new computers, cried Wiener with mingled alarm and triumph, are . . . harbingers of a whole new science of communication and control, which he promptly named "cybernetics." The newest machines, Wiener pointed out, already have an extraordinary resemblance to the human brain, both in structure and function. So far, they have no senses or "effectors" (arms and legs), but why shouldn't they have?

It was true, Gerard said, that his field was being profoundly affected by new ways of thought from communications engineering—helping them think of a nerve impulse not just as a "physical-chemical event" but as a sign or a signal. So it was helpful to take lessons from "calculating machines and communications systems," but it was dangerous, too.

> To say, as the public press says, that therefore these machines are brains, and that our brains are nothing but calculating machines, is presumptuous.

One might as well say that the telescope is an eye or that a bulldozer is a muscle.

Wiener felt he had to respond. "I have not been able to prevent these reports," he said, "but I have tried to make the publications exercise restraint. I still do not believe that the use of the word 'thinking' in them is entirely to be reprehended."*

Gerard's main purpose was to talk about whether the brain, with its mysterious architecture of neurons, branching dendrite trees, and complex interconnections alive within a chemical soup, could properly be described as analog or digital. Gregory Bateson instantly interrupted: he still found this distinction confusing. It was a basic question. Gerard owed his own understanding to "the expert tutelage that I have received here, primarily from John von Neumann"—who was sitting right there—but Gerard took a stab at it anyway. Analog is a slide rule, where number is represented as distance; digital is an abacus, where you either count a bead or you do not; there's nothing in between. A rheostat—light dimmer—is analog; a wall switch that snaps on or off, digital. Brain waves and neural chemistry, said Gerard, are analog.

Discussion ensued. Von Neumann had plenty to say. He had lately been developing a "game theory," which he viewed effectively as a mathematics of incomplete information. And he was taking the lead in designing an architecture for the new electronic computers. He wanted the more analog-minded of the group to think more abstractly—to recognize that digital processes take place in a messy, continuous world but are digital nonetheless. When a neuron snaps between two possible states—"the state of the nerve cell with no message in it and the state of the cell with a message in it"—the chemistry of this transition may

* As Jean-Pierre Dupuy remarks: "It was, at bottom, a perfectly ordinary situation, in which scientists blamed nonscientists for taking them at their word. Having planted the idea in the public mind that thinking machines were just around the corner, the cyberneticians hastened to dissociate themselves from anyone gullible enough to believe such a thing."

have intermediate shadings, but for theoretical purposes the shadings may be ignored. In the brain, he suggested, just as in a computer made of vacuum tubes, "these discrete actions are in reality simulated on the background of continuous processes." McCulloch had just put this neatly in a new paper called "Of Digital Computers Called Brains": "In this world it seems best to handle even apparent continuities as some numbers of some little steps." Remaining quiet in the audience was the new man in the group, Claude Shannon.

The next speaker was J. C. R. Licklider, an expert on speech and sound from the new Psycho-Acoustic Laboratory at Harvard, known to everyone as Lick. He was another young scientist with his feet in two different worlds—part psychologist and part electrical engineer. Later that year he moved to MIT, where he established a new psychology department within the department of electrical engineering. He was working on an idea for quantizing speech—taking speech waves and reducing them to the smallest quantities that could be reproduced by a "flip-flop circuit," a homemade gadget made from twenty-five dollars of vacuum tubes, resistors, and capacitors. It was surprising—even to people used to the crackling and hissing of telephones—how far speech could be reduced and still remain intelligible. Shannon listened closely, not just because he knew about the relevant telephone engineering but because he had dealt with the issues in his secret war work on audio scrambling. Wiener perked up, too, in part because of a special interest in prosthetic hearing aids.

When Licklider described some distortion as neither linear nor logarithmic but "halfway between," Wiener interrupted.

"What does 'halfway' mean? X plus S over N?"

Licklider sighed. "Mathematicians are always doing that, taking me up on inexact statements." But he had no problem with the math and later offered an estimate for how much information—using Shannon's new terminology—could be sent down a transmission line, given a certain bandwidth (5,000 cycles) and a certain signal-to-noise ratio (33 decibels), numbers that were realistic for commercial radio. "I think it appears that 100,000 bits of information can be transmitted through such a communication channel"—bits per second, he meant. That was a staggering number; by comparison, he calculated the rate of ordinary human speech this way: 10 phonemes per second, chosen from a vocabulary of 64 phonemes (2^6, "to make it easy"—the logarithm of the number of choices is 6), so a rate of 60 bits per second. "This assumes that the phonemes are all equally probable—"

"Yes!" interrupted Wiener.

"—and of course they are not."

Wiener wondered whether anyone had tried a similar calculation for "compression for the eye," for television. How much "real information" is necessary for intelligibility? Though he added, by the way: "I often wonder why people try to look at television."

Margaret Mead had a different issue to raise. She did not want the group to forget that meaning can exist quite apart from phonemes and dictionary definitions. "If you talk about another kind of information," she said, "if you are trying to communicate the fact that somebody is angry, what order of distortion might be introduced to take the anger out of a message that otherwise will carry exactly the same words?"

That evening Shannon took the floor. Never mind meaning, he said. He announced that, even though his topic was the redundancy of written English, he was not going to be interested in *meaning* at all.

He was talking about information as something transmitted from one point to another: "It might, for example, be a random sequence of digits, or it might be information for a guided missile or a television signal." What mattered was that he was going to represent the information source as a statistical process, generating messages with varying probabilities. He showed them the sample text strings he had used in *The Mathematical Theory of Communication*—which few of them had read—and described his "prediction experiment," in which the subject guesses text letter by letter. He told them that English has a specific *entropy*, a quantity correlated with redundancy, and that he could use these experiments to compute the number. His listeners were fascinated—Wiener, in particular, thinking of his own "prediction theory."

"My method has some parallelisms to this," Wiener interrupted. "Excuse me for interrupting."

There was a difference in emphasis between Shannon and Wiener. For Wiener, entropy was a measure of disorder; for Shannon, of uncertainty. Fundamentally, as they were realizing, these were the same. The more inherent order exists in a sample of English text—order in the form of statistical patterns, known consciously or unconsciously to speakers of the language—the more predictability there is, and in Shannon's terms, the less information is conveyed by each subsequent letter. When the subject guesses the next letter with confidence, it is redundant, and the arrival of the letter contributes no new information. Information is surprise.

The others brimmed with questions about different languages, different prose styles, ideographic writing, and phonemes. One psychologist asked whether newspaper writing would look different, statistically, from the work of James Joyce. Leonard Savage, a statistician who worked with von Neumann, asked how Shannon chose a book for his test: at random?

"I just walked over to the shelf and chose one."

"I wouldn't call that random, would you?" said Savage. "There is a danger that the book might be about engineering." Shannon did not tell them that in point of fact it had been a detective novel.

Someone else wanted to know if Shannon could say whether baby talk would be more or less predictable than the speech of an adult.

"I think more predictable," he replied, "if you are familiar with the baby."

English is actually many different languages—as many, perhaps, as there are English speakers—each with different statistics. It also spawns artificial dialects: the language of symbolic logic, with its restricted and precise alphabet, and the language one questioner called "Airplanese," employed by control towers and pilots. And language is in constant flux.

Heinz von Foerster, a young physicist from Vienna and an early acolyte of Wittgenstein, wondered how the degree of redundancy in a language might change as the language evolved, and especially in the transition from oral to written culture.

Von Foerster, like Margaret Mead and others, felt uncomfortable with the notion of information without meaning. "I wanted to call the whole of what they called information theory *signal* theory," he said later, "because information was not yet there. There were '*beep beeps*' but that was all, no information. The moment one transforms that set of signals into other signals our brain can make an understanding of, *then* information is born—it's not in the beeps." But he found himself thinking of the essence of language, its history in the mind and in the culture, in a new way. At first, he pointed out, no one is conscious of letters, or phonemes, as basic units of a language.

> I'm thinking of the old Maya texts, the hieroglyphics of the Egyptians or the Sumerian tables of the first period. During the development of writing it takes some considerable time—or an accident—to recognize that a language can be split into smaller units than words, e.g., syllables or letters. I have the feeling that there is a feedback between writing and speaking.

The discussion changed his mind about the centrality of information. He added an epigrammatic note to his transcript of the eighth conference: "Information can be considered as order wrenched from disorder."

Hard as Shannon tried to keep his listeners focused on his pure, meaning-free definition of information, this was a group that would not steer clear of semantic entanglements. They quickly grasped Shannon's essential ideas, and they speculated far afield. "If we could agree to define as information anything which changes probabilities or reduces uncertainties," remarked Alex Bavelas, a social psychologist, "changes in emotional security could be seen quite easily in this light." What about gestures or facial expressions, pats on the back or winks across the table? As the psychologists absorbed this artificial way of thinking about signals and the brain, their whole discipline stood on the brink of a radical transformation.

Ralph Gerard, the neuroscientist, was reminded of a story. A stranger is at a party of people who know one another well. One says, "72," and everyone laughs. Another says, "29," and the party roars. The stranger asks what is going on.

> His neighbor said, "We have many jokes and we have told them so often that now we just use a number." The guest thought he'd try it, and after a few words said, "63." The response was feeble. "What's the matter, isn't this a joke?"
> "Oh, yes, that is one of our very best jokes, but you did not tell it well."

The next year Shannon returned with a robot. It was not a very clever robot, nor lifelike in appearance, but it impressed the cybernetics group. It solved mazes. They called it Shannon's rat.

He wheeled out a cabinet with a five-by-five grid on its top panel. Partitions could be placed around and between any of the twenty-five squares to make mazes in different configurations. A pin could be placed in any square to serve as the goal, and moving around the maze was a sensing rod driven by a pair of little motors, one for east-west and one for north-south. Under the hood lay an array of electrical relays, about seventy-five of them, interconnected, switching on and off to form the robot's "memory." Shannon flipped the switch to power it up.

"When the machine was turned off," he said, "the relays essentially forgot everything they knew, so that they are now starting afresh, with no knowledge of the maze." His listeners were rapt. "You see the finger now exploring the maze, hunting for the goal. When it reaches the center of a square, the machine makes a new decision as to the next direction to try." When the rod hit a partition, the motors reversed and the relays recorded the event. The machine made each "decision" based on its previous "knowledge"—it was impossible to avoid these psychological words—according to a strategy Shannon had designed. It wandered about the space by trial and error, turning down blind alleys and bumping into walls. Finally, as they all watched, the rat found the goal, a bell rang, a lightbulb flashed on, and the motors stopped.

Then Shannon put the rat back at the starting point for a new run. This time it went directly to the goal without making any wrong turns or hitting any partitions. It had "learned." Placed in other, unexplored parts of the maze, it would revert to trial and error until, eventually, "it builds up a complete pattern of information and is able to reach the goal directly from any point."

To carry out the exploring and goal-seeking strategy, the machine had to store one piece of information for each square it visited: namely, the direction by which it last left the square. There were only four possibilities—north, west, south, east—so, as Shannon carefully explained, two relays were assigned as memory for each square. Two relays meant two bits of information, enough for a choice among four alternatives, because there were four possible states: off-off, off-on, on-off, and on-on.

Next Shannon rearranged the partitions so that the old solution would no longer work. The machine would then "fumble around" till it found a new solution. Sometimes, however, a particularly awkward combination of previous memory and a new maze would put the machine in an endless loop. He showed them: "When it arrives at A, it remembers that the old solution said to go to B, and so it goes around the circle, A, B, C, D, A, B, C, D. It has established a vicious circle, or a singing condition."

"A neurosis!" said Ralph Gerard.

Shannon added "an antineurotic circuit": a counter, set to break out of the loop when the machine repeated the same sequence six times. Leonard Savage saw that this was a bit of a cheat. "It doesn't have any way to recognize that it is 'psycho'—it just recognizes that it has been going too long?" he asked. Shannon agreed.

SHANNON AND HIS MAZE

"It is all too human," remarked Lawrence K. Frank.

"George Orwell should have seen this," said Henry Brosin, a psychiatrist.

A peculiarity of the way Shannon had organized the machine's memory—associating a single direction with each square—was that the path could not be reversed. Having reached the goal, the machine did not "know" how to return to its origin. The knowledge, such as it was, emerged from what Shannon called the vector field, the totality of the twenty-five directional vectors. "You can't say where the sensing finger came from by studying the memory," he explained.

"Like a man who knows the town," said McCulloch, "so he can go from any place to any other place, but doesn't always remember how he went."

Shannon's rat was kin to Babbage's silver dancer and the metal swans and fishes of Merlin's Mechanical Museum: automata performing a simulation of life. They never failed to amaze and entertain. The dawn of the information age brought a whole new generation of synthetic mice, beetles, and turtles, made with vacuum tubes and then transistors. They were crude, almost trivial, by the standards of just a few years later. In the case of the rat, the creature's total memory amounted to seventy-five bits. Yet Shannon could fairly claim that it solved a problem by trial and error; retained the solution and repeated it without the errors; integrated new information from further experience; and "forgot" the solution when circumstances changed. The machine was not only imitating lifelike behavior; it was performing functions previously reserved for brains.

One critic, Dennis Gabor, a Hungarian electrical engineer who later won the Nobel Prize for inventing holography, complained, "In reality it is the maze which remembers, not the mouse." This was true up to a point. After all, there was no mouse. The electrical relays could have been placed anywhere, and they held the memory. They became, in effect, a mental model of a maze—a *theory* of a maze.

The postwar United States was hardly the only place where biologists and neuroscientists were suddenly making common cause with mathematicians and electrical engineers—though Americans sometimes talked as though it was. Wiener, who recounted his travels to other countries at some length in his introduction to *Cybernetics*, wrote dismissively that in England he had found researchers to be "well-informed" but that not much progress had been made "in unifying the subject and in pulling the various threads of research together." New cadres of British scientists began coalescing in response to information theory and cybernetics in 1949—mostly young, with fresh experience in code breaking, radar, and gun control. One of their ideas was to form a dining club in the English fashion—"limited membership and a post-prandial situation," proposed John Bates, a pioneer in electroencephalography. This required considerable discussion of names, membership rules, venues, and emblems. Bates wanted electrically inclined biologists and biologically oriented engineers and suggested "about fifteen people who had Wiener's ideas before Wiener's book appeared." They met for the first time in the basement of the National Hospital for Nervous Diseases, in Bloomsbury, and decided to call themselves the Ratio Club—a name meaning whatever anyone wanted. (Their chroniclers Philip Husbands and Owen Holland, who interviewed many of the surviving members, report that half pronounced it RAY-she-oh and half RAT-ee-oh.) For their first meeting they invited Warren McCulloch.

They talked not just about understanding brains but "designing" them. A psychiatrist, W. Ross Ashby, announced that he was working on the idea that "a brain consisting of randomly connected impressional synapses would assume the required degree of orderliness as a result of experience"—in other words, that the mind is a self-organizing dynamical system. Others wanted to talk about pattern recognition, about noise in the nervous system, about robot chess and the possibility of mechanical self-awareness. McCulloch put it this way: "Think of the brain as a telegraphic relay, which, tripped by a signal, emits another signal." Relays had come a long way since Morse's time. "Of the molecular events of brains these signals are the atoms. Each goes or does not go." The fundamental unit is a choice, and it is binary. "It is the least event that can be true or false."

They also managed to attract Alan Turing, who published his own manifesto with a provocative opening statement—"I propose to consider the question, 'Can machines think?'"—followed by a sly admission that he would do so without even trying to define the terms *machine* and *think*. His idea was to replace the question with a test called the Imitation Game, destined to become famous as the "Turing Test." In its initial form the Imitation Game involves three people: a man, a woman, and an interrogator. The interrogator sits in a room apart and poses questions (ideally, Turing suggests, by way of a "teleprinter communicating between the two rooms"). The interrogator aims to determine which is the man and which is the woman. One of the two—say, the man—aims to trick the interrogator, while the other aims to help reveal the truth. "The best strategy for her is probably to give truthful answers," Turing suggests. "She can add such things as 'I am the woman, don't listen to him!' but it will avail nothing as the man can make similar remarks."

But what if the question is not which gender but which genus: human or machine?

It is understood that the essence of being human lies in one's "intellectual capacities"; hence this game of disembodied messages transmitted blindly between rooms. "We do not wish to penalise the machine for its inability to shine in beauty competitions," says Turing dryly, "nor to penalise a man for losing in a race against an aeroplane." Nor, for that matter, for slowness in arithmetic. Turing offers up some imagined questions and answers:

> Q: Please write me a sonnet on the subject of the Forth Bridge.
> A: Count me out on this one. I never could write poetry.

Before proceeding further, however, he finds it necessary to explain just what sort of machine he has in mind. "The present interest in 'thinking machines,'" he notes, "has been aroused by a particular kind of machine, usually called an 'electronic computer' or 'digital computer.'" These devices do the work of human computers, faster and more reliably. Turing spells out, as Shannon had not, the nature and properties of the digital computer. John von Neumann had done this, too, in constructing a successor machine to ENIAC. The digital computer comprises three parts: a "store of information," corresponding to the human computer's memory or paper; an "executive unit," which carries out individual operations; and a "control," which manages a list of instructions, making sure they are carried out in the right order. These instructions are encoded as numbers. They are sometimes called a "programme," Turing explains, and constructing such a list may be called "programming."

The idea is an old one, Turing says, and he cites Charles Babbage, whom he identifies as Lucasian Professor of Mathematics at Cambridge from 1828 to 1839—once so famous, now almost forgotten. Turing explains that Babbage "had all the essential ideas" and "planned such a machine, called the Analytical Engine, but it was never completed." It would have used wheels and cards—nothing to do with electricity. The existence (or nonexistence, but at least near existence) of Babbage's engine allows Turing to rebut a superstition he senses forming in the zeitgeist of 1950. People seem to feel that the magic of digital computers is essentially electrical; meanwhile, the nervous system is also electrical. But Turing is at pains to think of computation in a universal way, which means in an abstract way. He knows it is not about electricity at all:

> Since Babbage's machine was not electrical, and since all digital comput-
> ers are in a sense equivalent, we see that this use of electricity cannot be of
> theoretical importance. . . . The feature of using electricity is thus seen to
> be only a very superficial similarity.

Turing's famous computer was a machine made of logic: imaginary tape, arbitrary symbols. It had all the time in the world and unbounded memory, and it could do anything expressible in steps and operations. It could even judge the validity of a proof in the system of *Principia Mathematica*. "In the case that the formula is neither provable nor disprovable such a machine certainly does not behave in a very satisfactory manner, for it continues to work indefinitely without producing any result at all, but this cannot be regarded as very different from the reaction of the mathematicians." So Turing supposed it could play the Imitation Game.

He could not pretend to prove that, of course. He was mainly trying to change the terms of a debate he considered largely fatuous. He offered a few predictions for the half century to come: that computers would have a storage capacity of 10^9 bits (he imagined a few very large computers; he did not foresee our future of ubiquitous tiny computing devices with storage many magnitudes greater than that); and that they might be programmed to play the Imitation Game well enough to fool some interrogators for at least a few minutes (true, as far as it goes).

> The original question, "Can machines think?" I believe to be too meaningless to deserve discussion. Nevertheless I believe that at the end of the century the use of words and general educated opinion will have altered so much that one will be able to speak of machines thinking without expecting to be contradicted.

He did not live to see how apt his prophecy was. In 1952 he was arrested for the crime of homosexuality, tried, convicted, stripped of his security clearance, and subjected by the British authorities to a humiliating, emasculating program of estrogen injections. In 1954 he took his own life.

Until years later, few knew of Turing's crucial secret work for his country on the Enigma project at Bletchley Park. His ideas of thinking machines did attract attention, on both sides of the Atlantic. Some of the people who found the notion absurd or even frightening appealed to Shannon for his opinion; he stood squarely with Turing. "The idea of a machine thinking is by no means repugnant to all of us," Shannon told one engineer. "In fact, I find the converse idea, that the human brain may itself be a machine which could be duplicated functionally with inanimate objects, quite attractive." More useful, anyway, than "hypothecating intangible and unreachable 'vital forces,' 'souls' and the like."

Computer scientists wanted to know what their machines could do. Psychologists wanted to know whether brains are computers—or perhaps whether brains are *merely* computers. At midcentury computer scientists were new; but so, in their way, were psychologists.

Psychology at midcentury had grown moribund. Of all the sciences, it always had the most difficulty in saying what exactly it studied. Originally its object was the soul, as opposed to the body (somatology) and the blood (hematology). "*Psychologie* is a doctrine which searches out man's

Soul, and the effects of it; this is the part without which a man cannot consist," wrote James de Back in the seventeenth century. Almost by definition, though, the soul was ineffable—hardly a thing to be known. Complicating matters further was the entanglement (in psychology as in no other field) of the observer with the observed. In 1854, when it was still more likely to be called "mental philosophy," David Brewster lamented that no other department of knowledge had made so little progress as "the science of mind, if it can be called a science."

> Viewed as material by one inquirer, as spiritual by another, and by others as mysteriously compounded as both, the human mind escapes from the cognisance of sense and reason, and lies, a waste field with a northern exposure, upon which every passing speculator casts his mental tares.

The passing speculators were still looking mainly inward, and the limits of introspection were apparent. Looking for rigor, verifiability, and perhaps even mathematicization, students of the mind veered in radically different directions by the turn of the twentieth century. Sigmund Freud's path was only one. In the United States, William James constructed a discipline of psychology almost single-handed—professor of the first university courses, author of the first comprehensive textbook—and when he was done, he threw up his hands. His own *Principles of Psychology*, he wrote, was "a loathsome, distended, tumefied, bloated, dropsical mass, testifying to but two facts: *1st*, that there is no such thing as a *science* of psychology, and *2nd*, that WJ is an incapable."

In Russia, a new strain of psychology began with a physiologist, Ivan Petrovich Pavlov, known for his Nobel Prize–winning study of digestion, who scorned the word *psychology* and all its associated terminology. James, in his better moods, considered psychology the science of mental life, but for Pavlov there was no mind, only behavior. Mental states, thoughts, emotions, goals, and purpose—all these were intangible, subjective, and out of reach. They bore the taint of religion and superstition. What James had identified as central topics—"the stream of thought," "the consciousness of self," the perception of time and space, imagination, reasoning, and will—had no place in Pavlov's laboratory. All a scientist could observe was behavior, and this, at least, could be recorded and measured. The behaviorists, particularly John B. Watson in the United States and then, most famously, B. F. Skinner, made a science based on stimulus and response: food pellets, bells, electric shocks; salivation, lever pressing, maze running. Watson said that the whole purpose of psychology was to predict what responses would follow a given stimulus and what stimuli could produce a given behavior. Between stimulus and response lay a black box, known to be composed of sense organs, neural pathways, and motor functions, but fundamentally off limits. In effect, the behaviorists were saying yet again that the soul is ineffable. For a half century, their research program thrived because it produced results about conditioning reflexes and controlling behavior.

Behaviorists said, as the psychologist George Miller put it afterward: "You talk about memory; you talk about anticipation; you talk about your feelings; you talk about all these mentalistic things. That's moonshine. Show me one, point to one." They could teach pigeons to play ping-pong and rats to run mazes. But by midcentury, frustration had set in. The behaviorists' purity had become a dogma; their refusal to consider mental states became a cage, and psychologists still wanted to understand what the mind was.

Information theory gave them a way in. Scientists analyzed the processing of information and built machines to do it. The machines had memory. They simulated learning and goal seeking. A behaviorist running a rat through a maze would discuss the association between stimulus and response but would refuse to speculate in any way about the *mind* of the rat; now engineers were building mental models of rats out of a few electrical relays. They were not just prying open the black box; they were making their own. Signals were being transmitted, encoded, stored, and retrieved. Internal models of the external world were created and updated. Psychologists took note. From information theory and cybernetics, they received a set of useful metaphors and even a productive conceptual framework. Shannon's rat could be seen not only as a very crude model of the brain but also as a theory of behavior. Suddenly psychologists were free to talk about plans, algorithms, syntactic rules. They could investigate not just how living creatures react to the outside world but how they represent it to themselves.

Shannon's formulation of information theory seemed to invite researchers to look in a direction that he himself had not intended. He had declared, "The fundamental problem of communication is that of reproducing at one point either exactly or approximately a message selected at another point." A psychologist could hardly fail to consider the case where the source of the message is the outside world and the receiver is the mind.

Ears and eyes were to be understood as message channels, so why not test and measure them like microphones and cameras? "New concepts of the nature and measure of information," wrote Homer Jacobson, a chemist at Hunter College in New York, "have made it possible to specify quantitatively the informational capacity of the human ear," and he proceeded to do so. Then he did the same for the eye, arriving at an estimate four hundred times greater, in bits per second. Many more subtle kinds of experiments were suddenly fair game, some of them directly suggested by Shannon's work on noise and redundancy. A group in 1951 tested the likelihood that listeners would hear a word correctly when they knew it was one of just a few alternatives, as opposed to many alternatives. It seemed obvious but had never been done. Experimenters explored the effect of trying to understand two conversations at once. They began considering how much information an ensemble of items contained—digits or letters or words—and how much could be understood or remembered. In standard experiments, with speech and buzzers and key pressing and foot tapping, the language of stimulus and response began to give way to transmission and reception of information.

For a brief period, researchers discussed the transition explicitly; later it became invisible. Donald Broadbent, an English experimental psychologist exploring issues of attention and short-term memory, wrote of one experiment in 1958: "The difference between a description of the results in terms of stimulus and response, and a description in information theory terms, becomes most marked. . . . One could no doubt develop an adequate description of the results in S-R terms . . . but such a description is clumsy compared to the information theory description." Broadbent founded an applied psychology division at Cambridge University, and a flood of research followed, there and elsewhere, in the general realm of how people handle information: effects of noise on performance; selective attention and filtering of perception; short-term and long-term memory; pattern recognition; problem solving. And where did logic belong? To psychology or to computer science? Surely not just to philosophy.

An influential counterpart of Broadbent's in the United States was George Miller, who helped found the Center for Cognitive Studies at Harvard in 1960. He was already famous for a paper published in 1956 under the slightly whimsical title "The Magical Number Seven, Plus or Minus Two: Some Limits on Our Capacity for Processing Information." Seven seemed to be the number of items that most people could hold in working memory at any one time: seven digits (the typical American telephone number of the time), seven words, or seven objects displayed by an experimental psychologist. The number also kept popping up, Miller claimed, in other sorts of experiments. Laboratory subjects were fed sips of water with different amounts of salt, to see how many different levels of saltiness they could discriminate. They were asked to detect differences between tones of varying pitch or loudness. They were shown random patterns of dots, flashed on a screen, and asked how many (below seven, they almost always knew; above seven, they almost always estimated). In one way and another, the number seven kept recurring as a threshold. "This number assumes a variety of disguises," he wrote, "being sometimes a little larger and sometimes a little smaller than usual, but never changing so much as to be unrecognizable."

Clearly this was a crude simplification of some kind; as Miller noted, people can identify any of thousands of faces or words and can memorize long sequences of symbols. To see what kind of simplification, he turned to information theory, and especially to Shannon's understanding of information as a selection among possible alternatives. "The observer is considered to be a communication channel," he announced—a formulation sure to appall the behaviorists who dominated the profession. Information is being transmitted and stored—information about loudness, or saltiness, or number. He explained about bits:

> One bit of information is the amount of information that we need to make a decision between two equally likely alternatives. If we must decide whether a man is less than six feet tall or more than six feet tall and if we know that the chances are 50-50, then we need one bit of information. . . .
>
> Two bits of information enable us to decide among four equally likely alternatives. Three bits of information enable us to decide among eight equally likely alternatives . . . and so on. That is to say, if there are 32 equally likely alternatives, we must make five successive binary decisions, worth one bit each, before we know which alternative is correct. So the general rule is simple: every time the number of alternatives is increased by a factor of two, one bit of information is added.

The magical number seven is really just under three bits. Simple experiments measured discrimination, or channel capacity, in a single dimension; more complex measures arise from combinations of variables in multiple dimensions—for example, size, brightness, and hue. And people perform acts of what information theorists call "recoding," grouping information into larger and larger chunks—for example, organizing telegraph dots and dashes into letters, letters into words, and words into phrases. By now Miller's argument had become something in the nature of a manifesto. Recoding, he declared, "seems to me to be the very lifeblood of the thought processes."

> The concepts and measures provided by the theory of information provide a quantitative way of getting at some of these questions. The theory provides us with a yardstick for calibrating our stimulus materials and for measuring the performance of our subjects. . . . Informational concepts

have already proved valuable in the study of discrimination and of language; they promise a great deal in the study of learning and memory; and it has even been proposed that they can be useful in the study of concept formation. A lot of questions that seemed fruitless twenty or thirty years ago may now be worth another look.

This was the beginning of the movement called the cognitive revolution in psychology, and it laid the foundation for the discipline called cognitive science, combining psychology, computer science, and philosophy. Looking back, some philosophers have called this moment the informational turn. "Those who take the informational turn see information as the basic ingredient in building a mind," writes Frederick Adams. "Information has to contribute to the origin of the mental." As Miller himself liked to say, the mind came in on the back of the machine.

Shannon was hardly a household name—he never did become famous to the general public—but he had gained an iconic stature in his own academic communities, and sometimes he gave popular talks about "information" at universities and museums. He would explain the basic ideas; puckishly quote Matthew 5:37, "Let your communication be, Yea, yea; Nay, nay: for whatsoever is more than these cometh of evil" as a template for the notions of bits and of redundant encoding; and speculate about the future of computers and automata. "Well, to conclude," he said at the University of Pennsylvania, "I think that this present century in a sense will see a great upsurge and development of this whole information business; the business of collecting information and the business of transmitting it from one point to another, and perhaps most important of all, the business of processing it."

With psychologists, anthropologists, linguists, economists, and all sorts of social scientists climbing aboard the bandwagon of information theory, some mathematicians and engineers were uncomfortable.

Shannon himself called it a bandwagon. In 1956 he wrote a short warning notice—four paragraphs: "Our fellow scientists in many different fields, attracted by the fanfare and by the new avenues opened to scientific analysis, are using these ideas in their own problems. . . . Although this wave of popularity is certainly pleasant and exciting for those of us working in the field, it carries at the same time an element of danger." Information theory was in its hard core a branch of mathematics, he reminded them. He, personally, did believe that its concepts would prove useful in other fields, but not everywhere, and not easily: "The establishing of such applications is not a trivial matter of translating words to a new domain, but rather the slow tedious process of hypothesis and experimental verification." Furthermore, he felt the hard slogging had barely begun in "our own house." He urged more research and less exposition.

As for cybernetics, the word began to fade. The Macy cyberneticians held their last meeting in 1953, at the Nassau Inn in Princeton; Wiener had fallen out with several of the group, who were barely speaking to him. Given the task of summing up, McCulloch sounded wistful. "Our consensus has never been unanimous," he said. "Even had it been so, I see no reason why God should have agreed with us."

Throughout the 1950s, Shannon remained the intellectual leader of the field he had founded. His research produced dense, theorem-packed papers, pregnant with possibilities for development, laying foundations for broad fields of study. What Marshall McLuhan later called the "medium" was for Shannon the channel, and the channel was subject to rigorous mathematical treatment. The applications were immediate and the results fertile: broadcast channels and wiretap channels, noisy and noiseless channels, Gaussian channels, channels with input constraints and cost constraints, channels with feedback and channels with memory, multiuser channels and multiaccess channels. (When McLuhan announced that the medium was the message, he was being arch. The medium is both opposite to, and entwined with, the message.)

CLAUDE SHANNON (1963)

One of Shannon's essential results, the noisy coding theorem, grew in importance, showing that error correction can effectively counter noise and corruption. At first this was just a tantalizing theoretical nicety; error correction requires computation, which was not yet cheap. But during the 1950s, work on error-correcting methods began to fulfill Shannon's promise, and the need for them became apparent. One application was exploration of space with rockets and satellites; they needed to send messages very long distances with limited power. Coding theory became a crucial part of computer science, with error correction and data compression advancing side by side. Without it, modems, CDs, and digital television would not exist. For mathematicians interested in random processes, coding theorems are also measures of entropy.

Shannon, meanwhile, made other theoretical advances that planted seeds for future computer design. One discovery showed how to maximize flow through a network of many branches, where the network could be a communication channel or a railroad or a power grid or water pipes. Another was aptly titled "Reliable Circuits Using Crummy Relays" (though this was changed for publication to ". . . Less Reliable Relays"). He studied switching functions, rate-distortion theory, and differential entropy. All this was invisible to the public, but the seismic tremors that came with the dawn of computing were felt widely, and Shannon was part of that, too.

As early as 1948 he completed the first paper on a problem that he said, "of course, is of no importance in itself": how to program a machine to play chess. People had tried this before, beginning in the eighteenth and nineteenth centuries, when various chess automata toured Europe and were revealed every so often to have small humans hiding inside. In 1910 the Spanish mathematician and tinkerer Leonardo Torres y Quevedo built a real chess machine, entirely mechanical, called El Ajedrecista, that could play a simple three-piece endgame, king and rook against king.

Shannon now showed that computers performing numerical calculations could be made to play a full chess game. As he explained, these devices, "containing several thousand vacuum tubes, relays, and other elements," retained numbers in "memory," and a clever process of translation could make these numbers represent the squares and pieces of a chessboard. The principles he laid out have been employed in every chess program since. In these salad days of computing, many people immediately assumed that chess would be *solved:* fully known, in all its pathways and combinations. They thought a fast electronic computer would play perfect chess, just as they thought it would make reliable long-term weather forecasts. Shannon made a rough calculation, however, and suggested that the number of possible chess games was more than 10^{120}—a number that dwarfs the age of the universe in nanoseconds. So computers cannot play chess by brute force; they must reason, as Shannon saw, along something like human lines.

He visited the American champion Edward Lasker in his apartment on East Twenty-third Street in New York, and Lasker offered suggestions for improvement. When *Scientific American* published a simplified version of his paper in 1950, Shannon could not resist raising the question on everyone's minds: "Does a chess-playing machine of this type 'think'?"

> From a behavioristic point of view, the machine acts as though it were thinking. It has always been considered that skillful chess play requires the reasoning faculty. If we regard thinking as a property of external actions rather than internal method the machine is surely thinking.

Nonetheless, as of 1952 he estimated that it would take three programmers working six months to enable a large-scale computer to play even a tolerable amateur game. "The problem of a learning chess player is even farther in the future than a preprogrammed type. The methods which have been suggested are obviously extravagantly slow. The machine would wear out before winning a single game." The point, though, was to look in as many directions as possible for what a general-purpose computer could do.

He was exercising his sense of whimsy, too. He designed and actually built a machine to do arithmetic with Roman numerals: for example, IV times XII equals XLVIII. He dubbed this THROBAC I, an acronym for Thrifty Roman-numeral Backward-looking Computer. He created a "mind-reading machine" meant to play the child's guessing game of odds and evens. What all these flights of fancy had in common was an extension of algorithmic processes into new realms—the abstract mapping of ideas onto mathematical objects. Later, he wrote thousands of words on scientific aspects of juggling—with theorems and corollaries—and included from memory a quotation from E. E. Cummings: "Some son-of-a-bitch will invent a machine to measure Spring with."

In the 1950s Shannon was also trying to design a machine that would repair itself. If a relay failed, the machine would locate and replace it. He speculated on the possibility of a machine that could reproduce itself, collecting parts from the environment and assembling them. Bell Labs was happy for him to travel and give talks on such things, often demonstrating his maze-learning machine, but audiences were not universally delighted. The word "Frankenstein" was heard. "I wonder if you boys realize what you're toying around with there," wrote a newspaper columnist in Wyoming.

What happens if you switch on one of these mechanical computers but forget to turn it off before you leave for lunch? Well, I'll tell you. The same thing would happen in the way of computers in America that happened to Australia with jack rabbits. Before you could multiply 701,945,240 by 879,030,546, every family in the country would have a little computer of their own. . . .

Mr. Shannon, I don't mean to knock your experiments, but frankly I'm not remotely interested in even one computer, and I'm going to be pretty sore if a gang of them crowd in on me to multiply or divide or whatever they do best.

Two years after Shannon raised his warning flag about the bandwagon, a younger information theorist, Peter Elias, published a notice complaining about a paper titled "Information Theory, Photosynthesis, and Religion." There was, of course, no such paper. But there had been papers on information theory, life, and topology; information theory and the physics of tissue damage; and clerical systems; and psychopharmacology; and geophysical data interpretation; and crystal structure; and melody. Elias, whose father had worked for Edison as an engineer, was himself a serious specialist—a major contributor to coding theory. He mistrusted the softer, easier, platitudinous work flooding across disciplinary boundaries. The typical paper, he said, "discusses the surprisingly close relationship between the vocabulary and conceptual framework of information theory and that of psychology (or genetics, or linguistics, or psychiatry, or business organization). . . . The concepts of structure, pattern, entropy, noise, transmitter, receiver, and code are (when properly interpreted) central to both." He declared this to be larceny. "Having placed the discipline of psychology for the first time on a sound scientific basis, the author modestly leaves the filling in of the outline to the psychologists." He suggested his colleagues give up larceny for a life of honest toil.

These warnings from Shannon and Elias appeared in one of the growing number of new journals entirely devoted to information theory.

In these circles a notorious buzzword was *entropy*. Another researcher, Colin Cherry, complained, "We have heard of 'entropies' of languages, social systems, and economic systems and of its use in various method-starved studies. It is the kind of sweeping generality which people will clutch like a straw." He did not say, because it was not yet apparent, that information theory was beginning to change the course of theoretical physics and of the life sciences and that entropy was one of the reasons.

In the social sciences, the direct influence of information theorists had passed its peak. The specialized mathematics had less and less to contribute to psychology and more and more to computer science. But their contributions had been real. They had catalyzed the social sciences and prepared them for the new age under way. The work had begun; the informational turn could not be undone.

9 | ENTROPY AND ITS DEMONS

(You Cannot Stir Things Apart)

> *Thought interferes with the probability of events, and, in the long run therefore, with entropy.*
>
> —David L. Watson (1930)

IT WOULD BE AN EXAGGERATION TO SAY that no one knew what *entropy* meant. Still, it was one of those words. The rumor at Bell Labs was that Shannon had gotten it from John von Neumann, who advised him he would win every argument because no one would understand it. Untrue, but plausible. The word began by meaning the opposite of itself. It remains excruciatingly difficult to define. The *Oxford English Dictionary*, uncharacteristically, punts:

> 1. The name given to one of the quantitative elements which determine the thermodynamic condition of a portion of matter.

Rudolf Clausius coined the word in 1865, in the course of creating a science of thermodynamics. He needed to name a certain quantity that he had discovered—a quantity related to energy, but not energy.

Thermodynamics arose hand in hand with steam engines; it was at first nothing more than "the theoretical study of the steam engine." It concerned itself with the conversion of heat, or energy, into work. As this occurs—heat drives an engine—Clausius observed that the heat does not actually get lost; it merely passes from a hotter body into a cooler body. On its way, it accomplishes something. This is like a waterwheel, as Nicolas Sadi Carnot kept pointing out in France: water begins at the top and ends at the bottom, and no water is gained or lost, but the water performs work on the way down. Carnot imagined heat as just such a substance. The ability of a thermodynamic system to produce work depends not on the heat itself, but on the contrast between hot and cold. A hot stone plunged into cold water can generate work—for example, by creating steam that drives a turbine—but the total heat in the system (stone plus water) remains constant. Eventually, the stone and the water reach the same temperature. No matter how much energy a closed system contains, when everything is the same temperature, no work can be done.

It is the unavailability of this energy—its uselessness for work—that Clausius wanted to measure. He came up with the word *entropy*, formed from Greek to mean "transformation content." His English-speaking counterparts saw the point but decided Clausius had it backward in focusing on the negative. James Clerk Maxwell suggested in his *Theory of Heat* that it would be "more convenient" to make entropy mean the opposite: "the part which *can* be converted into mechanical work." Thus:

> When the pressure and temperature of the system have become uniform the entropy is exhausted.

Within a few years, though, Maxwell turned about-face and decided to follow Clausius. He rewrote his book and added an abashed footnote:

> In former editions of this book the meaning of the term Entropy, as introduced by Clausius, was erroneously stated to be that part of the energy which cannot be converted into work. The book then proceeded to use the term as equivalent to the available energy; thus introducing great confusion into the language of thermodynamics. In this edition I have endeavoured to use the word Entropy according to its original definition by Clausius.

The problem was not just in choosing between positive and negative. It was subtler than that. Maxwell had first considered entropy as a subtype of energy: the energy available for work. On reconsideration, he recognized that thermodynamics needed an entirely different measure. Entropy was not a kind of energy or an amount of energy; it was, as Clausius had said, the *unavailability* of energy. Abstract though this was, it turned out to be a quantity as measurable as temperature, volume, or pressure.

It became a totemic concept. With entropy, the "laws" of thermodynamics could be neatly expressed:

First law: The energy of the universe is constant.
Second law: The entropy of the universe always increases.

There are many other formulations of these laws, from the mathematical to the whimsical, e.g., "1. You can't win; 2. You can't break even either." But this is the cosmic, fateful one. The universe is running down. It is a degenerative one-way street. The final state of maximum entropy is our destiny.

William Thomson, Lord Kelvin, imprinted the second law on the popular imagination by reveling in its bleakness: "Although mechanical energy is *indestructible*," he declared in 1862, "there is a universal tendency to its dissipation, which produces gradual augmentation and diffusion of heat, cessation of motion, and exhaustion of potential energy through the material universe. The result of this would be a state of universal rest and death." Thus entropy dictated the universe's fate in H. G. Wells's novel *The Time Machine:* the life ebbing away, the dying sun, the "abominable desolation that hung over the world." Heat death is not cold; it is lukewarm and dull. Freud thought he saw something useful there in 1918, though he muddled it: "In considering the conversion of psychical energy no less than of physical, we must make use of the concept of an entropy, which opposes the undoing of what has already occurred."

Thomson liked the word *dissipation* for this. Energy is not lost, but it dissipates. Dissipated energy is present but useless. It was Maxwell, though, who began to focus on the confusion itself—the disorder—as entropy's essential quality. Disorder seemed strangely unphysical. It implied that a piece of the equation must be something like knowledge, or intelligence, or judgment. "The idea of dissipation of energy depends on the extent of our knowledge," Maxwell said. "Available energy is energy which we can direct into any desired channel. Dissipated energy is energy which we cannot lay hold of and direct at pleasure, such as the energy of the confused agitation of molecules which we call heat." What *we* can do, or know, became part of the definition. It seemed impossible to talk about order and disorder without involving an agent or an observer—without talking about the mind:

> Confusion, like the correlative term order, is not a property of material things in themselves, but only in relation to the mind which perceives them. A memorandum-book does not, provided it is neatly written, appear confused to an illiterate person, or to the owner who understands it thoroughly, but to any other person able to read it appears to be inextricably confused. Similarly the notion of dissipated energy could not occur to a being who could not turn any of the energies of nature to his own account, or to one who could trace the motion of every molecule and seize it at the right moment.

Order is subjective—in the eye of the beholder. Order and confusion are not the sorts of things a mathematician would try to define or measure. Or are they? If disorder corresponded to entropy, maybe it was ready for scientific treatment after all.

As an ideal case, the pioneers of thermodynamics considered a box of gas. Being made of atoms, it is far from simple or calm. It is a vast ensemble of agitating particles. Atoms were unseen and hypothetical, but these theorists—Clausius, Kelvin, Maxwell, Ludwig Boltzmann, Willard Gibbs—accepted the atomic nature of a fluid and tried to work out the consequences: mixing, violence, continuous motion. This motion constitutes heat, they now understood. Heat is no substance, no fluid, no "phlogiston"—just the motion of molecules.

Individually the molecules must be obeying Newton's laws—every action, every collision, measurable and calculable, in theory. But there were too many to measure and calculate individually. Probability entered the picture. The new science of statistical mechanics made a bridge between the microscopic details and the macroscopic behavior. Suppose the box of gas is divided by a diaphragm. The gas on side A is hotter than the gas on side B—that is, the A molecules are moving faster, with greater energy. As soon as the divider is removed, the molecules begin to mix; the fast collide with the slow; energy is exchanged; and after some time the gas reaches a uniform temperature. The mystery is this: Why can the process not be reversed? In Newton's equations of motion, time can have a plus sign or a minus sign; the mathematics works either way. In the real world past and future cannot be interchanged so easily.

"Time flows on, never comes back," said Léon Brillouin in 1949. "When the physicist is confronted with this fact he is greatly disturbed." Maxwell had been mildly disturbed. He wrote to Lord Rayleigh:

> If this world is a purely dynamical system, and if you accurately reverse the motion of every particle of it at the same instant, then all things will happen backwards to the beginning of things, the raindrops will collect themselves from the ground and fly up to the clouds, etc, etc, and men will see their friends passing from the grave to the cradle till we ourselves become the reverse of born, whatever that is.

His point was that in the microscopic details, if we watch the motions of individual molecules, their behavior is the same forward and backward in time. We can run the film backward. But pan out, watch the box of gas as an ensemble, and statistically the mixing process becomes a one-way street. We can watch the fluid for all eternity, and it will never divide itself into hot molecules on one side and cool on the other. The clever young Thomasina says in Tom Stoppard's *Arcadia*, "You cannot stir things apart," and this is precisely the same as "Time flows on, never comes back." Such processes run in one direction only. Probability is the reason. What is remarkable—physicists took a long time to accept it—is that every irreversible process must be explained the same way. Time itself depends on chance, or "the accidents of life," as Richard Feynman liked to say: "Well, you see that all there is to it is that the irreversibility is caused by the general accidents of life." For the box of gas to come unmixed is not physically impossible; it is just improbable in the extreme. So the second law is merely probabilistic. Statistically, everything tends toward maximum entropy.

Yet probability is enough: enough for the second law to stand as a pillar of science. As Maxwell put it:

> *Moral.* The 2nd law of Thermodynamics has the same degree of truth as the statement that if you throw a tumblerful of water into the sea, you cannot get the same tumblerful of water out again.

The improbability of heat passing from a colder to a warmer body (without help from elsewhere) is identical to the improbability of order arranging itself from disorder (without help from elsewhere). Both, fundamentally, are due only to statistics. Counting all the possible ways a system can be arranged, the disorderly ones far outnumber the orderly ones. There are many arrangements, or "states," in which molecules are all jumbled, and few in which they are neatly sorted. The orderly states have low probability and low entropy. For impressive degrees of orderliness, the probabilities may be *very* low. Alan Turing once whimsically proposed a number N, defined as "the odds against a piece of chalk leaping across the room and writing a line of Shakespeare on the board."

Eventually physicists began speaking of microstates and macrostates. A macrostate might be: all the gas in the top half of the box. The corresponding microstates would be all the possible arrangements of all particles—positions and velocities. Entropy thus became a physical equivalent of probability: the entropy of a given macrostate is the logarithm of the number of its possible microstates. The second law, then, is the tendency of the universe to flow from less likely (orderly) to more likely (disorderly) macrostates.

It was still puzzling, though, to hang so much of physics on a matter of mere probability. Can it be right to say that nothing in physics is stopping a gas from dividing itself into hot and cold—that it is only a matter of chance and statistics? Maxwell illustrated this conundrum with a thought experiment. Imagine, he suggested, "a finite being" who stands watch over a tiny hole in the diaphragm dividing the box of gas. This creature can see molecules coming, can tell whether they are fast or slow, and can choose whether or not to let them pass. Thus he could tilt the odds. By sorting fast from slow, he could make side A hotter and side B colder—"and yet no work has been done, only the intelligence of a very observant and neat-fingered being has been employed." The being defies ordinary probabilities. The chances are, things get mixed together. To sort them out requires information.

Thomson loved this idea. He dubbed the notional creature a demon:

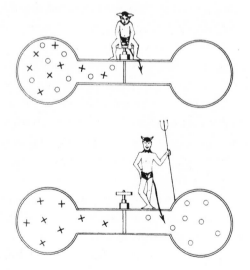

"Maxwell's intelligent demon," "Maxwell's sorting demon," and soon just "Maxwell's demon." Thomson waxed eloquent about the little fellow: "He differs from real living animals only [*only!*] in extreme smallness and agility." Lecturing to an evening crowd at the Royal Institution of Great Britain, with the help of tubes of liquid dyed two different colors, Thomson demonstrated the apparently irreversible process of diffusion and declared that only the demon can counteract it:

> He can cause one-half of a closed jar of air, or of a bar of iron, to become glowingly hot and the other ice cold; can direct the energy of the moving molecules of a basin of water to throw the water up to a height and leave it there proportionately cooled; can "sort" the molecules in a solution of salt or in a mixture of two gases, so as to reverse the natural process of diffusion, and produce concentration of the solution in one portion of the water, leaving pure water in the remainder of the space occupied; or, in the other case, separate the gases into different parts of the containing vessel.

The reporter for *The Popular Science Monthly* thought this was ridiculous. "All nature is supposed to be filled with infinite swarms of absurd little microscopic imps," he sniffed. "When men like Maxwell, of Cambridge, and Thomson, of Glasgow, lend their sanction to such a crude hypothetical fancy as that of little devils knocking and kicking the atoms this way and that . . . , we may well ask, What next?" He missed the point. Maxwell had not meant his demon to exist, except as a teaching device.

The demon sees what we cannot—because we are so gross and slow—namely, that the second law is statistical, not mechanical. At the level of molecules, it is violated all the time, here and there, purely by chance. The demon replaces chance with purpose. It uses information to reduce entropy. Maxwell never imagined how popular his demon would become, nor how long-lived. Henry Adams, who wanted to work some version of entropy into his theory of history, wrote to his brother Brooks in 1903, "Clerk Maxwell's demon who runs the second law of Thermo-dynamics ought to be made President." The demon presided over a gateway—at first, a magical gateway—from the world of physics to the world of information.

Scientists envied the demon's powers. It became a familiar character in cartoons enlivening physics journals. To be sure, the creature was a fantasy, but the atom itself had seemed fantastic, and the demon had helped tame it. Implacable as the laws of nature now seemed, the demon defied these laws. It was a burglar, picking the lock one molecule at a time. It had "infinitely subtle senses," wrote Henri Poincaré, and "could turn back the course of the universe." Was this not just what humans dreamed of doing?

Through their ever better microscopes, scientists of the early twentieth century examined the active, sorting processes of biological membranes. They discovered that living cells act as pumps, filters, and factories. Purposeful processes seemed to operate at tiny scales. Who or what was in

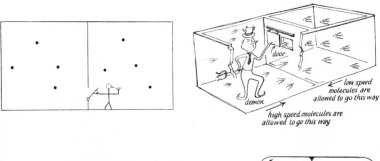

control? Life itself seemed an organizing force. "Now we must not introduce demonology into science," wrote the British biologist James Johnstone in 1914. In physics, he said, individual molecules must remain beyond our control. "These motions and paths are un-co-ordinated—'helter-skelter'—if we like so to term them. Physics considers only the statistical *mean* velocities." That is why the phenomena of physics are irreversible, "so that for the latter science Maxwell's demons do not exist." But what of life? What of physiology? The processes of terrestrial life *are* reversible, he argued. "We must therefore seek for evidence that the organism *can* control the, otherwise, un-co-ordinated motions of the individual molecules."

> Is it not strange that while we see that most of our human effort is that of *directing* natural agencies and energies into paths which they would not otherwise take, we should yet have failed to think of primitive organisms, or even of the tissue elements in the bodies of the higher organisms, as possessing also the power of directing physico-chemical processes?

When life remained so mysterious, maybe Maxwell's demon was not just a cartoon.

Then the demon began to haunt Leó Szilárd, a very young Hungarian physicist with a productive imagination who would later conceive the electron microscope and, not incidentally, the nuclear chain reaction. One of his more famous teachers, Albert Einstein, advised him out of avuncular protectiveness to take a paying job with the patent office, but Szilárd ignored the advice. He was thinking in the 1920s about how thermodynamics should deal with incessant molecular fluctuations. By definition, fluctuations ran counter to averages, like fish swimming momentarily upstream, and people naturally wondered: what if you could harness them? This irresistible idea led to a version of the perpetual motion machine, *perpetuum mobile*, holy grail of cranks and hucksters. It was another way of saying, "All that heat—why can't we use it?"

It was also another of the paradoxes engendered by Maxwell's demon.

In a closed system, a demon who could catch the fast molecules and let the slow molecules pass would have a source of useful energy, continually refreshed. Or, if not the chimerical imp, what about some other "intelligent being"? An experimental physicist, perhaps? A perpetual motion machine should be possible, declared Szilárd, "if we view the experimenting man as a sort of deus ex machina, one who is continuously informed of the existing state of nature." For his version of the thought experiment, Szilárd made clear that he did not wish to invoke a living demon, with, say, a brain—biology brought troubles of its own. "The very existence of a nervous system," he noted, "is dependent on continual dissipation of energy." (His friend Carl Eckart pithily rephrased this: "Thinking generates entropy.") Instead he proposed a "nonliving device," intervening in a model thermodynamic system, operating a piston in a cylinder of fluid. He pointed out that this device would need, in effect, "a sort of memory faculty." (Alan Turing was now, in 1929, a teenager. In Turing's terms, Szilárd was treating the mind of the demon as a computer with a two-state memory.)

Szilárd showed that even this perpetual motion machine would have to fail. What was the catch? Simply put: information is not free. Maxwell, Thomson, and the rest had implicitly talked as though knowledge was there for the taking—knowledge of the velocities and trajectories of molecules coming and going before the demon's eyes. They did not consider the cost of this information. They could not; for them, in a simpler time, it was as if the information belonged to a parallel universe, an astral plane, not linked to the universe of matter and energy, particles and forces, whose behavior they were learning to calculate.

But information is physical. Maxwell's demon makes the link. The demon performs a conversion between information and energy, one particle at a time. Szilárd—who did not yet use the word *information*— found that, if he accounted exactly for each measurement and memory, then the conversion could be computed exactly. So he computed it. He calculated that each unit of information brings a corresponding increase in entropy—specifically, by $k \log 2$ units. Every time the demon makes a choice between one particle and another, it costs one bit of information. The payback comes at the end of the cycle, when it has to clear its memory (Szilárd did not specify this last detail in words, but in mathematics). Accounting for this properly is the only way to eliminate the paradox of perpetual motion, to bring the universe back into harmony, to "restore concordance with the Second Law."

Szilárd had thus closed a loop leading to Shannon's conception of entropy as information. For his part, Shannon did not read German and did not follow *Zeitschrift für Physik*. "I think actually Szilárd was thinking of this," he said much later, "and he talked to von Neumann about it, and von Neumann may have talked to Wiener about it. But none of these people actually talked to me about it." Shannon reinvented the mathematics of entropy nonetheless.

To the physicist, entropy is a measure of uncertainty about the state of a physical system: one state among all the possible states it can be in. These microstates may not be equally likely, so the physicist writes $S = -\sum p_i \log p_i$.

To the information theorist, entropy is a measure of uncertainty about a message: one message among all the possible messages that a communications source can produce. The possible messages may not be equally likely, so Shannon wrote $H = -\sum p_i \log p_i$.

It is not just a coincidence of formalism: nature providing similar answers to similar problems. It is all one problem. To reduce entropy in a box of gas, to perform useful work, one pays a price in information. Likewise, a particular message reduces the entropy in the ensemble of possible messages—in terms of dynamical systems, a phase space.

That was how Shannon saw it. Wiener's version was slightly different. It was fitting—for a word that began by meaning the opposite of itself—that these colleagues and rivals placed opposite signs on their formulations of entropy. Where Shannon identified information with entropy, Wiener said it was *negative* entropy. Wiener was saying that information meant order, but an orderly thing does not necessarily embody much information. Shannon himself pointed out their difference and minimized it, calling it a sort of "mathematical pun." They get the same numerical answers, he noted:

> I consider how much information is *produced* when a choice is made from a set—the larger the set the *more* information. You consider the larger uncertainty in the case of a larger set to mean less knowledge of the situation and hence *less* information.

Put another way, H is a measure of surprise. Put yet another way, H is the average number of yes-no questions needed to guess the unknown message. Shannon had it right—at least, his approach proved fertile for mathematicians and physicists a generation later—but the confusion lingered for some years. Order and disorder still needed some sorting.

We all behave like Maxwell's demon. Organisms organize. In everyday experience lies the reason sober physicists across two centuries kept this cartoon fantasy alive. We sort the mail, build sand castles, solve jigsaw puzzles, separate wheat from chaff, rearrange chess pieces, collect stamps, alphabetize books, create symmetry, compose sonnets and sonatas, and put our rooms in order, and to do all this requires no great energy, as long as we can apply intelligence. We propagate structure (not just we humans but we who are alive). We disturb the tendency toward equilibrium. It would be absurd to attempt a thermodynamic accounting for such processes, but it is not absurd to say we are reducing entropy, piece by piece. Bit by bit. The original demon, discerning one molecule at a time, distinguishing fast from slow, and operating his little gateway, is sometimes described as "superintelligent," but compared to a real organism it is an idiot savant. Not only do living things lessen the disorder in their environments; they are in themselves, their skeletons and their flesh, vesicles and membranes, shells and carapaces, leaves and blossoms, circulatory systems and metabolic pathways—miracles of pattern and structure. It sometimes seems as if curbing entropy is our quixotic purpose in this universe.

In 1943 Erwin Schrödinger, the chain-smoking, bow-tied pioneer of quantum physics, asked to deliver the Statutory Public Lectures at Trinity College, Dublin, decided the time had come to answer one of the greatest of unanswerable questions: What is life? The equation bearing his name was the essential formulation of quantum mechanics. In looking beyond his field, as middle-aged Nobel laureates so often do, Schrödinger traded rigor for speculation and began by apologizing "that some of us should venture to embark on a synthesis of facts and theories, albeit with second-hand and incomplete knowledge of some of them—and at the risk of making fools of ourselves." Nonetheless, the little book he made from these lectures became influential. Without discovering or even stating anything new, it laid a foundation for a nascent science, as yet unnamed, combining genetics and biochemistry. "Schrödinger's book became a kind of *Uncle Tom's Cabin* of the revolution in biology that, when the dust had cleared, left molecular biology as its legacy," one of the discipline's founders wrote later. Biologists had not read anything like it before, and physicists took it as a signal that the next great problems might lie in biology.

Schrödinger began with what he called the enigma of biological stability. In notable contrast to a box of gas, with its vagaries of probability and fluctuation, and in seeming disregard of Schrödinger's own wave mechanics, where uncertainty is the rule, the structures of a living creature exhibit remarkable permanence. They persist, both in the life of the organism and across generations, through heredity. This struck Schrödinger as requiring explanation.

"When is a piece of matter said to be alive?" he asked. He skipped past the usual suggestions—growth, feeding, reproduction—and answered as simply as possible: "When it goes on 'doing something,' moving, exchanging material with its environment, and so forth, for a much longer period than we would expect an inanimate piece of matter to 'keep going' under similar circumstances." Ordinarily, a piece of matter comes to a standstill; a box of gas reaches a uniform temperature; a chemical system "fades away into a dead, inert lump of matter"—one way or another, the second law is obeyed and maximum entropy is reached. Living things manage to remain unstable. Norbert Wiener pursued this thought in *Cybernetics:* enzymes, he wrote, may be "metastable" Maxwell's demons—meaning not quite stable, or precariously stable. "The stable state of an enzyme is to be deconditioned," he noted, "and the stable state of a living organism is to be dead."

Schrödinger felt that evading the second law for a while, or seeming to, is exactly why a living creature "appears so enigmatic." The organism's ability to feign perpetual motion leads so many people to believe in a special, supernatural *life force*. He mocked this idea—*vis viva* or entelechy—and he also mocked the popular notion that organisms "feed upon energy." Energy and matter were just two sides of a coin, and anyway one calorie is as good as another. No, he said: the organism feeds upon negative entropy.

"To put it less paradoxically," he added paradoxically, "the essential thing in metabolism is that the organism succeeds in freeing itself from all the entropy it cannot help producing while alive."

In other words, the organism sucks orderliness from its surroundings. Herbivores and carnivores dine on a smorgasbord of structure; they feed on organic compounds, matter in a well-ordered state, and return it "in a very much degraded form—not entirely degraded, however, for plants can make use of it." Plants meanwhile draw not just energy but negative entropy from sunlight. In terms of energy, the accounting can be more or less rigorously performed. In terms of order, calculations are not so simple. The mathematical reckoning of order and chaos remains more ticklish, the relevant definitions being subject to feedback loops of their own.

Much more remained to be learned, Schrödinger said, about how life stores and perpetuates the orderliness it draws from nature. Biologists with their microscopes had learned a great deal about cells. They could see gametes—sperm cells and egg cells. Inside them were the rodlike fibers called chromosomes, arranged in pairs, with consistent numbers from species to species, and known to be carriers of hereditary features. As Schrödinger put it now, they hold within them, somehow, the "pattern" of the organism: "It is these chromosomes, or probably only an axial skeleton fibre of what we actually see under the microscope as the chromosome, that contain in some kind of code-script the entire pattern of the individual's future development." He considered it amazing—mysterious, but surely crucial in some way as yet unknown—that every single cell of an organism "should be in possession of a complete (double) copy of the code-script." He compared this to an army in which every soldier knows every detail of the general's plans.

These details were the many discrete "properties" of an organism, though it remained far from clear what a property entailed. ("It seems neither adequate nor possible to dissect into discrete 'properties' the pattern of an organism which is essentially a unity, a 'whole,'" Schrödinger mused.) The color of an animal's eyes, blue or brown, might be a property, but it is more useful to focus on the *difference* from one individual to another, and this difference was understood to be controlled by something conveyed in the chromosomes. He used the term *gene:* "the hypothetical material carrier of a definite hereditary feature." No one could yet see these hypothetical genes, but surely the time was not far off. Microscopic observations made it possible to estimate their size: perhaps 100 or 150 atomic distances; perhaps one thousand atoms or fewer. Yet somehow these tiny entities must encapsulate the entire pattern of a living creature—a fly or a rhododendron, a mouse or a human. And we must understand this pattern as a four-dimensional object: the structure of the organism through the whole of its ontogenetic development, every stage from embryo to adult.

In seeking a clue to the gene's molecular structure, it seemed natural to look to the most organized forms of matter, crystals. Solids in crystalline form have a relative permanence; they can begin with a tiny germ and build up larger and larger structures; and quantum mechanics was beginning to give deep insight into the forces involved in their bonding. But Schrödinger felt something was missing. Crystals are *too* orderly—built up in "the comparatively dull way of repeating the same structure in three directions again and again." Elaborate though they seem, crystalline solids contain just a few types of atoms. Life must depend on a higher level of complexity, structure without predictable repetition, he argued. He invented a term: *aperiodic crystals*. This was his hypothesis: *We believe a gene—or perhaps the whole chromosome fiber—to be an aperiodic solid.* He could hardly emphasize enough the glory of this difference, between periodic and aperiodic:

> The difference in structure is of the same kind as that between an ordinary wallpaper in which the same pattern is repeated again and again in regular periodicity and a masterpiece of embroidery, say a Raphael tapestry, which shows no dull repetition, but an elaborate, coherent, *meaningful* design.

Some of his most admiring readers, such as Léon Brillouin, the French physicist recently decamped to the United States, said that Schrödinger was too clever to be completely convincing, even as they demonstrated in their own work just how convinced they were. Brillouin was particularly taken with the comparison to crystals, with their elaborate but inanimate structures. Crystals have some capacity for self-repair, he noted; under stress, their atoms may shift to new positions for the sake of equilibrium. That may be understood in terms of thermodynamics and now quantum mechanics. How much more exalted, then, is self-repair in the organism: "The living organism heals its own wounds, cures its sicknesses, and may rebuild large portions of its structure when they have been destroyed by some accident. This is the most striking and unexpected behavior." He followed Schrödinger, too, in using entropy to connect the smallest and largest scales.

The earth is not a closed system, and life feeds upon energy and negative entropy leaking into the earth system. . . . The cycle reads: first, creation of unstable equilibriums (fuels, food, waterfalls, etc.); then use of these reserves by all living creatures.

Living creatures confound the usual computation of entropy. More generally, so does information. "Take an issue of *The New York Times*, the book on cybernetics, and an equal weight of scrap paper," suggested Brillouin. "Do they have the same entropy?" If you are feeding the furnace, yes. But not if you are a reader. There is entropy in the arrangement of the ink spots.

For that matter, physicists themselves go around transforming negative entropy into information, said Brillouin. From observations and measurements, the physicist derives scientific laws; with these laws, people create machines never seen in nature, with the most improbable structures. He wrote this in 1950, as he was leaving Harvard to join the IBM Corporation in Poughkeepsie.

That was not the end for Maxwell's demon—far from it. The problem could not truly be solved, the demon effectively banished without a deeper understanding of a realm far removed from thermodynamics: mechanical computing. Later, Peter Landsberg wrote its obituary this way: "Maxwell's demon died at the age of 62 (when a paper by Leó Szilárd appeared), but it continues to haunt the castles of physics as a restless and lovable poltergeist."

LIFE'S OWN CODE

(The Organism Is Written in the Egg)

> *What lies at the heart of every living thing is not a fire, not warm breath, not a "spark of life." It is information, words, instructions. If you want a metaphor, don't think of fires and sparks and breath. Think, instead, of a billion discrete, digital characters carved in tablets of crystal.*
>
> —Richard Dawkins (1986)

SCIENTISTS LOVE THEIR FUNDAMENTAL PARTICLES. If traits are handed down from one generation to the next, these traits must take some primal form or have some carrier. Hence the putative particle of protoplasm. "The biologist must be allowed as much scientific use of the imagination as the physicist," *The Popular Science Monthly* explained in 1875. "If the one must have his atoms and molecules, the other must have his physiological units, his plastic molecules, his 'plasticules.'"

Plasticule did not catch on, and almost everyone had the wrong idea about heredity anyway. So in 1910 a Danish botanist, Wilhelm Johannsen, self-consciously invented the word *gene*. He was at pains to correct the common mythology and thought a word might help. The myth was this: that "personal qualities" are transmitted from parent to progeny. This is "the most naïve and oldest conception of heredity," Johannsen said in a speech to the American Society of Naturalists. It was understandable. If father and daughter are fat, people might be tempted to think that his fatness caused hers, or that he passed it on to her. But that is wrong. As Johannsen declared, "The *personal qualities* of any individual organism do not at all cause the qualities of its offspring; but the qualities of both ancestor and descendent are in quite the same manner determined by the nature of the 'sexual substances'—i.e., the gametes—from which they have developed." What is inherited is more abstract, more in the nature of potentiality.

To banish the fallacious thinking, he proposed a new terminology, beginning with *gene*: "nothing but a very applicable little word, easily combined with others."* It hardly mattered that neither he nor anyone else knew what a gene actually was; "it may be useful as an expression for the 'unit-factors,' 'elements,' or 'allelomorphs.' . . . As to the nature of the 'genes' it is as yet of no value to propose a hypothesis." Gregor Mendel's years of research with green and yellow peas showed that such a thing must exist. Colors and other traits vary depending on many factors, such as temperature and soil content, but *something* is preserved whole; it does not blend or diffuse; it must be quantized. Mendel had discovered the gene, though he did not name it. For him it was more an algebraic convenience than a physical entity.

When Schrödinger contemplated the gene, he faced a problem. How could such a "tiny speck of material" contain the entire complex code-script that determines the elaborate development of the organism? To resolve the difficulty Schrödinger summoned an example not from wave mechanics or theoretical physics but from telegraphy: Morse code. He noted that two signs, dot and dash, could be combined in well-ordered groups to generate all human language. Genes, too, he suggested, must employ a code: "The miniature code should precisely correspond with a highly complicated and specified plan of development and should somehow contain the means to put it into action."

Codes, instructions, signals—all this language, redolent of machinery and engineering, pressed in on biologists like Norman French invading

* He added: "Old terms are mostly compromised by their application in antiquated or erroneous theories and systems, from which they carry splinters of inadequate ideas, not always harmless to the developing insight."

medieval English. In the 1940s the jargon had a precious, artificial feeling, but that soon passed. The new molecular biology began to examine information storage and information transfer. Biologists could count in terms of "bits." Some of the physicists now turning to biology saw information as exactly the concept needed to discuss and measure biological qualities for which tools had not been available: complexity and order, organization and specificity. Henry Quastler, an early radiologist from Vienna, then at the University of Illinois, was applying information theory to both biology and psychology; he estimated that an amino acid has the information content of a written word and a protein molecule the information content of a paragraph. His colleague Sidney Dancoff suggested to him in 1950 that a chromosomal thread is "a linear coded tape of information":

> The entire thread constitutes a "message." This message can be broken down into sub-units which may be called "paragraphs," "words," etc. The smallest message unit is perhaps some flip-flop which can make a yes-no decision.

In 1952 Quastler organized a symposium on information theory in biology, with no purpose but to deploy these new ideas—entropy, noise, messaging, differentiating—in areas from cell structure and enzyme catalysis to large-scale "biosystems." One researcher constructed an estimate of the number of bits represented by a single bacterium: as much as 10^{13}. (But that was the number needed to describe its entire molecular structure in three dimensions—perhaps there was a more economical description.) The growth of the bacterium could be analyzed as a reduction in the entropy of its part of the universe. Quastler himself wanted to take the measure of higher organisms in terms of information content: not in terms of atoms ("this would be extremely wasteful") but in terms of "hypothetical instructions to build an organism." This brought him, of course, to genes.

The whole set of instructions—situated "somewhere in the chromosomes"—is the genome. This is a "catalogue," he said, containing, if not all, then at least "a substantial fraction of all information about an adult organism." He emphasized, though, how little was known about genes. Were they discrete physical entities, or did they overlap? Were they "independent sources of information" or did they affect one another? How many were there? Multiplying all these unknowns, he arrived at a result:

> that the essential complexity of a single cell and of a whole man are both not more than 10^{12} nor less than 10^5 bits; this is an extremely coarse estimate, but is better than no estimate at all.

These crude efforts led to nothing, directly. Shannon's information theory could not be grafted onto biology whole. It hardly mattered. A seismic shift was already under way: from thinking about energy to thinking about information.

Across the Atlantic, an odd little letter arrived at the offices of the journal *Nature* in London in the spring of 1953, with a list of signatories from Paris, Zurich, Cambridge, and Geneva, most notably Boris Ephrussi, France's first professor of genetics. The scientists complained of "what seems to us a rather chaotic growth in technical vocabulary." In particular, they had seen genetic recombination in bacteria described as "transformation," "induction," "transduction," and even "infection." They proposed to simplify matters:

> As a solution to this confusing situation, we would like to suggest the use of the term "interbacterial information" to replace those above. It does not imply necessarily the transfer of material substances, and recognizes the possible future importance of cybernetics at the bacterial level.

This was the product of a wine-flushed lakeside lunch at Locarno, Switzerland—meant as a joke, but entirely plausible to the editors of

Nature, who published it forthwith. The youngest of the lunchers and signers was a twenty-five-year-old American named James Watson.

The very next issue of *Nature* carried another letter from Watson, along with his collaborator, Francis Crick. It made them famous. They had found the gene.

A consensus had emerged that whatever genes were, however they functioned, they would probably be proteins: giant organic molecules made of long chains of amino acids. Alternatively, a few geneticists in the 1940s focused instead on simple viruses—phages. Then again, experiments on heredity in bacteria had persuaded a few researchers, Watson and Crick among them, that genes might lie in a different substance, which, for no known reason, was found within the nucleus of every cell, plant and animal, phages included. This substance was a nucleic acid, particularly deoxyribonucleic acid, or DNA. The people working with nucleic acids, mainly chemists, had not been able to learn much about it, except that the molecules were built up from smaller units, called nucleotides. Watson and Crick thought this must be the secret, and they raced to figure out its structure at the Cavendish Laboratory in Cambridge. They could not see these molecules; they could only seek clues in the shadows cast by X-ray diffraction. But they knew a great deal about the subunits. Each nucleotide contained a "base," and there were just four different bases, designated as A, C, G, and T. They came in strictly predictable proportions. They must be the letters of the code. The rest was trial and error, fired by imagination.

What they discovered became an icon: the double helix, heralded on magazine covers, emulated in sculpture. DNA is formed of two long sequences of bases, like ciphers coded in a four-letter alphabet, each sequence complementary to the other, coiled together. Unzipped, each strand may serve as a template for replication. (Was it Schrödinger's "aperiodic crystal"? In terms of physical structure, X-ray diffraction showed DNA to be entirely regular. The aperiodicity lies at the abstract level of language—the sequence of "letters.") In the local pub, Crick, ebullient, announced to anyone who would listen that they had discovered "the secret of life"; in their one-page note in *Nature* they were more circumspect. They ended with a remark that has been called "one of the most coy statements in the literature of science":

> It has not escaped our notice that the specific pairing we have postulated immediately suggests a possible copying mechanism for the genetic material.

They dispensed with the timidity in another paper a few weeks later. In each chain the sequence of bases appeared to be irregular—any sequence was possible, they observed. "It follows that in a long molecule many different permutations are possible." Many permutations—many possible messages. Their next remark set alarms sounding on both sides of the Atlantic: "It therefore seems likely that the precise sequence of the bases is the code which carries the genetical information." In using these terms, *code* and *information*, they were no longer speaking figuratively.

The macromolecules of organic life embody information in an intricate structure. A single hemoglobin molecule comprises four chains of polypeptides, two with 141 amino acids and two with 146, in strict linear sequence, bonded and folded together. Atoms of hydrogen, oxygen, carbon, and iron could mingle randomly for the lifetime of the universe and be no more likely to form hemoglobin than the proverbial chimpanzees to type the works of Shakespeare. Their genesis requires energy; they are built up from simpler, less patterned parts, and the law of entropy applies. For earthly life, the energy comes as photons from the sun. The information comes via evolution.

The DNA molecule was special: the information it bears is its only function. Having recognized this, microbiologists turned to the problem of deciphering the code. Crick, who had been inspired to leave physics for biology when he read Schrödinger's *What Is Life?*, sent Schrödinger a copy of the paper but did not receive a reply.

On the other hand, George Gamow saw the Watson-Crick report when he was visiting the Radiation Laboratory at Berkeley. Gamow was a Ukrainian-born cosmologist—an originator of the Big Bang theory—and he knew a big idea when he saw one. He sent off a letter:

> Dear Drs. Watson & Crick,
>
> I am a physicist, not a biologist. . . . But I am very much excited by your article in May 30th *Nature*, and think that brings Biology over into the group of "exact" sciences. . . . If your point of view is correct each organism will be characterized by a long number written in quadrucal (?) system with figures 1, 2, 3, 4 standing for different bases. . . . This would open a very exciting possibility of theoretical research based on combinatorix and the theory of numbers! . . . I have a feeling this can be done. What do you think?

For the next decade, the struggle to understand the genetic code consumed a motley assortment of the world's great minds, many of them, like Gamow, lacking any useful knowledge of biochemistry. For Watson and Crick, the initial problem had depended on a morass of specialized particulars: hydrogen bonds, salt linkages, phosphate-sugar chains with deoxyribofuranose residues. They had to learn how inorganic ions could be organized in three dimensions; they had to calculate exact angles of chemical bonds. They made models out of cardboard and tin plates. But now the problem was being transformed into an abstract game of symbol manipulation. Closely linked to DNA, its single-stranded cousin, RNA, appeared to play the role of messenger or translator. Gamow said explicitly that the underlying chemistry hardly mattered. He and others who followed him understood this as a puzzle in mathematics—a mapping between messages in different alphabets. If this was a coding problem, the tools they needed came from combinatorics and information theory. Along with physicists, they consulted cryptanalysts.

Gamow himself began impulsively by designing a combinatorial code. As he saw it, the problem was to get from the four bases in DNA to the twenty known amino acids in proteins—a code, therefore, with four letters and twenty words.* Pure combinatorics made him think of nucleotide triplets: three-letter words. He had a detailed solution—soon known as his "diamond code"—published in *Nature* within a few months. A few months after that, Crick showed this to be utterly wrong: experimental data on protein sequences ruled out the diamond code. But Gamow was not giving up. The triplet idea was seductive. An unexpected cast of scientists joined the hunt: Max Delbrück, an ex-physicist now at Caltech in biology; his friend Richard Feynman, the quantum theorist; Edward Teller, the famous bomb maker; another Los Alamos alumnus, the mathematician Nicholas Metropolis; and Sydney Brenner, who joined Crick at the Cavendish.

They all had different coding ideas. Mathematically the problem seemed daunting even to Gamow. "As in the breaking of enemy messages during the war," he wrote in 1954, "the success depends on the available length of the coded text. As every intelligence officer will tell you, the work is very hard, and the success depends mostly on luck. . . . I am afraid that the problem cannot be solved without the help of electronic computer." Gamow and Watson decided to make it a club: the RNA Tie Club, with exactly twenty members. Each member received a woolen tie in black and green, made to Gamow's design by a haberdasher in Los Angeles. The game playing aside, Gamow wanted to create a communication channel to bypass journal publication. News in science had never moved so fast. "Many of the essential concepts were first proposed in informal discussions on both sides of the Atlantic and were then quickly broadcast to the cognoscenti," said another member, Gunther Stent, "by private international bush telegraph." There were false starts, wild guesses, and dead ends, and the established biochemistry community did not always go along willingly.

"People didn't necessarily *believe* in the code," Crick said later. "The

---

* In listing twenty amino acids, Gamow was getting ahead of what was actually known. The number twenty turned out to be correct, though Gamow's list was not.

majority of biochemists simply weren't thinking along those lines. It was a completely novel idea, and moreover they were inclined to think it was oversimplified." They thought the way to understand proteins would be to study enzyme systems and the coupling of peptide units. Which was reasonable enough.

> They thought protein synthesis couldn't be a simple matter of coding from one thing to another; that sounded too much like something a *physicist* had invented. It didn't sound like biochemistry to *them*. . . . So there was a certain resistance to simple ideas like three nucleotides' coding an amino acid; people thought it was rather like cheating.

Gamow, at the other extreme, was bypassing the biochemical details to put forward an idea of shocking simplicity: that any living organism is determined by "a long number written in a four-digital system." He called this "the number of the beast" (from Revelation). If two beasts have the same number, they are identical twins.

By now the word *code* was so deeply embedded in the conversation that people seldom paused to notice how extraordinary it was to find such a thing—abstract symbols representing arbitrarily different abstract symbols—at work in chemistry, at the level of molecules. The genetic code performed a function with uncanny similarities to the metamathematical code invented by Gödel for his philosophical purposes. Gödel's code substitutes plain numbers for mathematical expressions and operations; the genetic code uses triplets of nucleotides to represent amino acids. Douglas Hofstadter was the first to make this connection explicitly, in the 1980s: "between the complex machinery in a living cell that enables a DNA molecule to replicate itself and the clever machinery in a mathematical system that enables a formula to say things about itself." In both cases he saw a twisty feedback loop. "Nobody had ever in the least suspected that one set of chemicals could *code* for another set," Hofstadter wrote.

Indeed, the very idea is somewhat baffling: If there is a code, then who invented it? What kinds of messages are written in it? Who writes them? Who reads them?

The Tie Club recognized that the problem was not just information storage but information transfer. DNA serves two different functions. First, it preserves information. It does this by copying itself, from generation to generation, spanning eons—a Library of Alexandria that keeps its data safe by copying itself billions of times. Notwithstanding the beautiful double helix, this information store is essentially one-dimensional: a string of elements arrayed in a line. In human DNA, the nucleotide units number more than a billion, and this detailed gigabit message must be conserved perfectly, or almost perfectly. Second, however, DNA also sends that information outward for use in the making of the organism. The data stored in a one-dimensional strand has to flower forth in three dimensions. This information transfer occurs via messages passing from the nucleic acids to proteins. So DNA not only replicates itself; separately, it dictates the manufacture of something entirely different. These proteins, with their own enormous complexity, serve as the material of a body, the mortar and bricks, and also as the control system, the plumbing and wiring and the chemical signals that control growth.

The replication of DNA is a copying of information. The manufacture of proteins is a transfer of information: the sending of a message. Biologists could see this clearly now, because the *message* was now well defined and abstracted from any particular substrate. If messages could be borne upon sound waves or electrical pulses, why not by chemical processes?

Gamow framed the issue simply: "The nucleus of a living cell is a storehouse of information." Furthermore, he said, it is a transmitter of information. The continuity of all life stems from this "information system"; the proper study of genetics is "the language of the cells."

When Gamow's diamond code proved wrong, he tried a "triangle code," and more variations followed—also wrong. Triplet codons remained central, and a solution seemed tantalizingly close but out of reach. A problem was how nature punctuated the seemingly unbroken DNA and RNA strands. No one could see a biological equivalent for the pauses that separate letters in Morse code, or the spaces that separate words. Perhaps every fourth base was a comma. Or maybe (Crick suggested) commas would be unnecessary if some triplets made "sense" and others made "nonsense." Then again, maybe a sort of tape reader just needed to start at a certain point and count off the nucleotides three by three. Among the mathematicians drawn to this problem were a group at the new Jet Propulsion Laboratory in Pasadena, California, meant to be working on aerospace research. To them it looked like a classic problem in Shannon coding theory: "the sequence of nucleotides as an infinite message, written without punctuation, from which any finite portion must be decodable into a sequence of amino acids by suitable insertion of commas." They constructed a *dictionary* of codes. They considered the problem of *misprints*.

Biochemistry did matter. All the world's cryptanalysts, lacking petri dishes and laboratory kitchens, would not have been able to guess from among the universe of possible answers. When the genetic code was solved, in the early 1960s, it turned out to be full of redundancy. Much of the mapping from nucleotides to amino acids seemed arbitrary—not as neatly patterned as any of Gamow's proposals. Some amino acids correspond to just one codon, others to two, four, or six. Particles called ribosomes ratchet along the RNA strand and translate it, three bases at a time. Some codons are redundant; some actually serve as start signals and stop signals. The redundancy serves exactly the purpose that an information theorist would expect. It provides tolerance for errors. Noise affects biological messages like any other. Errors in DNA—misprints—are mutations.

Even before the exact answer was reached, Crick crystallized its fundamental principles in a statement that he called (and is called to this day) the Central Dogma. It is a hypothesis about the direction of evolution and the origin of life; it is provable in terms of Shannon entropy in the possible chemical alphabets:

Once "information" has passed into protein it *cannot get out again*. In more detail, the transfer of information from nucleic acid to nucleic acid, or from nucleic acid to protein may be possible, but transfer from protein to protein, or from protein to nucleic acid is impossible. Information means here the *precise* determination of sequence.

The genetic message is independent and impenetrable: no information from events outside can change it.

Information had never been writ so small. Here is scripture at angstrom scale, published where no one can see, the Book of Life in the eye of a needle.

Omne vivum ex ovo. "The complete description of the organism is already written in the egg," said Sydney Brenner to Horace Freeland Judson, molecular biology's great chronicler, at Cambridge in the winter of 1971. "Inside every animal there is an internal description of that animal. . . . What is going to be difficult is the immense amount of detail that will have to be subsumed. The most economical language of description is the molecular, genetic description that is already there. We do not yet know, in that language, what the *names* are. What does the organism name *to itself*? We cannot say that an organism has, for example, a name for a finger. There's no guarantee that in making a hand, the explanation can be couched in the terms we use for making a glove."

Brenner was in a thoughtful mood, drinking sherry before dinner at King's College. When he began working with Crick, less than two decades before, molecular biology did not even have a name. Two decades later, in the 1990s, scientists worldwide would undertake the mapping of the entire human genome: perhaps 20,000 genes, 3 billion base pairs. What was the most fundamental change? It was a shift of the frame, from energy and matter to information.

"All of biochemistry up to the fifties was concerned with where you get the energy and the materials for cell function," Brenner said.

"Biochemists only thought about the flux of energy and the flow of matter. Molecular biologists started to talk about the flux of information. Looking back, one can see that the double helix brought the realization that information in biological systems could be studied in much the same way as energy and matter. . . .

"Look," he told Judson, "let me give you an example. If you went to a biologist twenty years ago and asked him, How do you make a protein, he would have said, Well, that's a horrible problem, I don't know . . . but the important question is where do you get the energy to make the peptide bond. Whereas the molecular biologist would have said, That's not the problem, the important problem is where do you get the instructions to assemble the sequence of amino acids, and to hell with the energy; the energy will look after itself."

By this time, the technical jargon of biologists included the words *alphabet*, *library*, *editing*, *proofreading*, *transcription*, *translation*, *nonsense*, *synonym*, and *redundancy*. Genetics and DNA had drawn the attention not just of cryptographers but of classical linguists. Certain proteins, capable of flipping from one relatively stable state to another, were found to act as relays, accepting ciphered commands and passing them to their neighbors—switching stations in three-dimensional communications networks. Brenner, looking forward, thought the focus would turn to computer science as well. He envisioned a science—though it did not yet have a name—of chaos and complexity. "I think in the next twenty-five years we are going to have to teach biologists another language still," he said. "I don't know what it's called yet; nobody knows. But what one is aiming at, I think, is the fundamental problem of the theory of elaborate systems." He recalled John von Neumann, at the dawn of information theory and cybernetics, proposing to understand biological processes and mental processes in terms of how a computing machine might operate. "In other words," said Brenner, "where a science like physics works in terms of laws, or a science like molecular biology, to now, is stated in terms of mechanisms, maybe now what one has to begin to think of is algorithms. Recipes. Procedures."

If you want to know what a mouse is, ask instead how you could build a mouse. How does the mouse build itself? The mouse's genes switch one another on and off and perform computation, in steps. "I feel that this new molecular biology has to go in this direction—to explore the high-level logical computers, the programs, the algorithms of development. . . .

"One would like to be able to fuse the two—to be able to move between the molecular hardware and the *logical* software of how it's all organized, without feeling they are different sciences."

Even now—or especially now—the gene was not what it seemed. Having begun as a botanist's hunch and an algebraic convenience, it had been tracked down to the chromosome and revealed as molecular coiled strands. It was decoded, enumerated, and catalogued. And then, in the heyday of molecular biology, the idea of the gene broke free of its moorings once again.

The more was known, the harder it was to define. Is a gene nothing more or less than DNA? Is it *made* of DNA, or is it something *carried* in DNA? Is it properly pinned down as a material thing at all?

Not everyone agreed there was a problem. Gunther Stent declared in 1977 that one of the field's great triumphs was the "unambiguous identification" of the Mendelian gene as a particular length of DNA. "It is in this sense that all working geneticists now employ the term 'gene,'" he wrote. To put it technically but succinctly: "The gene is, in fact, a linear array of DNA nucleotides which determines a linear array of protein amino acids." It was Seymour Benzer, said Stent, who established that definitively.

Yet Benzer himself had not been quite so sanguine. He argued as early as 1957 that the classical gene was dead. It was a concept trying to serve three purposes at once—as a unit of recombination, of mutation, and of function—and already he had strong reason to suspect that these were incompatible. A strand of DNA carries many base pairs, like beads on a string or letters in a sentence; as a physical object it could not be called an elementary unit. Benzer offered a batch of new particle names: "recon," for the smallest unit that can be interchanged by recombination; "muton," for the smallest unit of mutational change (a single base pair); and "cistron" for the unit of function—which in turn, he admitted, was difficult to define. "It depends upon what level of function is meant," he wrote—perhaps just the specification of an amino acid, or perhaps a whole ensemble of steps "leading to *one* particular physiological end-effect." *Gene* was not going away, but that was a lot of weight for one little word to bear.

Part of what was happening was a collision between molecular biology and evolutionary biology, as studied in fields from botany to paleontology. It was as fruitful a collision as any in the history of science—before long, neither side could move forward without the other—but on the way some sparks flared. Quite of few of them were set off by a young zoologist at Oxford, Richard Dawkins. It seemed to Dawkins that many of his colleagues were looking at life the wrong way round.

As molecular biology perfected its knowledge of the details of DNA and grew more skillful in manipulating these molecular prodigies, it was natural to see them as the answer to the great question of life: how do organisms reproduce themselves? We use DNA, just as we use lungs to breathe and eyes to see. We *use* it. "This attitude is an error of great profundity," Dawkins wrote. "It is the truth turned crashingly on its head." DNA came first—by billions of years—and DNA *comes* first, he argued, when life is viewed from the proper perspective. From that perspective, genes are the focus, the sine qua non, the star of the show. In his first book—published in 1976, meant for a broad audience, provocatively titled *The Selfish Gene*—he set off decades of debate by declaring: "We are survival machines—robot vehicles blindly programmed to preserve the selfish molecules known as genes." He said this was a truth he had known for years.

Genes, not organisms, are the true units of natural selection. They began as "replicators"—molecules formed accidentally in the primordial soup, with the unusual property of making copies of themselves.

> They are past masters of the survival arts. But do not look for them floating loose in the sea; they gave up that cavalier freedom long ago. Now they swarm in huge colonies, safe inside gigantic lumbering robots, sealed off from the outside world, communicating with it by tortuous indirect routes, manipulating it by remote control. They are in you and in me; they created us, body and mind; and their preservation is the ultimate rationale for our existence. They have come a long way, those replicators. Now they go by the name of genes, and we are their survival machines.

This was guaranteed to raise the hackles of organisms who thought of themselves as more than robots. "English biologist Richard Dawkins has recently raised my hackles," wrote Stephen Jay Gould in 1977, "with his claim that genes themselves are units of selection, and individuals merely their temporary receptacles." Gould had plenty of company. Speaking for many molecular biologists, Gunther Stent dismissed Dawkins as "a thirty-six-year-old student of animal behavior" and filed him under "the old prescientific tradition of animism, under which natural objects are endowed with souls."

Yet Dawkins's book was brilliant and transformative. It established a new, multilayered understanding of the gene. At first, the idea of the selfish gene seemed like a trick of perspective, or a joke. Samuel Butler had said a century earlier—and did not claim to be the first—that a hen is only an egg's way of making another egg. Butler was quite serious, in his way:

> Every creature must be allowed to "run" its own development in its own way; the egg's way may seem a very roundabout manner of doing things; but it *is* its way, and it is one of which man, upon the whole, has no great reason to complain. Why the fowl should be considered more alive than the egg, and why it should be said that the hen lays the egg, and not that the egg lays the hen, these are questions which lie beyond the power of

philosophic explanation, but are perhaps most answerable by considering the conceit of man, and his habit, persisted in during many ages, of ignoring all that does not remind him of himself.

He added, "But, perhaps, after all, the real reason is, that the egg does not cackle when it has laid the hen." Some time later, Butler's template, *X is just a Y's way of making another Y*, began reappearing in many forms. "A scholar," said Daniel Dennett in 1995, "is just a library's way of making another library." Dennett, too, was not entirely joking.

It was prescient of Butler in 1878 to mock a man-centered view of life, but he had read Darwin and could see that all creation had not been designed in behalf of *Homo sapiens*. "Anthropocentrism is a disabling vice of the intellect," Edward O. Wilson said a century later, but Dawkins was purveying an even more radical shift of perspective. He was not just nudging aside the human (and the hen) but the organism, in all its multifarious glory. How could biology *not* be the study of organisms? If anything, he understated the difficulty when he wrote, "It requires a deliberate mental effort to turn biology the right way up again, and remind ourselves that the replicators come first, in importance as well as in history."

A part of Dawkins's purpose was to explain altruism: behavior in individuals that goes against their own best interests. Nature is full of examples of animals risking their own lives in behalf of their progeny, their cousins, or just fellow members of their genetic club. Furthermore, they share food; they cooperate in building hives and dams; they doggedly protect their eggs. To explain such behavior—to explain any adaptation, for that matter—one asks the forensic detective's question, *cui bono*? Who benefits when a bird spots a predator and cries out, warning the flock but also calling attention to itself? It is tempting to think in terms of the good of the group—the family, tribe, or species—but most theorists agree that evolution does not work that way. Natural selection can seldom operate at the level of groups. It turns out, however, that many explanations fall neatly into place if one thinks of the individual as trying to propagate its particular assortment of genes down through the future. Its species shares most of those genes, of course, and its kin share even more. Of course, the individual does not know about its genes. It is not consciously *trying* to do any such thing. Nor, of course, would anyone impute intention to the gene itself—tiny brainless entity. But it works quite well, as Dawkins showed, to flip perspectives and say that the gene works to maximize its own replication. For example, a gene "might ensure its survival by tending to endow the successive bodies with long legs, which help those bodies escape from predators." A gene might maximize its own numbers by giving an organism the instinctive impulse to sacrifice its life to save its offspring: the gene itself, the particular clump of DNA, dies with its creature, but copies of the gene live on. The process is blind. It has no foresight, no intention, no knowledge. The genes, too, are blind: "They do not plan ahead," says Dawkins. "Genes just *are*, some genes more so than others, and that is all there is to it."

The history of life begins with the accidental appearance of molecules complex enough to serve as building blocks—replicators. The replicator is an information carrier. It survives and spreads by copying itself. The copies must be coherent and reliable but need not be perfect; on the contrary, for evolution to proceed, errors must appear. Replicators could exist long before DNA, even before proteins. In one scenario, proposed by the Scots biologist Alexander Cairns-Smith, replicators appeared in sticky layers of clay crystals: complex molecules of silicate minerals. In other models the evolutionary playground is the more traditional "primordial soup." Either way, some of these information-bearing macromolecules disintegrate more quickly than others; some make more or better copies; some have the chemical effect of breaking up competing molecules. Absorbing photon energy like the miniature Maxwell's demons they are, molecules of ribonucleic acid, RNA, catalyze the formation of bigger and more information-rich molecules. DNA, ever so slightly more stable, possesses the dual capability of copying itself while also manufacturing another sort of molecule, and this provides a special advantage. It can protect itself by building a shell of proteins around it. This is Dawkins's "survival machine"—first cells, then larger and larger bodies, with growing inventories of membranes and tissues and limbs and organs and skills. They are the genes' fancy vehicles, racing against other vehicles, converting energy, and even processing information. In the game of survival some vehicles outplay, outmaneuver, and outpropagate others.

It took some time, but the gene-centered, information-based perspective led to a new kind of detective work in tracing the history of life. Where paleontologists look back through the fossil record for skeletal precursors of wings and tails, molecular biologists and biophysicists look for telltale relics of DNA in hemoglobin, oncogenes, and all the rest of the library of proteins and enzymes. "There is a molecular archeology in the making," says Werner Loewenstein. The history of life is written in terms of negative entropy. "What actually evolves is information in all its forms or transforms. If there were something like a guidebook for living creatures, I think, the first line would read like a biblical commandment, *Make thy information larger.*"

No one gene makes an organism. Insects and plants and animals are collectives, communal vehicles, cooperative assemblies of a multitude of genes, each playing its part in the organism's development. It is a complex ensemble in which each gene interacts with thousands of others in a hierarchy of effects extending through both space and time. The body is a colony of genes. Of course, it acts and moves and procreates as a unit, and furthermore, in the case of at least one species, it feels itself, with impressive certainty, to be a unit. The gene-centered perspective has helped biologists appreciate that the genes composing the human genome are only a fraction of the genes carried around in any one person, because humans (like other species) host an entire ecosystem of microbes—bacteria, especially, from our skin to our digestive systems.

Our "microbiomes" help us digest food and fight disease, all the while evolving fast and flexibly in service of their own interests. All these genes engage in a grand process of mutual co-evolution—competing with one another, and with their alternative alleles, in nature's vast gene pool, but no longer competing on their own. Their success or failure comes through interaction. "Selection favors those genes which succeed *in the presence of other genes*," says Dawkins, "*which in turn succeed in the presence of them.*"

The effect of any one gene depends on these interactions with the ensemble and depends, too, on effects of the environment and on raw chance. Indeed, just to speak of a gene's *effect* became a complex business. It was not enough simply to say that the effect of a gene is the protein it synthesizes. One might want to say that a sheep or a crow has a gene for blackness. This might be a gene that manufactures a protein for black pigment in wool or feathers. But sheep and crows and all the other creatures capable of blackness exhibit it in varying circumstances and degrees; even so simple-seeming a quality seldom has a biological on-off switch. Dawkins suggests the case of a gene that synthesizes a protein that acts as an enzyme with many indirect and distant effects, one of which is to facilitate the synthesis of black pigment. Even more remotely, suppose a gene encourages an organism to seek sunlight, which is in turn necessary for the black pigment. Such a gene serves as a mere co-conspirator but its role may be indispensable. To call it a gene *for* blackness, however, becomes difficult. And it is harder still to specify genes for more complex qualities—genes for obesity or aggression or nest building or braininess or homosexuality.

Are there genes for such things? Not if a gene is a particular strand of DNA that expresses a protein. Strictly speaking, one cannot say there are genes *for* almost anything—not even eye color. Instead, one should say that differences in genes tend to cause differences in phenotype (the actualized organism). But from the earliest days of the study of heredity, scientists have spoken of genes more broadly. If a population varies in some trait—say, tallness—and if the variation is subject to natural selection, then by definition it is at least partly genetic. There is a genetic component to the variation in tallness. There is no gene for long legs; there is no gene for a leg at all. To build a leg requires many genes, each issuing instructions in the form of proteins, some making raw materials, some making timers and on-off switches. Some of these genes surely have the effect of making legs longer than they would otherwise be, and it is those genes that we may call, for short, genes *for* long legs—as long as we remember that long-leggedness is not directly represented or encoded directly in the gene.

So geneticists and zoologists and ethologists and paleontologists all got into the habit of saying "a gene for X" instead of "a genetic contribution to the variation in X." Dawkins was forcing them to face the logical consequences. If there is any genetic variation in a trait—eye color or obesity—then there must be a gene or genes for that trait. It doesn't matter that the actual appearance of the trait may depend on an unfathomable array of other factors, which may be environmental or even accidental. By way of illustration, he offered a deliberately extreme example: a gene for reading.

The idea seems absurd, for several reasons. Reading is learned behavior. No one is born able to read. If ever a skill depends on environmental factors, such as education, it is reading. Until a few millennia ago, the behavior did not exist, so it could not have been subject to natural selection. You might as well say (as the geneticist John Maynard Smith did, mockingly) that there is a gene for tying shoelaces. But Dawkins was undaunted. He pointed out that genes are about *differences*, after all. So he began with a simple counterpoint: might there not be a gene for dyslexia?

All we would need in order to establish the existence of a gene for reading is to discover a gene for not reading, say a gene which induced a brain lesion causing specific dyslexia. Such a dyslexic person might be normal and intelligent in all respects except that he could not read. No geneticist would be particularly surprised if this type of dyslexia turned out to breed true in some Mendelian fashion. Obviously, in this event the gene would only exhibit its effect in an environment which included normal education. In a prehistoric environment it might have had no detectable effect, or it might have had some different effect and have been known to cave-dwelling geneticists as, say, a gene for inability to read animal footprints. . . .

It follows from the ordinary conventions of genetic terminology that the wild-type gene at the same locus, the gene that the rest of the population has in double dose, would properly be called a gene "for reading." If you object to that, you must also object to our speaking of a gene for tallness in Mendel's peas. . . . In both cases the character of interest is a *difference*, and in both cases the difference only shows itself in some specified environment. The reason why something so simple as a one gene difference can have such a complex effect . . . is basically as follows. However complex a given state of the world may be, the *difference* between that state of the world and some alternative state of the world may be caused by something extremely simple.

Can there be a gene for altruism? Yes, says Dawkins, if this means "any gene that influences the development of nervous systems in such a way as to make them likely to behave altruistically." Such genes—these replicators, these survivors—know nothing about altruism and nothing about reading, of course. Whatever and wherever they are, their phenotypic effects matter only insofar as they help the genes propagate.

Molecular biology, in its signal achievement, had pinpointed the gene in a protein-encoding piece of DNA. This was the hardware definition. The software definition was older and fuzzier: the unit of heredity; the bearer of a phenotypic difference. With the two definitions uneasily coexisting, Dawkins looked past them both.

If genes are meant to be masters of survival, they can hardly be fragments of nucleic acid. Such things are fleeting. To say that a replicator manages to survive for eons is to define the replicator as *all the copies considered as one*. Thus the gene does not "grow senile," Dawkins declared.

It is no more likely to die when it is a million years old than when it is only a hundred. It leaps from body to body down the generations, manipulating body after body in its own way and for its own ends, abandoning a succession of mortal bodies before they sink in senility and death.

"What I am doing," he says, "is emphasizing the potential near-immortality of a gene, in the form of copies, as its defining property." This is where life breaks free from its material moorings. (Unless you already believed in the immortal soul.) The gene is not an information-carrying macromolecule. The gene is the information. The physicist Max Delbrück wrote in 1949, "Today the tendency is to say 'genes are just molecules, or hereditary particles,' and thus to do away with the abstractions." Now the abstractions returned.

Where, then, is any particular gene—say, the gene for long legs in humans? This is a little like asking where is Beethoven's Piano Sonata in E minor. Is it in the original handwritten score? The printed sheet music? Any one performance—or perhaps the sum of all performances, historical and potential, real and imagined?

The quavers and crotchets inked on paper are not the music. Music is not a series of pressure waves sounding through the air; nor grooves etched in vinyl or pits burned in CDs; nor even the neuronal symphonies stirred up in the brain of the listener. The music is the information. Likewise, the base pairs of DNA are not genes. They encode genes. Genes themselves are made of bits.

11 | INTO THE MEME POOL

(You Parasitize My Brain)

> *When I muse about memes, I often find myself picturing an*
> *ephemeral flickering pattern of sparks leaping from brain to brain,*
> *screaming "Me, me!"*
>
> —Douglas Hofstadter (1983)

"NOW THROUGH THE VERY UNIVERSALITY of its structures, starting with the code, the biosphere looks like the product of a unique event," Jacques Monod wrote in 1970. "The universe was not pregnant with life, nor the biosphere with man. Our number came up in the Monte Carlo game. Is it any wonder if, like a person who has just made a million at the casino, we feel a little strange and a little unreal?"

Monod, the Parisian biologist who shared the Nobel Prize for working out the role of messenger RNA in the transfer of genetic information, was not alone in thinking of the biosphere as more than a notional place: an entity, composed of all the earth's life-forms, simple and complex, teeming with information, replicating and evolving, coding from one level of abstraction to the next. This view of life was more abstract—more mathematical—than anything Darwin had imagined, but he would have recognized its basic principles. Natural selection directs the whole show. Now biologists, having absorbed the methods and vocabulary of communications science, went further to make their own contributions to the understanding of information itself. Monod proposed an analogy: Just as the biosphere stands above the world of nonliving matter, so an "abstract kingdom" rises above the biosphere. The denizens of this kingdom? Ideas.

> Ideas have retained some of the properties of organisms. Like them, they tend to perpetuate their structure and to breed; they too can fuse, recombine, segregate their content; indeed they too can evolve, and in this evolution selection must surely play an important role.

Ideas have "spreading power," he noted—"infectivity, as it were"—and some more than others. An example of an infectious idea might be a religious ideology that gains sway over a large group of people. The American neurophysiologist Roger Sperry had put forward a similar notion several years earlier, arguing that ideas are "just as real" as the neurons they inhabit. Ideas have power, he said.

> Ideas cause ideas and help evolve new ideas. They interact with each other and with other mental forces in the same brain, in neighboring brains, and thanks to global communication, in far distant, foreign brains. And they also interact with the external surroundings to produce in toto a burstwise advance in evolution that is far beyond anything to hit the evolutionary scene yet. . . .

Monod added, "I shall not hazard a theory of the selection of ideas." No need. Others were willing.

Richard Dawkins made his own connection between the evolution of genes and the evolution of ideas. His essential actor was the replicator, and it scarcely mattered whether replicators were made of nucleic acid. His rule is "All life evolves by the differential survival of replicating entities." Wherever there is life, there must be replicators. Perhaps on other worlds replicators could arise in a silicon-based chemistry—or in no chemistry at all.

What would it mean for a replicator to exist without chemistry? "I think that a new kind of replicator has recently emerged on this planet," he proclaimed at the end of his first book, in 1976. "It is staring us in the face. It is still in its infancy, still drifting clumsily about in its primeval soup, but already it is achieving evolutionary change at a rate that leaves the old gene panting far behind." That "soup" is human culture; the vector of transmission is language; and the spawning ground is the brain.

For this bodiless replicator itself, Dawkins proposed a name. He called it the *meme*, and it became his most memorable invention, far more influential than his selfish genes or his later proselytizing against religiosity. "Memes propagate themselves in the meme pool by leaping from brain to brain via a process which, in the broad sense, can be called imitation," he wrote. They compete with one another for limited resources: brain time or bandwidth. They compete most of all for *attention*. For example:

Ideas. Whether an idea arises uniquely or reappears many times, it may thrive in the meme pool or it may dwindle and vanish. The belief in God is an example Dawkins offers—an ancient idea, replicating itself not just in words but in music and art. The belief that the earth orbits the sun is no less a meme, competing with others for survival. (Truth may be a helpful quality for a meme, but it is only one among many.)

Tunes. This tune

has spread for centuries across several continents. This one

a notorious though shorter-lived invader of brains, overran an immense population many times faster.

Catchphrases. One text snippet, "What hath God wrought?" appeared early and spread rapidly in more than one medium. Another, "Read my lips," charted a peculiar path through late twentieth-century America.

"Survival of the fittest" is a meme that, like other memes, mutates wildly ("survival of the fattest"; "survival of the sickest"; "survival of the fakest"; "survival of the twittest"; . . .).

Images. In Isaac Newton's lifetime, no more than a few thousand people had any idea what he looked like, though he was one of England's most famous men, yet now millions of people have quite a clear idea—based on replicas of copies of rather poorly painted portraits. Even more pervasive and indelible are the smile of *Mona Lisa*, *The Scream* of Edvard Munch, and the silhouettes of various fictional extraterrestrials. These are memes, living a life of their own, independent of any physical reality. "This may not be what George Washington looked like then," a tour guide was overheard saying of the Gilbert Stuart painting at the Metropolitan Museum of Art, "but this is what he looks like now." Exactly.

Memes emerge in brains and travel outward, establishing beachheads on paper and celluloid and silicon and anywhere else information can go. They are not to be thought of as elementary particles but as organisms. The number three is not a meme; nor is the color blue, nor any simple thought, any more than a single nucleotide can be a gene. Memes are complex units, distinct and memorable—units with staying power. Also, an object is not a meme. The hula hoop is not a meme; it is made of plastic, not of bits. When this species of toy spread worldwide in a mad epidemic in 1958, it was the product, the physical manifestation of a meme, or memes: the craving for hula hoops; the swaying, swinging, twirling skill set of hula-hooping. The hula hoop itself is a meme vehicle. So, for that matter, is each human hula hooper—a strikingly effective meme vehicle, in the sense neatly explained by the philosopher Daniel Dennett: "A wagon with spoked wheels carries not only grain or freight from place to place; it carries the brilliant idea of a wagon with spoked wheels from mind to mind." Hula hoopers did that for the hula hoop's memes—and in 1958 they found a new transmission vector, broadcast television, sending its messages immeasurably faster and farther than any wagon. The moving image of the hula hooper seduced new minds by hundreds, and then by thousands, and then by millions. The meme is not the dancer but the dance.

We are their vehicles and their enablers. For most of our biological history they existed fleetingly; their main mode of transmission was the one called "word of mouth." Lately, however, they have managed to adhere in solid substance: clay tablets, cave walls, paper sheets. They achieve longevity through our pens and printing presses, magnetic tapes and optical disks. They spread via broadcast towers and digital networks. Memes may be stories, recipes, skills, legends, and fashions. We copy them, one person at a time. Alternatively, in Dawkins's meme-centered perspective, they copy themselves. At first some of Dawkins's readers wondered how literally to take that. Did he mean to give memes anthropomorphic desires, intentions, and goals? It was the selfish gene all over again. (Typical salvo: "Genes cannot be selfish or unselfish, any more than atoms can be jealous, elephants abstract or biscuits teleological." Typical rebuttal: a reminder that *selfishness* is defined by the geneticist as the tendency to increase one's chances of survival relative to its competitors.)

Dawkins's way of speaking was not meant to suggest that memes are conscious actors, only that they are entities with interests that can be furthered by natural selection. Their interests are not our interests. "A meme," Dennett says, "is an information packet with attitude." When we speak of *fighting for a principle* or *dying for an idea*, we may be more literal than we know. "To die for an idea; it is unquestionably noble," H. L. Mencken wrote. "But how much nobler it would be if men died for ideas that were true!"

Tinker, tailor, soldier, sailor . . . Rhyme and rhythm help people remember bits of text. Or: rhyme and rhythm help bits of text get remembered. Rhyme and rhythm are qualities that aid a meme's survival, just as strength and speed aid an animal's. Patterned language has an evolutionary advantage. Rhyme, rhythm, and reason—for reason, too, is a form of pattern. *I was promised on a time to have reason for my rhyme; from that time unto this season, I received nor rhyme nor reason.*

Like genes, memes have effects on the wide world beyond themselves: phenotypic effects. In some cases (the meme for making fire; for wearing clothes; for the resurrection of Jesus) the effects can be powerful indeed. As they broadcast their influence on the world, memes thus influence the conditions affecting their own chances of survival. The meme or memes composing Morse code had strong positive feedback effects. "I believe that, given the right conditions, replicators automatically band together to create systems, or machines, that carry them around and work to favour their continued replication," wrote Dawkins. Some memes have evident benefits for their human hosts ("look before you leap," knowledge of CPR, belief in hand washing before cooking), but memetic success and genetic success are not the same. Memes can replicate with impressive virulence while leaving swaths of collateral damage—patent medicines and psychic surgery, astrology and satanism, racist myths, superstitions, and (a special case) computer viruses. In a way, these are the most interesting—the memes that thrive to their hosts' detriment, such as the idea that suicide bombers will find their reward in heaven.

When Dawkins first floated the *meme* meme, Nicholas Humphrey, an evolutionary psychologist, said immediately that these entities should be considered "living structures, not just metaphorically but technically":

> When you plant a fertile meme in my mind you literally parasitize my brain, turning it into a vehicle for the meme's propagation in just the way that a virus may parasitize the genetic mechanism of a host cell. And this isn't just a way of talking—the meme for, say, "belief in life after death" is actually realized physically, millions of times over, as a structure in the nervous systems of individual men the world over.

Most early readers of *The Selfish Gene* passed over memes as a fanciful afterthought, but the pioneering ethologist W. D. Hamilton, reviewing the book for *Science*, ventured this prediction:

> Hard as this term may be to delimit—it surely must be harder than gene, which is bad enough—I suspect that it will soon be in common use by

biologists and, one hopes, by philosophers, linguists, and others as well and that it may become absorbed as far as the word "gene" has been into everyday speech.

Memes could travel wordlessly even before language was born. Plain mimicry is enough to replicate knowledge—how to chip an arrowhead or start a fire. Among animals, chimpanzees and gorillas are known to acquire behaviors by imitation. Some species of songbirds *learn* their songs, or at least song variants, after hearing them from neighboring birds (or, more recently, from ornithologists with audio players). Birds develop song repertoires and song dialects—in short, they exhibit a bird-song *culture* that predates human culture by eons. These special cases notwithstanding, for most of human history memes and language have gone hand in glove. (Clichés are memes.) Language serves as culture's first catalyst. It supersedes mere imitation, spreading knowledge by abstraction and encoding.

Perhaps the analogy with disease was inevitable. Before anyone understood anything of epidemiology, its language was applied to species of information. An emotion can be *infectious*, a tune *catchy*, a habit *contagious*. "From look to look, contagious through the crowd / The panic runs," wrote the poet James Thomson in 1730. Lust, likewise, according to Milton: "Eve, whose eye darted contagious fire." But only in the new millennium, in the time of global electronic transmission, has the identification become second nature. Ours is the age of virality: viral education, viral marketing, viral e-mail and video and networking. Researchers studying the Internet itself as a medium—crowdsourcing, collective attention, social networking, and resource allocation—employ not only the language but also the mathematical principles of epidemiology.

One of the first to use the terms *viral text* and *viral sentences* seems to have been a reader of Dawkins named Stephen Walton of New York City, corresponding in 1981 with Douglas Hofstadter. Thinking logically—perhaps in the mode of a computer—Walton proposed simple self-replicating sentences along the lines of "Say me!" "Copy me!" and "If you copy me, I'll grant you three wishes!" Hofstadter, then a columnist for *Scientific American*, found the term *viral text* itself to be even catchier.

> Well, now, Walton's own viral text, as you can see here before your eyes, has managed to commandeer the facilities of a very powerful host—an entire magazine and printing press and distribution service. It has leapt aboard and is now—even as you read this viral sentence—propagating itself madly throughout the ideosphere!

(In the early 1980s, a magazine with a print circulation of 700,000 still seemed like a powerful communications platform.) Hofstadter gaily declared himself infected by the *meme* meme.

One source of resistance—or at least unease—was the shoving of us humans toward the wings. It was bad enough to say that a person is merely a gene's way of making more genes. Now humans are to be considered as vehicles for the propagation of memes, too. No one likes to be called a puppet. Dennett summed up the problem this way: "I don't know about you, but I am not initially attracted by the idea of my brain as a sort of dung heap in which the larvae of other people's ideas renew themselves, before sending out copies of themselves in an informational diaspora. . . . Who's in charge, according to this vision—we or our memes?"

He answered his own question by reminding us that, like it or not, we are seldom "in charge" of our own minds. He might have quoted Freud; instead he quoted Mozart (or so he thought):

> In the night when I cannot sleep, thoughts crowd into my mind. . . . Whence and how do they come? I do not know and I have nothing to do with it. Those which please me I keep in my head and hum them.

Later Dennett was informed that this well-known quotation was not Mozart's after all. It had taken on a life of its own; it was a fairly successful meme.

For anyone taken with the idea of memes, the landscape was changing faster than Dawkins had imagined possible in 1976, when he wrote, "The computers in which memes live are human brains." By 1989, the time of the second edition of *The Selfish Gene*, having become an adept programmer himself, he had to amend that: "It was obviously predictable that manufactured electronic computers, too, would eventually play host to self-replicating patterns of information." Information was passing from one computer to another "when their owners pass floppy discs around," and he could see another phenomenon on the near horizon: computers connected in networks. "Many of them," he wrote, "are literally wired up together in electronic mail exchange. . . . It is a perfect milieu for self-replicating programs to flourish." Indeed, the Internet was in its birth throes. Not only did it provide memes with a nutrient-rich culture medium; it also gave wings to the *idea* of memes. *Meme* itself quickly became an Internet buzzword. Awareness of memes fostered their spread.

A notorious example of a meme that could not have emerged in pre-Internet culture was the phrase "jumped the shark." Loopy self-reference characterized every phase of its existence. To jump the shark means to pass a peak of quality or popularity and begin an irreversible decline. The phrase was thought to have been used first in 1985 by a college student named Sean J. Connolly, in reference to a certain television series. The origin of the phrase requires a certain amount of explanation without which it could not have been initially understood. Perhaps for that reason, there is no recorded usage until 1997, when Connolly's roommate, Jon Hein, registered the domain name *jumptheshark.com* and created a web site devoted to its promotion. The web site soon featured a list of frequently asked questions:

> Q. Did "jump the shark" originate from this web site, or did you create the site to capitalize on the phrase?
> A. This site went up December 24, 1997 and gave birth to the phrase "jump the shark." As the site continues to grow in popularity, the term has become more commonplace. The site is the chicken, the egg, and now a Catch-22.

It spread to more traditional media in the next year; Maureen Dowd devoted a column to explaining it in *The New York Times* in 2001; in 2003 the same newspaper's "On Language" columnist, William Safire, called it "the popular culture's phrase of the year"; soon after that, people were using the phrase in speech and in print without self-consciousness—no quotation marks or explanation—and eventually, inevitably, various cultural observers asked, "Has 'jump the shark' jumped the shark?" ("Granted, Jump the Shark is a brilliant cultural concept. . . . But now the damn thing is everywhere.") Like any good meme, it spawned mutations. The "jumping the shark" entry in Wikipedia advised in 2009, "See also: jumping the couch; nuking the fridge."

Is this science? In his 1983 column, Hofstadter proposed the obvious memetic label for such a discipline: *memetics*. The study of memes has attracted researchers from fields as far apart as computer science and microbiology. In bioinformatics, chain letters are an object of study. They are memes; they have evolutionary histories. The very purpose of a chain letter is replication; whatever else a chain letter may say, it embodies one message: *Copy me*. One student of chain-letter evolution, Daniel W. VanArsdale, listed many variants, in chain letters and even earlier texts: "Make seven copies of it exactly as it is written" [1902]; "Copy this in full and send to nine friends" [1923]; "And if any man shall take away from the words of the book of this prophecy, God shall take away his part out of the book of life" [Revelation 22:19]. Chain letters flourished with the help of a new nineteenth-century technology: "carbonic paper," sandwiched between sheets of writing paper in stacks. Then carbon paper made a symbiotic partnership with another technology, the typewriter. Viral outbreaks of chain letters occurred all through the early twentieth century.

"An unusual chain-letter reached Quincy during the latter part of 1933," wrote a local Illinois historian. "So rapidly did the chain-letter fad develop symptoms of mass hysteria and spread throughout the United States, that by 1935–1936 the Post Office Department, as well as agencies of public opinion, had to take a hand in suppressing the movement." He provided a sample—a meme motivating its human carriers with promises and threats:

> We trust in God. He supplies our needs.
>> Mrs. F. Streuzel . . . . Mich.
>> Mrs. A. Ford . . . . Chicago, Ill.
>> Mrs. K. Adkins . . . . Chicago, Ill.
>> etc.
>
> Copy the above names, omitting the first. Add your name last. Mail it to five persons who you wish prosperity to. The chain was started by an American Colonel and must be mailed 24 hours after receiving it. This will bring prosperity within 9 days after mailing it.
>> Mrs. Sanford won $3,000. Mrs. Andres won $1,000.
>> Mrs. Howe who broke the chain lost everything she possessed.
>> The chain grows a definite power over the expected word.
>> DO NOT BREAK THE CHAIN.

Two subsequent technologies, when their use became widespread, provided orders-of-magnitude boosts in chain-letter fecundity: photocopying (c. 1950) and e-mail (c. 1995). One team of information scientists—Charles H. Bennett from IBM in New York and Ming Li and Bin Ma from Ontario, Canada—inspired by a chance conversation on a hike in the Hong Kong mountains, began an analysis of a set of chain letters collected during the photocopier era. They had thirty-three, all variants of a single letter, with mutations in the form of misspellings, omissions, and transposed words and phrases. "These letters have passed from host to host, mutating and evolving," they reported.

> Like a gene, their average length is about 2,000 characters. Like a potent virus, the letter threatens to kill you and induces you to pass it on to your "friends and associates"—some variation of this letter has probably reached millions of people. Like an inheritable trait, it promises

benefits for you and the people you pass it on to. Like genomes, chain letters undergo natural selection and sometimes parts even get transferred between coexisting "species."

Reaching beyond these appealing metaphors, they set out to use the letters as a "test bed" for algorithms used in evolutionary biology. The algorithms were designed to take the genomes of various modern creatures and work backward, by inference and deduction, to reconstruct their phylogeny—their evolutionary trees. If these mathematical methods worked with genes, the scientists suggested, they should work with chain letters, too. In both cases the researchers were able to verify mutation rates and relatedness measures.

Still, most of the elements of culture change and blur too easily to qualify as stable replicators. They are rarely as neatly fixed as a sequence of DNA. Dawkins himself emphasized that he had never imagined founding anything like a new science of memetics. A peer-reviewed *Journal of Memetics* came to life in 1997—published online, naturally—and then faded away after eight years partly spent in self-conscious debate over status, mission, and terminology. Even compared with genes, memes are hard to mathematize or even to define rigorously. So the gene-meme analogy causes uneasiness and the genetics-memetics analogy even more.

Genes at least have a grounding in physical substance. Memes are abstract, intangible, and unmeasurable. Genes replicate with near-perfect fidelity, and evolution depends on that: some variation is essential, but mutations need to be rare. Memes are seldom copied exactly; their boundaries are always fuzzy, and they mutate with a wild flexibility that would be fatal in biology. The term *meme* could be applied to a suspicious cornucopia of entities, from small to large. For Dennett, the first four notes of Beethoven's Fifth Symphony were "clearly" a meme, along with Homer's *Odyssey* (or at least the *idea* of the *Odyssey*), the wheel, anti-Semitism, and writing. "Memes have not yet found their Watson and Crick," said Dawkins; "they even lack their Mendel."

Yet here they are. As the arc of information flow bends toward ever greater connectivity, memes evolve faster and spread farther. Their presence is felt if not seen in herd behavior, bank runs, informational cascades, and financial bubbles. Diets rise and fall in popularity, their very names becoming catchphrases—the South Beach Diet and the Atkins Diet, the Scarsdale Diet, the Cookie Diet and the Drinking Man's Diet all replicating according to a dynamic about which the science of nutrition has nothing to say. Medical practice, too, experiences "surgical fads" and "iatroepidemics"—epidemics caused by fashions in treatment—like the iatroepidemic of children's tonsillectomies that swept the United States and parts of Europe in the mid-twentieth century, with no more medical benefit than ritual circumcision. Memes were seen through car windows when yellow diamond-shaped BABY ON BOARD signs appeared as if in an instant of mass panic in 1984, in the United States and then Europe and Japan, followed an instant later by a spawn of ironic mutations (BABY I'M BOARD, EX IN TRUNK). Memes were felt when global discourse was dominated in the last year of the millennium by the belief that the world's computers would stammer or choke when their internal clocks reached a special round number.

In the competition for space in our brains and in the culture, the effective combatants are the messages. The new, oblique, looping views of genes and memes have enriched us. They give us paradoxes to write on Möbius strips. "The human world is made of stories, not people," writes David Mitchell. "The people the stories use to tell themselves are not to be blamed." Margaret Atwood writes: "As with all knowledge, once you knew it, you couldn't imagine how it was that you hadn't known it before. Like stage magic, knowledge before you knew it took place before your very eyes, but you were looking elsewhere." Nearing death, John Updike reflects on

A life poured into words—apparent waste
intended to preserve the thing consumed.

Fred Dretske, a philosopher of mind and knowledge, wrote in 1981: "In the beginning there was information. The word came later." He added this explanation: "The transition was achieved by the development of organisms with the capacity for selectively exploiting this information in order to survive and perpetuate their kind." Now we might add, thanks to Dawkins, that the transition was achieved by the information itself, surviving and perpetuating its kind and selectively exploiting organisms.

Most of the biosphere cannot see the infosphere; it is invisible, a parallel universe humming with ghostly inhabitants. But they are not ghosts to us—not anymore. We humans, alone among the earth's organic creatures, live in both worlds at once. It is as though, having long coexisted with the unseen, we have begun to develop the needed extrasensory perception. We are aware of the many species of information. We name their types sardonically, as though to reassure ourselves that we understand: *urban myths* and *zombie lies*. We keep them alive in air-conditioned server farms. But we cannot own them. When a jingle lingers in our ears, or a fad turns fashion upside down, or a hoax dominates the global chatter for months and vanishes as swiftly as it came, who is master and who is slave?

12 | THE SENSE OF RANDOMNESS

(In a State of Sin)

> *"I wonder," she said. "It's getting harder to* see *the patterns, don't you think?"*
>
> —Michael Cunningham (2005)

IN 1958, GREGORY CHAITIN, a precocious eleven-year-old New Yorker, the son of Argentine émigrés, found a magical little book in the library and carried it around with him for a while trying to explain it to other children—and then, he had to admit, trying to understand it himself. It was *Gödel's Proof*, by Ernest Nagel and James R. Newman. Expanded from an article in *Scientific American*, it reviewed the renaissance in logic that began with George Boole; the process of "mapping," encoding statements about mathematics in the form of symbols and even integers; and the idea of metamathematics, systematized language *about* mathematics and therefore *beyond* mathematics. This was heady stuff for the boy, who followed the authors through their simplified but rigorous exposition of Gödel's "astounding and melancholy" demonstration that formal mathematics can never be free of self-contradiction.

The vast bulk of mathematics as practiced at this time cared not at all for Gödel's proof. Startling though incompleteness surely was, it seemed incidental somehow—contributing nothing to the useful work of mathematicians, who went on making discoveries and proving theorems. But philosophically minded souls remained deeply disturbed by it, and these were the sorts of people Chaitin liked to read. One was John von Neumann—who had been there at the start, in Königsberg, 1930, and then in the United States took the central role in the development of computation and computing theory. For von Neumann, Gödel's proof was a point of no return:

> It was a very serious conceptual crisis, dealing with rigor and the proper way to carry out a correct mathematical proof. In view of the earlier notions of the absolute rigor of mathematics, it is surprising that such a thing could have happened, and even more surprising that it could have happened in these latter days when miracles are not supposed to take place. Yet it did happen.

Why? Chaitin asked. He wondered if at some level Gödel's incompleteness could be connected to that new principle of quantum physics, uncertainty, which smelled similar somehow. Later, the adult Chaitin had a chance to put this question to the oracular John Archibald Wheeler. Was Gödel incompleteness related to Heisenberg uncertainty? Wheeler answered by saying he had once posed that very question to Gödel himself, in his office at the Institute for Advanced Study—Gödel with his legs wrapped in a blanket, an electric heater glowing warm against the wintry drafts. Gödel refused to answer. In this way, Wheeler refused to answer Chaitin.

When Chaitin came upon Turing's proof of uncomputability, he thought this must be the key. He also found Shannon and Weaver's book, *The Mathematical Theory of Communication*, and was struck by its upside-down seeming reformulation of entropy: an entropy of bits, measuring information on the one hand and disorder on the other. The common element was randomness, Chaitin suddenly thought. Shannon linked randomness, perversely, to information. Physicists had found randomness inside the atom—the kind of randomness that Einstein deplored by complaining about God and dice. All these heroes of science were talking about or around randomness.

It is a simple word, *random*, and everyone knows what it means. Everyone, that is, and no one. Philosophers and mathematicians struggled endlessly. Wheeler said this much, at least: "Probability, like time, is a concept invented by humans, and humans have to bear the responsibility for the obscurities that attend it." The toss of a fair coin is random, though every detail of the coin's trajectory may be determined à la Newton. Whether the population of France is an even or odd number at any given instant is random, but the population of France itself is surely *not* random: it is a definite fact, even if not knowable. John Maynard Keynes tackled randomness in terms of its opposites, and he chose three: knowledge, causality, and design. What is known in advance, determined by a cause, or organized according to plan cannot be random.

"Chance is only the measure of our ignorance," Henri Poincaré famously said. "Fortuitous phenomena are by definition those whose laws we do not know." Immediately he recanted: "Is this definition very satisfactory? When the first Chaldean shepherds watched the movements of the stars, they did not yet know the laws of astronomy, but would they have dreamed of saying that the stars move at random?" For Poincaré, who understood chaos long before it became a science, examples of randomness included such phenomena as the scattering of raindrops, their causes physically determined but so numerous and complex as to be unpredictable. In physics—or wherever natural processes seem unpredictable—apparent randomness may be noise or may arise from deeply complex dynamics.

Ignorance is subjective. It is a quality of the observer. Presumably randomness—if it exists at all—should be a quality of the thing itself. Leaving humans out of the picture, one would like to say that an event, a choice, a distribution, a game, or, most simply, a number is random.

The notion of a random number is full of difficulties. Can there be such thing as a *particular* random number; a *certain* random number? This number is arguably random:

100973253337652013586346735487680959091173929274945 . . .

Then again, it is special. It begins a book published in 1955 with the title *A Million Random Digits*. The RAND Corporation generated the digits by means of what it described as an electronic roulette wheel: a pulse generator, emitting 100,000 pulses per second, gated through a five-place binary counter, then passed through a binary-to-decimal converter, fed into an IBM punch, and printed by an IBM model 856 Cardatype. The process took years. When the first batch of digits was tested, statisticians discovered significant biases: digits, or groups of digits, or patterns of digits that appeared too frequently or not frequently enough. Finally, however, the tables were published. "Because of the very nature of the tables," the editors said wryly, "it did not seem necessary to proofread every page of the final manuscript in order to catch random errors of the Cardatype."

The book had a market because scientists had a working need for random numbers in bulk, to use in designing statistically fair experiments and building realistic models of complex systems. The new method of Monte Carlo simulation employed random sampling to model phenomena that could not be solved analytically; Monte Carlo simulation was invented and named by von Neumann's team at the atomic-bomb project, desperately trying to generate random numbers to help them calculate neutron diffusion. Von Neumann realized that a mechanical computer, with its deterministic algorithms and finite storage capacity, could never generate truly random numbers. He would have to settle for *pseudorandom* numbers: deterministically generated numbers that behaved as if random. They were random enough for practical purposes. "Any one who considers arithmetical methods of producing random digits is, of course, in a state of sin," said von Neumann.

Randomness might be defined in terms of order—its absence, that is. This orderly little number sequence can hardly be called "random":

00000

Yet it makes a cameo appearance in the middle of the famous million random digits. In terms of probability, that is to be expected: "00000" is as likely to occur as any of the other 99,999 possible five-digit strings. Elsewhere in the million random digits we find:

010101

This, too, appears patterned.

To pick out fragments of pattern in this jungle of digits requires work by an intelligent observer. Given a long enough random string, every possible short-enough substring will appear somewhere. One of them will be the combination to the bank vault. Another will be the encoded complete works of Shakespeare. But they will not do any good, because no one can find them.

Perhaps we may say that numbers like 00000 and 010101 can be random in a particular context. If a person flips a fair coin (one of the simplest mechanical random-number generators) long enough, at some point the coin is bound to come up heads ten times in a row. When that happens, the random-number seeker will typically discard the result and go for a coffee break. This is one of the ways humans do poorly at generating random numbers, even with mechanical assistance. Researchers have established that human intuition is useless both in predicting randomness and in recognizing it. Humans drift toward pattern willy-nilly. The New York Public Library bought *A Million Random Digits* and shelved it under Psychology. In 2010 it was still available from Amazon for eighty-one dollars.

A number is (we now understand) information. When we modern people, Shannon's heirs, think about information in its purest form, we may imagine a string of 0s and 1s, a binary number. Here are two binary strings, fifty digits long:

A: 01010101010101010101010101010101010101010101010101
B: 10001010111110101110100110101000011000100111101111

If Alice (A) and Bob (B) both say they generated their strings by flipping a coin, no one will ever believe Alice. The strings are surely not equally random. Classical probability theory offers no solid reason for claiming that B is more random than A, because a random process *could* produce either string. Probability is about ensembles, not individual events. Probability theory treats events statistically. It does not like questions in the form "How likely was that to happen?" If it happened, it happened.

To Claude Shannon, these strings would look like messages. He would ask, *How much information* does each string contain? On their face, they both contain fifty bits. A telegraph operator charging by the digit would measure the length of the messages and give Alice and Bob the same bill. Then again, the two messages seem to differ profoundly. Message A immediately becomes boring: once you see the pattern, further repetitions provide no new information. In message B, every bit is as valuable as every other. Shannon's first formulation of information theory treated messages statistically, as choices from the ensemble of all possible messages—in the case of A and B, 2^{50} of them. But Shannon also considered redundancy within a message: the pattern, the regularity, the order that makes a message compressible. The more regularity in a message, the more predictable it is. The more predictable, the more redundant. The more redundant a message is, the less information it contains.

The telegraph operator sending message A has a shortcut: he can transmit something like "Repeat '01' twenty-five times." For longer messages with easy patterns, the savings in keystrokes becomes enormous. Once the pattern is clear, the extra characters are free. The operator for message B must soldier on the hard way, sending every character, because every character is a complete surprise; every character costs one bit. This pair of questions—*how random* and *how much information*—turn out to be one and the same. They have a single answer.

Chaitin was not thinking about telegraphs. The device he could not get out of his head was the Turing machine—that impossibly elegant abstraction, marching back and forth along its infinite paper tape, reading and writing symbols. Free from all the real world's messiness, free from creaking wheel-work and finical electricity, free from any need for speed, the Turing machine was the ideal computer. Von Neumann, too, had kept coming back to Turing machines. They were the ever-handy lab mice of computer theory. Turing's U had a transcendent power: a universal Turing machine can simulate any other digital computer, so computer scientists can disregard the messy details of any particular make or model. This is liberating.

Claude Shannon, having moved from Bell Labs to MIT, reanalyzed the Turing machine in 1956. He stripped it down to the smallest possible skeleton, proving that the universal computer could be constructed with just two internal states, or with just two symbols, 0 and 1, or blank and nonblank. He wrote his proof in words more pragmatic than mathematical: he described exactly how the two-state Turing machine would step left and right, "bouncing" back and forth to keep track of the larger numbers of states in a more complex computer. It was all very intricate and specific, redolent of Babbage. For example:

> When the reading head moves, the state information must be transferred to the next cell of the tape to be visited using only two internal states in machine B. If the next state in machine A is to be (say) state 17 (according to some arbitrary numbering system) this is transferred in machine B by "bouncing" the reading head back and forth between the old cell and the new one 17 times (actually 18 trips to the new cell and 17 back to the old one).

The "bouncing operation" carries the information from cell to cell, and the cells act as "transmitters" and "controllers."

Turing had titled his great paper "On Computable Numbers," but of course the real focus was on *un*computable numbers. Could uncomputable numbers and random numbers be related? In 1965 Chaitin was an undergraduate at the City College of New York, writing up a discovery he hoped to submit to a journal; it would be his first publication. He began, "In this paper the Turing machine is regarded as a general purpose computer and some practical questions are asked about programming it." Chaitin, as a high-school student in the Columbia Science Honors Program, had the opportunity to practice programming in machine language on giant IBM mainframes, using decks of punched cards—one card for each line of a program. He would leave his card deck in the computer center and come back the next day for the program's output. He could run Turing machines in his head, too: *write 0, write 1, write blank, shift tape left, shift tape right.* . . . The universal computer gave him a nice way to distinguish between numbers like Alice and Bob's A and B. He could write a program to make a Turing machine print out "010101 . . ." a million times, and he could write down the length of that program—quite short. But given a million random digits—no pattern, no regularity, nothing special at all—there could be no shortcut. The computer program would have to incorporate the entire number. To make the IBM mainframe print out those million digits, he would have to put the whole million digits into the punched cards. To make the Turing machine do it, he would still need the million digits for input.

Here is another number (in decimal this time):

C: 3.1415926535897932384626433832795028841971693993751 . . .

This looks random. Statistically each digit appears with the expected frequency (one in ten); likewise each pair of digits (one in a hundred), each triplet, and so on. A statistician would say it appears to be "normal," as far as anyone can tell. The next digit is always a surprise. The works of Shakespeare will be in there, eventually. But someone might recognize this as a familiar number, π. So it is not random after all.

But why do we say π is not random? Chaitin proposed a clear answer: a number is not random if it is computable—if a definable computer program will generate it. Thus computability is a measure of randomness.

For Turing computability was a yes-or-no quality—a given number either is or is not. But we would like to say that some numbers are more random than others—they are less patterned, less orderly. Chaitin said the patterns and the order express computability. Algorithms generate patterns. So we can gauge computability by looking at *the size of the algorithm*. Given a number—represented as a string of any length—we ask, what is the length of the shortest program that will generate it? Using the language of a Turing machine, that question can have a definite answer, measured in bits.

Chaitin's algorithmic definition of randomness also provides an algorithmic definition of information: the size of the algorithm measures how much information a given string contains.

Looking for patterns—seeking the order amid chaos—is what scientists do, too. The eighteen-year-old Chaitin felt this was no accident. He ended this first paper by applying algorithmic information theory to the process of science itself. "Consider a scientist," he proposed, "who has been observing a closed system that once every second either emits a ray of light or does not."

He summarizes his observations in a sequence of 0s and 1s in which a 0 represents "ray not emitted" and a 1 represents "ray emitted." The sequence may start

0110101110 . . .

and continue for a few thousand more bits. The scientist then examines the sequence in the hope of observing some kind of pattern or law. What does he mean by this? It seems plausible that a sequence of 0s and 1s is patternless if there is no better way to calculate it than just by writing it all out at once from a table giving the whole sequence.

But if the scientist could discover a way to produce the same sequence with an algorithm, a computer program significantly shorter than the sequence, then he would surely know the events were not random. He would say that he had hit upon a theory. This is what science always seeks: a simple theory that accounts for a large set of facts and allows for prediction of events still to come. It is the famous Occam's razor. "We are to admit no more causes of natural things than such as are both true and sufficient to explain their appearances," said Newton, "for nature is pleased with simplicity." Newton quantified *mass* and *force*, but *simplicity* had to wait.

Chaitin sent his paper to the *Journal of the Association for Computing Machinery*. They were happy to publish it, but one referee mentioned that he had heard rumors of similar work coming from the Soviet Union. Sure enough, the first issue of a new journal arrived (after a journey of months) in early 1966: *Проблемы Передачи Информации*, *Problems of Information Transmission*. It contained a paper titled "Three Approaches to the Definition of the Concept 'Amount of Information,'" by A. N. Kolmogorov. Chaitin, who did not read Russian, had just time to add a footnote.

Andrei Nikolaevich Kolmogorov was the outstanding mathematician of the Soviet era. He was born in Tambov, three hundred miles southeast of Moscow, in 1903; his unwed mother, one of three sisters Kolmogorova, died in childbirth, and his aunt Vera raised him in a village near the river Volga. In the waning years of tsarist Russia, this independent-minded woman ran a village school and operated a clandestine printing press in her home, sometimes hiding forbidden documents under baby Andrei's cradle.

Moscow University accepted Andrei Nikolaevich as a student of mathematics soon after the revolution of 1917. Within ten years he was proving a collection of influential results that took form in what became the theory of probability. His *Foundations of the Theory of Probability*, published in Russian in 1933 and in English in 1950, remains the modern classic. But his interests ranged widely, to physics and linguistics as well as other fast-growing branches of mathematics. Once he made a foray into genetics but drew back after a dangerous run-in with Stalin's favorite pseudoscientist, Trofim Lysenko. During World War II

Kolmogorov applied his efforts to statistical theory in artillery fire and devised a scheme of stochastic distribution of barrage balloons to protect Moscow from Nazi bombers. Apart from his war work, he studied turbulence and random processes. He was a Hero of Socialist Labor and seven times received the Order of Lenin.

He first saw Claude Shannon's *Mathematical Theory of Communication* rendered into Russian in 1953, purged of its most interesting features by a translator working in Stalin's heavy shadow. The title became *Statistical Theory of Electrical Signal Transmission*. The word *information*, информация, was everywhere replaced with данные, *data*. The word *entropy* was placed in quotation marks to warn the reader against inferring a connection with entropy in physics. The section applying information theory to the statistics of natural language was omitted entirely. The result was technical, neutral, juiceless, and thus unlikely to attract interpretation in the terms of Marxist ideology. These were serious concerns; "cybernetics" was initially defined in the *Short Philosophical Dictionary* (standard reference of ideological orthodoxy) as a "reactionary pseudoscience" and "an ideological weapon of imperialist reaction." Kolmogorov leapt upon Shannon's paper nonetheless; he, at least, was unafraid to use the word *information*. Working with his students in Moscow, he put forth a rigorous mathematical formulation of information theory, with definitions of the fundamental concepts, careful proofs, and new discoveries—some of which, he soon learned to his sorrow, had appeared in Shannon's original paper but had been omitted from the Russian version.

In the Soviet Union, still moderately isolated from the rest of the world's science, Kolmogorov was well placed to carry the banner of information. He was in charge of all mathematics in the *Great Soviet Encyclopedia*, choosing the authors, editing the articles, and writing much of it himself. In 1956 he delivered a long plenary report on the theory of information transmission to the Soviet Academy of Sciences. His colleagues thought this was a bit "addled"—that Shannon's work was "more technology than mathematics," as Kolmogorov recalled it afterward. "It is true," he said, "that Shannon left to his successors the rigorous 'justification' of his ideas in some difficult cases. However, his mathematical intuition was amazingly precise." Kolmogorov was not as enthusiastic about cybernetics. Norbert Wiener felt a kinship with him—they had both done early work on stochastic processes and Brownian motion. On a visit to Moscow, Wiener said, "When I read the works of Academician Kolmogorov, I feel that these are my thoughts as well, this is what I wanted to say. And I know that Academician Kolmogorov has the same feeling when reading my works." But the feeling was evidently not shared. Kolmogorov steered his colleagues toward Shannon instead. "It is easy to understand that as a mathematical discipline cybernetics in Wiener's understanding lacks unity," he said, "and it is difficult to imagine productive work in training a specialist, say a postgraduate student, in cybernetics in this sense." He already had real results to back up his instincts: a useful generalized formulation of Shannon entropy, and an extension of his information measure to processes in both discrete and continuous time.

Prestige in Russia was finally beginning to flow toward any work that promised to aid electronic communication and computing. Such work began almost in a void. Pragmatic electrical engineering barely existed; Soviet telephony was notoriously dismal, a subject for eternally bitter Russian humor. As of 1965, there was still no such thing as direct long-distance dialing. The number of toll calls nationally had yet to surpass the number of telegrams, a milestone that had been reached in the United States before the end of the previous century. Moscow had fewer telephones per capita than any major world city. Nonetheless, Kolmogorov and his students generated enough activity to justify a new quarterly journal, *Problems of Information Transmission*, devoted to information theory, coding theory, theory of networks, and even information in living organisms. The inaugural issue opened with Kolmogorov's "Three Approaches to the Definition of the Concept 'Amount of Information'"—almost a manifesto—which then began its slow journey toward the awareness of mathematicians in the West.

"At each given moment there is only a fine layer between the 'trivial' and the impossible," Kolmogorov mused in his diary. "Mathematical discoveries are made in this layer." In the new, quantitative view of information he saw a way to attack a problem that had eluded probability theory, the problem of randomness. How much information is contained in a given "finite object"? An object could be a number (a series of digits) or a message or a set of data.

He described three approaches: the combinatorial, the probabilistic, and the algorithmic. The first and second were Shannon's, with refinements. They focused on the probability of one object among an ensemble of objects—one particular message, say, chosen from a set of possible messages. How would this work, Kolmogorov wondered, when the object was not just a symbol in an alphabet or a lantern in a church window but something big and complicated—a genetic organism, or a work of art? How would one measure the amount of information in Tolstoy's *War and Peace*? "Is it possible to include this novel in a reasonable way in the set of 'all possible novels' and further to postulate the existence of a certain probability distribution in this set?" he asked. Or could one measure the amount of genetic information in, say, the cuckoo bird by considering a probability distribution in the set of all possible species?

His third approach to measuring information—the algorithmic—avoided the difficulties of starting with ensembles of possible objects. It focused on the object itself.* Kolmogorov introduced a new word for the thing he was trying to measure: *complexity*. As he defined this term, the complexity of a number, or message, or set of data is the inverse of simplicity and order and, once again, it corresponds to information. The simpler an object is, the less information it conveys. The more

---

* "Our definition of the quantity of information has the advantage that it refers to individual objects and not to objects treated as members of a set of objects with a probability distribution given on it. The probabilistic definition can be convincingly applied to the information contained, for example, in a stream of congratulatory telegrams. But it would not be clear how to apply it, for example, to an estimate of the quantity of information contained in a novel or in the translation of a novel into another language relative to the original."

complexity, the more information. And, just as Gregory Chaitin did, Kolmogorov put this idea on a solid mathematical footing by calculating complexity in terms of algorithms. The complexity of an object is the size of the smallest computer program needed to generate it. An object that can be produced by a short algorithm has little complexity. On the other hand, an object needing an algorithm every bit as long as the object itself has maximal complexity.

A simple object can be generated—or computed, or described—with just a few bits. A complex object requires an algorithm of many bits. Put this way, it seemed obvious. But until now it had not been understood mathematically. Kolmogorov put it this way:

> The intuitive difference between "simple" and "complicated" objects has apparently been perceived a long time ago. On the way to its formalization, an obvious difficulty arises: something that can be described simply in one language may not have a simple description in another and it is not clear what method of description should be chosen.

That difficulty is solved by using computer language. It does not matter which computer language, because they are all equivalent, reducible to the language of a universal Turing machine. The Kolmogorov complexity of an object is the size, in bits, of the shortest algorithm needed to generate it. This is also the amount of information. And it is also the degree of randomness—Kolmogorov declared "a new conception of the notion 'random' corresponding to the natural assumption that randomness is the absence of regularity." The three are fundamentally equivalent: information, randomness, and complexity—three powerful abstractions, bound all along like secret lovers.

For Kolmogorov, these ideas belonged not only to probability theory but also to physics. To measure the complexity of an orderly crystal or a helter-skelter box of gas, one could measure the shortest algorithm needed to describe the state of the crystal or gas. Once again entropy was the key. Kolmogorov had a useful background in difficult physical problems to which these new methods could be applied. In 1941 he had produced the first useful, though flawed, understanding of the local structure of turbulent flows—equations to predict the distribution of whorls and eddies. He had also worked on perturbations in planetary orbits, another problem surprisingly intractable for classical Newtonian physics. Now he began laying the groundwork for the renaissance in chaos theory to come in the 1970s: analyzing dynamical systems in terms of entropy and information dimension. It made sense now to say that a dynamical system produces information. If it is unpredictable, it produces a great deal of information.

Kolmogorov knew nothing of Gregory Chaitin, nor did either man know of an American probability theorist named Ray Solomonoff, who had developed some of the same ideas. The world was changing. Time, distance, and language still divided mathematicians in Russia from their Western counterparts, but the gulf narrowed every year. Kolmogorov often said that no one should do mathematics after the age of sixty. He dreamed of spending his last years as a buoy keeper on the Volga, making a watery circuit in a boat with oars and a small sail. When the time came, buoy keepers had switched to motorboats, and for Kolmogorov, this ruined the dream.

Now the paradoxes returned.

Zero is an interesting number. Books have been written about it. One is certainly an interesting number—it is the first and the foremost (not counting zero), the singular and unique. Two is interesting in all kinds of ways: the smallest prime, the definitive even number, the number needed for a successful marriage, the atomic number of helium, the number of candles to light on Finnish Independence Day. *Interesting* is an everyday word, not mathematicians' jargon. It seems safe to say that any small number is interesting. All the two-digit numbers and many of the three-digit numbers have their own Wikipedia entries.

Number theorists name entire classes of interesting numbers: prime numbers, perfect numbers, squares and cubes, Fibonacci numbers, factorials. The number 593 is more interesting than it looks; it happens to be the sum of nine squared and two to the ninth—thus a "Leyland number" (any number that can be expressed as $x^y + y^x$). Wikipedia also devotes an article to the number 9,814,072,356. It is the largest holodigital square—which is to say, the largest square number containing each decimal digit exactly once.

What would be an uninteresting number? Presumably a random number. The English number theorist G. H. Hardy randomly rode in taxi No. 1729 on his way to visit the ailing Srinivasa Ramanujan in 1917 and remarked to his colleague that, as numbers go, 1,729 was "rather a dull one." On the contrary, replied Ramanujan (according to a standard anecdote of mathematicians), it is the smallest number expressible as the sum of two cubes in two different ways.* "Every positive integer is one of Ramanujan's personal friends," remarked J. E. Littlewood. Due to the anecdote, 1,729 is known nowadays as the Hardy-Ramanujan number. Nor is that all; 1,729 also happens to be a Carmichael number, an Euler pseudoprime, and a Zeisel number.

But even the mind of Ramanujan was finite, as is Wikipedia, as is the aggregate sum of human knowledge, so the list of interesting numbers must end somewhere. Surely there must be a number about which there is nothing special to say. Wherever it is, there stands a paradox: the number we may describe, interestingly, as "the smallest uninteresting number."

This is none other than Berry's paradox reborn, the one described by Bertrand Russell in *Principia Mathematica*. Berry and Russell had devilishly asked, What is the least integer not nameable in fewer than nineteen syllables? Whatever this number is, it can be named in eighteen syllables: *the least integer not nameable in fewer than nineteen syllables*. Explanations for why a number is interesting are ways of naming the number: "the square of eleven," for example, or "the number of stars in the American

* $1729 = 1^3 + 12^3 = 9^3 + 10^3$

flag." Some of these names do not seem particularly helpful, and some are rather fuzzy. Some are pure mathematical facts: whether, for example, a number is expressible as the sum of two cubes in two different ways. But some are facts about the world, or about language, or about human beings, and they may be accidental and ephemeral—for example, whether a number corresponds to a subway stop or a date in history.

Chaitin and Kolmogorov revived Berry's paradox in inventing algorithmic information theory. An algorithm names a number. "The paradox originally talks about English, but that's much too vague," Chaitin says. "I pick a computer-programming language instead." Naturally he picks the language of a universal Turing machine.

> And then what does it mean, how do you name an integer? Well, you name an integer by giving a way to calculate it. A program names an integer if its output is that integer—you know, it outputs that integer, just one, and then it stops.

Asking whether a number is interesting is the inverse of asking whether it is random. If the number n can be computed by an algorithm that is relatively short, then n is interesting. If not, it is random. The algorithm PRINT I AND THEN PRINT 100 ZEROES generates an interesting number (a googol). Similarly, FIND THE FIRST PRIME NUMBER, ADD THE NEXT PRIME NUMBER, AND REPEAT A MILLION TIMES generates a number that is interesting: the sum of the first million primes. It would take a Turing machine a long time to compute that particular number, but a finite time nonetheless. The number is computable.

But if the most concise algorithm for n is "PRINT [n]"—an algorithm incorporating the entire number, with no shorthand—then we may say that there is nothing interesting about n. In Kolmogorov's terms, this number is random—maximally complex. It will have to be patternless, because any pattern would provide a way to devise a shorthand algorithm. "If there is a small, concise computer program that calculates the number, that means it has some quality or characteristic that enables you to pick it out and to compress it into a smaller algorithmic description," Chaitin says. "So that's unusual; that's an interesting number."

But *is* it unusual? Looking generally at all the numbers, how can a mathematician know whether the interesting ones are rare or common? For that matter, looking at any one number, can a mathematician ever know for sure whether a smaller algorithm might be found? For Chaitin, these were the critical questions.

He answered the first with a counting argument. The vast majority of numbers have to be uninteresting because there cannot possibly be enough concise computer programs to go around. Count them. Given 1,000 bits (say), one has 2^{1000} numbers; but not nearly that many useful computer programs can be written in 1,000 bits. "There are a lot of positive integers," Chaitin says. "If the programs have to be smaller, then there just aren't enough of them to name all those different positive integers." So most n's of any given length are random.

The next question was far more troubling. Knowing that most numbers are random, and given any particular number n, can mathematicians prove it to be random? They cannot tell by looking at it. They can often prove the opposite, that n is interesting: in that case they just have to find a short algorithm that generates n. (Technically, it must be shorter than $\log_2 n$ bits, the number needed to write n in binary.) Proving the negative is a different story. "Even though most positive integers are uninteresting," Chaitin declared, "you can never be sure. . . . You can only prove it in a small number of cases." One could imagine trying to do it by brute force, writing down every possible algorithm and testing them one by one. But a computer will have to perform the tests—an algorithm testing other algorithms—and soon, Chaitin demonstrated, a new version of Berry's paradox appears. Instead of "the smallest uninteresting number," one inevitably encounters a statement in the form of "the smallest number that we can prove cannot be named in fewer than n syllables." (We are not really talking about syllables any more, of course, but Turing-machine states.)* It is another recursive, self-looping twist. This was Chaitin's version of Gödel's incompleteness. Complexity, defined in terms of program size, is generally uncomputable. Given an arbitrary string of a million digits, a mathematician knows that it is almost certainly random, complex, and patternless—but cannot be absolutely sure.

Chaitin did this work in Buenos Aires. When he was still a teenager, before he could graduate from City College, his parents moved back to their home in Argentina, and he got a job there with IBM World Trade. He continued to nurse his obsession with Gödel and incompleteness and to send papers to the American Mathematical Society and the Association for Computing Machinery. Eight years later, Chaitin returned to the United States to visit IBM's research center in Yorktown Heights, New York, and placed a telephone call to his hero, then nearing seventy at the Institute for Advanced Study in Princeton. Gödel answered, and Chaitin introduced himself and said he had a new approach to incompleteness, based on Berry's paradox instead of the liar paradox.

"It doesn't make any difference which paradox you use," said Gödel.

"Yes, but . . ." Chaitin said he was on the trail of a new "information-theoretic" view of incompleteness and asked if he could call on Gödel in Princeton. He was staying in the YMCA in White Plains and would take the train, changing in New York City. Gödel agreed, but when the day came, he canceled. It was snowing, and he was fearful for his health. Chaitin never did meet him. Gödel, increasingly unstable, afraid of poisoning, died in the winter of 1978 of self-starvation.

Chaitin spent the rest of his career at the IBM Watson Research Center, one of the last great scientists to be so well supported in work of no plausible use to his corporate patron. He sometimes said he was "hiding" in a physics department; he felt that more conventional

* More precisely, it looked like this: "The finite binary sequence S with the first proof that S cannot be described by a Turing machine with n states or less" is a $(\log_2 n + c_F)$–state description of S.

mathematicians dismissed him as "a closet physicist" anyway. His work treated mathematics as a sort of empirical science—not a Platonic pipeline to absolute truth, but a research program subject to the world's contingencies and uncertainties. "In spite of incompleteness and uncomputability and even algorithmic randomness," he said, "mathematicians don't want to give up absolute certainty. Why? Well, absolute certainty is like God."

In quantum physics and later in chaos, scientists found the limits to their knowledge. They explored the fruitful uncertainty that at first so vexed Einstein, who did not want to believe that God plays dice with the universe. Algorithmic information theory applies the same limitations to the universe of whole numbers—an ideal, mental universe. As Chaitin put it, "God not only plays dice in quantum mechanics and nonlinear dynamics, but even in elementary number theory."

Among its lessons were these:

- Most numbers are random. Yet very few of them can be *proved* random.
- A chaotic stream of information may yet hide a simple algorithm. Working backward from the chaos to the algorithm may be impossible.
- Kolmogorov-Chaitin (KC) complexity is to mathematics what entropy is to thermodynamics: the antidote to perfection. Just as we can have no perpetual-motion machines, there can be no complete formal axiomatic systems.
- Some mathematical facts are true for no reason. They are accidental, lacking a cause or deeper meaning.

Joseph Ford, a physicist studying the behavior of unpredictable dynamical systems in the 1980s, said that Chaitin had "charmingly captured the essence of the matter" by showing the path from Gödel's incompleteness to chaos. This was the "deeper meaning of chaos," Ford declared:

Chaotic orbits exist but they are Gödel's children, so complex, so over-laden with information that humans can never comprehend them. But chaos is ubiquitous in nature; therefore the universe is filled with count-less mysteries that man can never understand.

Yet one still tries to take their measure.

How much information . . . ?

When an object (a number or a bitstream or a dynamical system) can be expressed a different way in fewer bits, it is compressible. A frugal telegraph operator prefers to send the compressed version. Because the spirit of fru-gal telegraph operators kept the lights on at Bell Labs, it was natural for Claude Shannon to explore data compression, both theory and practice. Compression was fundamental to his vision: his war work on cryptogra-phy analyzed the disguising of information at one end and the recovery of the information at the other; data compression likewise encodes the information, with a different motivation—the efficient use of bandwidth. Satellite television channels, pocket music players, efficient cameras and telephones and countless other modern appurtenances depend on coding algorithms to compress numbers—sequences of bits—and those algo-rithms trace their lineage to Shannon's original 1948 paper.

The first of these, now called Shannon-Fano coding, came from his colleague Robert M. Fano. It began with the simple idea of assigning short codes to frequent symbols, as in Morse code. They knew their method was not optimal, however: it could not be relied on to produce the shortest possible messages. Within three years it was surpassed by work of a graduate student of Fano's at MIT, David Huffman. In the decades since, versions of the Huffman coding algorithm have squeezed many, many bytes.

Ray Solomonoff, a child of Russian immigrants who studied at the University of Chicago, encountered Shannon's work in the early

1950s and began thinking about what he called the Information Packing Problem: how much information could one "pack" into a given number of bits, or conversely, given some information, how could one pack it into the fewest possible bits. He had majored in physics, studied mathematical biology and probability and logic on the side, and gotten to know Marvin Minsky and John McCarthy, pioneers in what would soon be called artificial intelligence. Then he read Noam Chomsky's offbeat and original paper "Three Models for the Description of Language," applying the new information-theoretic ideas to the formalization of structure in language. All this was bouncing around in Solomonoff's mind; he was not sure where it led, but he found himself focusing on the problem of *induction*. How do people create theories to account for their experience of the world? They have to make generalizations, find patterns in data that are always influenced by randomness and noise. Could one enable a machine to do that? In other words, could a computer be made to learn from experience?

He worked out an elaborate answer and published it in 1964. It was idiosyncratic, and hardly anyone noticed until the 1970s, when both Chaitin and Kolmogorov discovered that Solomonoff had anticipated the essential features of what by then was called algorithmic information theory. In effect, Solomonoff, too, had been figuring out how a computer might look at sequences of data—number sequences or bit strings—and measure their randomness and their hidden patterns. When humans or computers learn from experience, they are using induction: recognizing regularities amid irregular streams of information. From this point of view, the laws of science represent data compression in action. A theoretical physicist acts like a very clever coding algorithm. "The laws of science that have been discovered can be viewed as summaries of large amounts of empirical data about the universe," wrote Solomonoff. "In the present context, each such law can be transformed into a method of compactly coding the empirical data that gave rise to that law." A good scientific theory is economical. This was yet another way of saying so.

Solomonoff, Kolmogorov, and Chaitin tackled three different problems and came up with the same answer. Solomonoff was interested in inductive inference: given a sequence of observations, how can one make the best predictions about what will come next? Kolmogorov was looking for a mathematical definition of randomness: what does it mean to say that one sequence is more random than another, when they have the same probability of emerging from a series of coin flips? And Chaitin was trying to find a deep path into Gödel incompleteness by way of Turing and Shannon—as he said later, "putting Shannon's information theory and Turing's computability theory into a cocktail shaker and shaking vigorously." They all arrived at minimal program size. And they all ended up talking about complexity.

The following bitstream (or number) is not very complex, because it is rational:

D: 142857142857142857142857142857142857142857142857142857714 . . .

It may be rephrased concisely as "PRINT 142857 AND REPEAT," or even more concisely as "1/7." If it is a message, the compression saves keystrokes. If it is an incoming stream of data, the observer may recognize a pattern, grow more and more confident, and settle on *one-seventh* as a theory for the data.

In contrast, this sequence contains a late surprise:

E: 1010101010101010101010101010101010101010101010101010101013

The telegraph operator (or theorist, or compression algorithm) must pay attention to the whole message. Nonetheless, the extra information is minimal; the message can still be compressed, wherever pattern exists. We may say it contains a redundant part and an arbitrary part.

It was Shannon who first showed that anything nonrandom in a message allows compression:

F: 1011010111101101101011101011101111010011101101001111101110

Heavy on ones, light on zeroes, this might be emitted by the flip of a biased coin. Huffman coding and other such algorithms exploit statistical regularities to compress the data. Photographs are compressible because of their subjects' natural structure: light pixels and dark pixels come in clusters; statistically, nearby pixels are likely to be similar; distant pixels are not. Video is even more compressible, because the differences between one frame and the next are relatively slight, except when the subject is in fast and turbulent motion. Natural language is compressible because of redundancies and regularities of the kind Shannon analyzed. Only a wholly random sequence remains incompressible: nothing but one surprise after another.

Random sequences are "normal"—a term of art meaning that on average, in the long run, each digit appears exactly as often as the others, one time in ten; and each pair of digits, from 00 to 99, appears one time in a hundred; and each triplet likewise, and so on. No string of any particular length is more likely to appear than any other string of that length. Normality is one of those simple-seeming ideas that, when mathematicians look closely, turn out to be covered with thorns. Even though a truly random sequence must be normal, the reverse is not necessarily the case. A number can be statistically normal yet not random at all. David Champernowne, a young Cambridge friend of Turing's, invented (or discovered) such a creature in 1933—a construction made of all the integers, chained together in order:

G: 123456789101112131415161718192021222324252627282 93 . . .

It is easy to see that each digit, and each combination of digits, occurs equally often in the long run. Yet the sequence could not be less random. It is rigidly structured and completely predictable. If you know where you are, you know what comes next.

Even apart from freaks like Champernowne's, it turns out that normal numbers are difficult to recognize. In the universe of numbers, normality is the rule; mathematicians know for sure that almost all numbers are normal. The rational numbers are not normal, and there are infinitely many rational numbers, but they are infinitely outnumbered by normal numbers. Yet, having settled the great and general question, mathematicians can almost never prove that any particular number is normal. This in itself is one of the more remarkable oddities of mathematics.

Even π retains some mysteries:

C: 3.14159265358979323846264338327950288419716939937751 . . .

The world's computers have spent many cycles analyzing the first trillion or so known decimal digits of this cosmic message, and as far as anyone can tell, they appear normal. No statistical features have been discovered—no biases or correlations, local or remote. It is a quintessentially nonrandom number that seems to behave randomly. Given the nth digit, there is no shortcut for guessing the nth plus one. Once again, the next bit is always a surprise.

How much information, then, is represented by this string of digits? Is it information rich, like a random number? Or information poor, like an ordered sequence?

The telegraph operator could, of course, save many keystrokes—infinitely many, in the long run—by simply sending the message "π." But this is a cheat. It presumes knowledge previously shared by the sender and the receiver. The sender has to recognize this special sequence to begin with, and then the receiver has to know what π is, and how to look up its decimal expansion, or else how to compute it. In effect, they need to share a code book.

This does not mean, however, that π contains a lot of information. The essential message can be sent in fewer keystrokes. The telegraph operator has several strategies available. For example, he could say, "Take 4, subtract 4/3, add 4/5, subtract 4/7, and so on." The telegraph operator sends an algorithm, that is. This infinite series of fractions converges slowly upon π, so the recipient has a lot of work to do, but the message itself is economical: the total information content is the same no matter how many decimal digits are required.

The issue of shared knowledge at the far ends of the line brings complications. Sometimes people like to frame this sort of problem—the problem of information content in messages—in terms of communicating with an alien life-form in a faraway galaxy. What could we tell them? What would we want to say? The laws of mathematics being universal, we tend to think that π would be one message any intelligent race would recognize. Only, they could hardly be expected to know the Greek letter. Nor would they be likely to recognize the decimal digits "3.1415926535 . . ." unless they happened to have ten fingers.

The sender of a message can never fully know his recipient's mental code book. Two lights in a window might mean nothing or might mean "The British come by sea." Every poem is a message, different for every reader. There is a way to make the fuzziness of this line of thinking go away. Chaitin expressed it this way:

> It is preferable to consider communication not with a distant friend but with a digital computer. The friend might have the wit to make inferences about numbers or to construct a series from partial information or from vague instructions. The computer does not have that capacity, and for our purposes that deficiency is an advantage. Instructions given the computer must be complete and explicit, and they must enable it to proceed step by step.

In other words: the message is an algorithm. The recipient is a machine; it has no creativity, no uncertainty, and no knowledge, except whatever "knowledge" is inherent in the machine's structure. By the 1960s, digital computers were already getting their instructions in a form measured in bits, so it was natural to think about how much information was contained in any algorithm.

A different sort of message would be this:

Even to the eye this sequence of notes seems nonrandom. It happens that the message they represent is already making its way through interstellar space, 10 billion miles from its origin, at a tiny fraction of light speed. The message is not encoded in this print-based notation, nor in any digital form, but as microscopic waves in a single long groove winding in a spiral engraved on a disc twelve inches in diameter and one-fiftieth of an inch in thickness. The disc might have been vinyl, but in this case it was copper, plated with gold. This analog means of capturing, preserving, and reproducing sound was invented in 1877 by Thomas Edison, who called it phonography. It remained the most popular audio technology a hundred years later—though not for much longer—and in 1977 a committee led by the astronomer Carl Sagan created a particular phonograph record and stowed copies in a pair of spacecraft named *Voyager 1* and *Voyager 2*, each the size of a small automobile, launched that summer from Cape Canaveral, Florida.

So it is a message in an interstellar bottle. The message has no meaning, apart from its patterns, which is to say that it is abstract art: the first prelude of Johann Sebastian Bach's *Well-Tempered Clavier*, as played on the piano by Glenn Gould. More generally, perhaps the meaning is "There is intelligent life here." Besides the Bach prelude, the record includes music samples from several different cultures and a selection of earthly sounds: wind, surf, and thunder; spoken greetings in fifty-five languages; the voices of crickets, frogs, and whales; a ship's horn, the clatter of a horse-drawn cart, and a tapping in Morse code. Along with the phonograph record are a cartridge and needle and a brief pictographic instruction manual. The committee did not bother with a phonograph player or a source of electrical power. Maybe the aliens will find a way to convert those analog metallic grooves into waves in whatever fluid serves as their atmosphere— or into some other suitable input for their alien senses.

Would they recognize the intricate patterned structure of the Bach prelude (say), as distinct from the less interesting, more

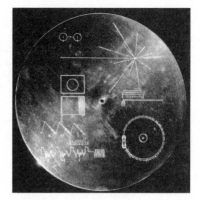

THE "GOLDEN RECORD" STOWED
ABOARD THE VOYAGER SPACECRAFT

random chatter of crickets? Would the sheet music convey a clearer message—the written notes containing, after all, the essence of Bach's creation? And, more generally, what kind of knowledge would be needed at the far end of the line—what kind of code book—to decipher the message? An appreciation of counterpoint and voice leading? A sense of the tonal context and performance practices of the European Baroque? The sounds—the notes—come in groups; they form shapes, called melodies; they obey the rules of an implicit grammar. Does the music carry its own logic with it, independent of geography and history? On earth, meanwhile, within a few years, even before the *Voyagers* had sailed past the solar system's edge, music was seldom recorded in analog form anymore. Better to store the sounds of the *Well-Tempered Clavier* as bits: the waveforms discretized without loss as per the Shannon sampling theorem, and the information preserved in dozens of plausible media.

In terms of bits, a Bach prelude might not seem to have much information at all. As penned by Bach on two manuscript pages, this one amounts to six hundred notes, characters in a small alphabet. As Glenn Gould played it on a piano in 1964—adding the performer's layers of nuance and variation to the bare instructions—it lasts a minute and thirty-six seconds. The sound of that performance, recorded onto a CD, microscopic pits burned by a laser onto a slim disc of polycarbonate plastic, comprises 135 million bits. But this bitstream can be compressed considerably with no loss of information. Alternatively, the prelude fits on a small player-piano roll (descendant of Jacquard's loom, predecessor of punched-card computing); encoded electronically with the MIDI protocol, it uses a few thousands bits. Even the basic six-hundred-character message has tremendous redundancy: unvarying tempo, uniform timbre, just a brief melodic pattern, a word, repeated over and over with slight variations till the final bars. It is famously, deceptively simple. The very repetition creates expectations and breaks them. Hardly anything happens, and everything is a surprise. "Immortal broken chords of radiantly white harmonies," said Wanda Landowska. It is simple the way a Rembrandt drawing is simple. It does a lot with a little. Is it then rich in information? Certain music could be considered information poor. At one extreme John Cage's composition titled *4'33"* contains no "notes" at all: just four minutes and thirty-three seconds of near silence, as the piece absorbs the ambient sounds around the still pianist—the listeners' shifting in their seats, rustling clothes, breathing, sighing.

How much information in the Bach C-major Prelude? As a set of patterns, in time and frequency, it can be analyzed, traced, and understood, but only up to a point. In music, as in poetry, as in any art, perfect understanding is meant to remain elusive. If one could find the bottom it would be a bore.

In a way, then, the use of minimal program size to define complexity seems perfect—a fitting apogee for Shannon information theory. In another way it remains deeply unsatisfying. This is particularly so when turning to the big questions—one might say, the human questions—of art, of biology, of intelligence.

According to this measure, a million zeroes and a million coin tosses lie at opposite ends of the spectrum. The empty string is as simple as can be; the random string is maximally complex. The zeroes convey no information; coin tosses produce the most information possible. Yet these extremes have something in common. They are dull. They have no value. If either one were a message from another galaxy, we would attribute no intelligence to the sender. If they were music, they would be equally worthless.

Everything we care about lies somewhere in the middle, where pattern and randomness interlace.

Chaitin and a colleague, Charles H. Bennett, sometimes discussed these matters at IBM's research center in Yorktown Heights, New York. Over a period of years, Bennett developed a new measure of value, which he called "logical depth." Bennett's idea of depth is connected to complexity but orthogonal to it. It is meant to capture the usefulness of a message, whatever usefulness might mean in any particular domain. "From the earliest days of information theory it has been appreciated that information per se is not a good measure of message value," he wrote, finally publishing his scheme in 1988.

A typical sequence of coin tosses has high information content but little value; an ephemeris, giving the positions of the moon and planets every day for a hundred years, has no more information than the equations of motion and initial conditions from which it was calculated, but saves its owner the effort of recalculating these positions.

The amount of work it takes to compute something had been mostly disregarded—set aside—in all the theorizing based on Turing machines, which work, after all, so ploddingly. Bennett brought it back. There is no logical depth in the parts of a message that are sheer randomness and unpredictability, nor is there logical depth in obvious redundancy— plain repetition and copying. Rather, he proposed, the value of a message lies in "what might be called its buried redundancy—parts predictable only with difficulty, things the receiver could in principle have figured out without being told, but only at considerable cost in money, time, or computation." When we value an object's complexity, or its information content, we are sensing a lengthy hidden computation. This might be true of music or a poem or a scientific theory or a crossword puzzle, which gives its solver pleasure when it is neither too cryptic nor too shallow, but somewhere in between.

Mathematicians and logicians had developed a tendency to think of information processing as free—not like pumping water or carrying stones. In our time, it certainly has gotten cheap. But it embodies work after all, and Bennett suggests that we recognize this work, reckon its expense in understanding complexity. "The more subtle something is, the harder it is to discover," Bennett says. He applied the idea of logical depth to the problem of self-organization: the question of how complex structures develop in nature. Evolution starts with simple initial conditions; complexity arises, apparently building on itself. Whatever the basic processes involved, physical or biological, something is under way that begins to resemble computation.

3 | INFORMATION IS PHYSICAL

(It from Bit)

The more energy, the faster the bits flip. Earth, air, fire, and water in the end are all made of energy, but the different forms they take are determined by information. To do anything requires energy. To specify what is done requires information.

—Seth Lloyd (2006)

QUANTUM MECHANICS HAS WEATHERED in its short history more crises, controversies, interpretations (the Copenhagen, the Bohm, the Many Worlds, the Many Minds), factional implosions, and general philosophical breast-beating than any other science. It is happily riddled with mysteries. It blithely disregards human intuition. Albert Einstein died unreconciled to its consequences, and Richard Feynman was not joking when he said no one understands it. Perhaps arguments about the nature of reality are to be expected; quantum physics, so uncannily successful in practice, deals in theory with the foundations of all things, and its own foundations are continually being rebuilt. Even so, the ferment sometimes seems more religious than scientific.

"How did this come about?" asks Christopher Fuchs, a quantum theorist at Bell Labs and then the Perimeter Institute in Canada.

Go to any meeting, and it is like being in a holy city in great tumult. You will find all the religions with all their priests pitted in holy war— the Bohmians, the Consistent Historians, the Transactionalists, the

Spontaneous Collapseans, the Einselectionists, the Contextual Objectivists, the outright Everettics, and many more beyond that. They all declare to see the light, the ultimate light. Each tells us that if we will accept their solution as our savior, then we too will see the light.

It is time, he says, to start fresh. Throw away the existing quantum axioms, exquisite and mathematical as they are, and turn to deep physical principles. "Those principles should be crisp; they should be compelling. They should stir the soul." And where should these physical principles be found? Fuchs answers his own question: in quantum *information* theory.

"The reason is simple, and I think inescapable," he declares. "Quantum mechanics has always been about information; it is just that the physics community has forgotten this."

One who did not forget—or who rediscovered it—was John Archibald Wheeler, pioneer of nuclear fission, student of Bohr and teacher of Feynman, namer of black holes, the last giant of twentieth-century physics. Wheeler was given to epigrams and gnomic utterances. *A black hole has no hair* was his famous way of stating that nothing but mass, charge, and spin can be perceived from outside. "It teaches us," he wrote, "that space can be crumpled like a piece of paper into an infinitesimal dot, that time can be extinguished like a blown-out flame, and that the laws of physics that we regard as 'sacred,' as immutable, are anything but." In 1989 he offered his final catchphrase: *It from Bit*. His view was extreme. It was immaterialist: information first, everything else later. "Otherwise put," he said,

QM is about
   Information.

$$H(X)$$

Plain old ordinary Shannon
information :

   ignorance
   lack of predictability

**VISUAL AID BY
CHRISTOPHER FUCHS**

every it—every particle, every field of force, even the space-time continuum itself—derives its function, its meaning, its very existence . . . from *bits*.

Why does nature appear quantized? Because information is quantized. The bit is the ultimate unsplittable particle.

Among the physics phenomena that pushed information front and center, none were more spectacular than black holes. At first, of course, they had not seemed to involve information at all.

Black holes were the brainchild of Einstein, though he did not live to know about them. He established by 1915 that light must submit to the pull of gravity; that gravity curves the fabric of spacetime; and that a sufficient mass, compacted together, as in a dense star, would collapse utterly, intensifying its own gravity and contracting without limit. It took almost a half century more to face up to the consequences, because they are strange. Anything goes in, nothing comes out. At the center lies the singularity. Density becomes infinite; gravity becomes infinite; spacetime curves infinitely. Time and space are interchanged. Because no light, no signal of any kind, can escape the interior, such things are quintessentially invisible. Wheeler began calling them "black holes" in 1967. Astronomers are sure they have found some, by gravitational inference, and no one can ever know what is inside.

At first astrophysicists focused on matter and energy falling in. Later they began to worry about the information. A problem arose when Stephen Hawking, adding quantum effects to the usual calculations of general relativity, argued in 1974 that black holes should, after all, radiate particles—a consequence of quantum fluctuations near the event horizon. Black holes slowly evaporate, in other words. The problem was that Hawking radiation is featureless and dull. It is thermal radiation— heat. But matter falling *into* the black hole carries information, in its very structure, its organization, its quantum states—in terms of statistical mechanics, its accessible microstates. As long as the missing information stayed out of reach beyond the event horizon, physicists did not have to worry about it. They could say it was inaccessible but not obliterated. "All colours will agree in the dark," as Francis Bacon said in 1625.

The outbound Hawking radiation carries no information, however. If the black hole evaporates, where does the information go? According to quantum mechanics, information may never be destroyed. The deterministic laws of physics require the states of a physical system at one instant to determine the states at the next instant; in microscopic detail, the laws are reversible, and information must be preserved. Hawking was the first to state firmly—even alarmingly—that this was a problem challenging the very foundations of quantum mechanics. The loss of information would violate unitarity, the principle that probabilities must add up to one. "God not only plays dice, He sometimes throws the dice where they cannot be seen," Hawking said. In the summer of 1975, he submitted a paper to the *Physical Review* with a dramatic headline, "The Breakdown of Physics in Gravitational Collapse." The journal held it for more than a year before publishing it with a milder title.

As Hawking expected, other physicists objected vehemently. Among them was John Preskill at the California Institute of Technology, who continued to believe in the principle that information cannot be lost: even when a book goes up in flames, in physicists' terms, if you could track every photon and every fragment of ash, you should be able to integrate backward and reconstruct the book. "Information loss is highly infectious," warned Preskill at a Caltech Theory Seminar. "It is very hard to modify quantum theory so as to accommodate a little bit of information loss without it leaking into all processes." In 1997 he made a much-publicized wager with Hawking that the information must be escaping the black hole somehow. They bet an encyclopedia of the winner's choice. "Some physicists feel the question of what happens in a black hole is academic or even theological, like counting angels on pinheads," said Leonard Susskind of Stanford, siding with Preskill. "But it is not so at all: at stake are the future rules of physics." Over the next few years a cornucopia of solutions was proposed. Hawking himself said at one point: "I think the information probably goes off into another universe. I have not been able to show it yet mathematically."

It was not until 2004 that Hawking, then sixty-two, reversed himself and conceded the bet. He announced that he had found a way to show that quantum gravity is unitary after all and that information is preserved. He applied a formalism of quantum indeterminacy—the "sum over histories" path integrals of Richard Feynman—to the very topology of spacetime and declared, in effect, that black holes are never unambiguously black. "The confusion and paradox arose because people thought classically in terms of a single topology for space-time," he wrote.* His new formulation struck some physicists as cloudy and left many questions unanswered, but he was firm on one point. "There is no baby universe branching off, as I once thought," he wrote. "The information remains firmly in our universe. I'm sorry to disappoint science fiction fans." He gave Preskill a copy of *Total Baseball: The Ultimate Baseball Encyclopedia*, weighing in at 2,688 pages—"from which information can be recovered with ease," he said. "But maybe I should have just given him the ashes."

Charles Bennett came to quantum information theory by a very different route. Long before he developed his idea of logical depth, he was thinking about the "thermodynamics of computation"—a peculiar topic, because information processing was mostly treated as disembodied. "The thermodynamics of computation, if anyone had stopped to wonder about it, would probably have seemed no more urgent as a topic of scientific inquiry than, say, the thermodynamics of love," says Bennett. It is like the energy of thought. Calories may be expended, but no one is counting.

Stranger still, Bennett tried investigating the thermodynamics of the least thermodynamic computer of all—the nonexistent, abstract, idealized Turing machine. Turing himself never worried about his thought

---

* "It was either R^4 or a black hole. But the Feynman sum over histories allows it to be both at once."

experiment consuming any energy or radiating any heat as it goes about its business of marching up and down imaginary paper tapes. Yet in the early 1980s Bennett was talking about using Turing-machine tapes for fuel, their caloric content to be measured in bits. Still a thought experiment, of course, meant to focus on a very real question: What is the physical cost of logical work? "Computers," he wrote provocatively, "may be thought of as engines for transforming free energy into waste heat and mathematical work." Entropy surfaced again. A tape full of zeroes, or a tape encoding the works of Shakespeare, or a tape rehearsing the digits of π, has "fuel value." A random tape has none.

Bennett, the son of two music teachers, grew up in the Westchester suburbs of New York; he studied chemistry at Brandeis and then Harvard in the 1960s. James Watson was at Harvard then, teaching about the genetic code, and Bennett worked for him one year as a teaching assistant. He got his doctorate in molecular dynamics, doing computer simulations that ran overnight on a machine with a memory of about twenty thousand decimal digits and generated output on pages and pages of fan-fold paper. Looking for more computing power to continue his molecular-motion research, he went to the Lawrence Livermore Laboratory in Berkeley, California, and Argonne National Laboratory in Illinois, and then joined IBM Research in 1972.

IBM did not manufacture Turing machines, of course. But at some point it dawned on Bennett that a special-purpose Turing machine had already been found in nature: namely RNA polymerase. He had learned about polymerase directly from Watson; it is the enzyme that crawls along a gene—its "tape"—transcribing the DNA. It steps left and right; its logical state changes according to the chemical information written in sequence; and its thermodynamic behavior can be measured.

In the real world of 1970s computing, hardware had rapidly grown thousands of times more energy-efficient than during the early vacuum-tube era. Nonetheless, electronic computers dissipate considerable energy in the form of waste heat. The closer they come to their theoretical minimum of energy use, the more urgently scientists want to know just what that theoretical minimum is. Von Neumann, working with his big computers, made a back-of-the-envelope calculation as early as 1949, proposing an amount of heat that must be dissipated "per elementary act of information, that is per elementary decision of a two-way alternative and per elementary transmittal of one unit of information." He based it on the molecular work done in a model thermodynamic system by Maxwell's demon, as reimagined by Leó Szilárd.* Von Neumann said the price is paid by every elementary act of information processing, every choice between two alternatives. By the 1970s this was generally accepted. But it was wrong.

Von Neumann's error was discovered by the scientist who became Bennett's mentor at IBM, Rolf Landauer, an exile from Nazi Germany. Landauer devoted his career to establishing the physical basis of information. "Information Is Physical" was the title of one famous paper, meant to remind the community that computation requires physical objects and obeys the laws of physics. Lest anyone forget, he titled a later essay—his last, it turned out—"Information Is Inevitably Physical." Whether a bit is a mark on a stone tablet or a hole in a punched card or a particle with spin up or down, he insisted that it could not exist without *some* embodiment. Landauer tried in 1961 to prove von Neumann's formula for the cost of information processing and discovered that he could not. On the contrary, it seemed that most logical operations have no entropy cost at all. When a bit flips from zero to one, or vice-versa, the information is preserved. The process is reversible. Entropy is unchanged; no heat needs to be dissipated. Only an irreversible operation, he argued, increases entropy.

Landauer and Bennett were a double act: a straight and narrow old IBM type and a scruffy hippie (in Bennett's view, anyway). The younger man pursued Landauer's principle by analyzing every kind of computer he could imagine, real and abstract, from Turing machines and

---

* Von Neumann's formula for the theoretical energy cost of every logical operation was $kT \ln 2$ joules per bit, where T is the computer's operating temperature and k is the Boltzman constant. Szilárd had proved that the demon in his engine can get $kT \ln 2$ of work out of every molecule it selects, so that energy cost must be paid somewhere in the cycle.

messenger RNA to "ballistic" computers, carrying signals via something like billiard balls. He confirmed that a great deal of computation can be done with no energy cost at all. In every case, Bennett found, heat dissipation occurs only when information is *erased*. Erasure is the irreversible logical operation. When the head on a Turing machine erases one square of the tape, or when an electronic computer clears a capacitor, a bit is lost, and *then* heat must be dissipated. In Szilárd's thought experiment, the demon does not incur an entropy cost when it observes or chooses a molecule. The payback comes at the moment of clearing the record, when the demon erases one observation to make room for the next.

Forgetting takes work.

"You might say this is the revenge of information theory on quantum mechanics," Bennett says. Sometimes a successful idea in one field can impede progress in another. In this case the successful idea was the uncertainty principle, which brought home the central role played by the measurement process itself. One can no longer talk simply about "looking" at a molecule; the observer needs to employ photons, and the photons must be more energetic than the thermal background, and complications ensue. In quantum mechanics the act of observation has consequences of its own, whether performed by a laboratory scientist or by Maxwell's demon. Nature is sensitive to our experiments.

"The quantum theory of radiation helped people come to the incorrect conclusion that computing had an irreducible thermodynamic cost per step," Bennett says. "In the other case, the success of Shannon's theory of information processing led people to abstract away all of the physics from information processing and think of it as a totally mathematical thing." As communications engineers and chip designers came closer and closer to atomic levels, they worried increasingly about quantum limitations interfering with their clean, classical ability to distinguish zero and one states. But now they looked again—and this, finally, is where quantum information science is born. Bennett and others began to think differently: that quantum effects, rather than being a nuisance, might be turned to advantage.

Wedged like a hope chest against a wall of his office at IBM's research laboratory in the wooded hills of Westchester is a light-sealed device called Aunt Martha (short for Aunt Martha's coffin). Bennett and his research assistant John Smolin jury-rigged it in 1988 and 1989 with a little help from the machine shop: an aluminum box spray-painted dull black on the inside and further sealed with rubber stoppers and black velvet. With a helium-neon laser for alignment and high-voltage cells to polarize the photons, they sent the first message ever to be encoded by quantum cryptography. It was a demonstration of an information-processing task that could be effectively accomplished only via a quantum system. Quantum error correction, quantum teleportation, and quantum computers followed shortly behind.

The quantum message passed between Alice and Bob, a ubiquitous mythical pair. Alice and Bob got their start in cryptography, but the quantum people own them now. Occasionally they are joined by Charlie. They are constantly walking into different rooms and flipping quarters and sending each other sealed envelopes. They choose states and perform Pauli rotations. "We say things such as 'Alice sends Bob a qubit and forgets what she did,' 'Bob does a measurement and tells Alice,'" explains Barbara Terhal, a colleague of Bennett's and one of the next generation of quantum information theorists. Terhal herself has investigated whether Alice and Bob are *monogamous*—another term of art, naturally.

In the Aunt Martha experiment, Alice sends information to Bob, encrypted so that it cannot be read by a malevolent third party (Eve the eavesdropper). If they both know their private key, Bob can decipher the message. But how is Alice to send Bob the key in the first place? Bennett and Gilles Brassard, a computer scientist in Montreal, began by encoding each bit of information as a single quantum object, such as a photon. The information resides in the photon's quantum states—for example, its horizontal or vertical polarization. Whereas an object in classical physics, typically composed of billions of particles, can be intercepted, monitored, observed, and passed along, a quantum object cannot. Nor can it be copied or cloned. The act of observation inevitably disrupts the message. No matter how delicately eavesdroppers try to listen in, they can be detected. Following an intricate and complex protocol worked out by Bennett and Brassard, Alice generates a sequence of random bits to use as the key, and Bob is able to establish an identical sequence at his end of the line.

The first experiments with Aunt Martha's coffin managed to send quantum bits across thirty-two centimeters of free air. It was not *Mr. Watson, come here, I want to see you*, but it was a first in the history of cryptography: an absolutely unbreakable cryptographic key. Later experimenters moved on to optical fiber. Bennett, meanwhile, moved on to quantum teleportation.

He regretted that name soon enough, when the IBM marketing department featured his work in an advertisement with the line "Stand by: I'll teleport you some goulash." But the name stuck, because teleportation worked. Alice does not send goulash; she sends qubits.*

The qubit is the smallest nontrivial quantum system. Like a classical bit, a qubit has two possible values, zero or one—which is to say, two states that can be reliably distinguished. In a classical system, *all* states are distinguishable in principle. (If you cannot tell one color from another, you merely have an imperfect measuring device.) But in a quantum system, imperfect distinguishability is everywhere, thanks to Heisenberg's uncertainty principle. When you measure any property of a quantum object, you thereby lose the ability to measure a complementary property. You can discover a particle's momentum or its position but not both. Other complementary properties include directions of spin and,

---

* This word is not universally accepted, though the *OED* recognized it as of December 2007. David Mermin wrote that same year: "Unfortunately the preposterous spelling *qubit* currently holds sway . . . . Although "qubit" honors the English (German, Italian, . . .) rule that *q* should be followed by *u*, it ignores the equally powerful requirement that *qu* should be followed by a vowel. My guess is that "qubit" has gained acceptance because it visually resembles an obsolete English unit of distance, the homonymic *cubit*. To see its ungainliness with fresh eyes, it suffices to imagine . . . that one erased transparencies and cleaned ones ears with *Qutips*."

as in Aunt Martha's coffin, polarization. Physicists think of these quantum states in a geometrical way—the states of a system corresponding to directions in space (a space of many possible dimensions), and their distinguishability depending on whether those directions are perpendicular (or "orthogonal").

This imperfect distinguishability is what gives quantum physics its dreamlike character: the inability to observe systems without disturbing them; the inability to clone quantum objects or broadcast them to many listeners. The qubit has this dreamlike character, too. It is not just either-or. Its 0 and 1 values are represented by quantum states that can be reliably distinguished—for example, horizontal and vertical polarizations—but coexisting with these are the whole continuum of intermediate states, such as diagonal polarizations, that lean toward 0 or 1 with different probabilities. So a physicist says that a qubit is a *superposition* of states; a combination of probability amplitudes. It is a determinate thing with a cloud of indeterminacy living inside. But the qubit is not a muddle; a superposition is not a hodgepodge but a combining of probabilistic elements according to clear and elegant mathematical rules.

"A nonrandom whole can have random parts," says Bennett. "This is the most counterintuitive part of quantum mechanics, yet it follows from the superposition principle and is the way nature works, as far as we know. People may not like it at first, but after a while you get used to it, and the alternatives are far worse."

The key to teleportation and to so much of the quantum information science that followed is the phenomenon known as entanglement. Entanglement takes the superposition principle and extends it across space, to a pair of qubits far apart from each other. They have a definite state *as a pair* even while neither has a measurable state on its own. Before entanglement could be discovered, it had to be invented, in this case by Einstein. Then it had to be named, not by Einstein but by Schrödinger. Einstein invented it for a thought experiment designed to illuminate what he considered flaws in quantum mechanics as it stood

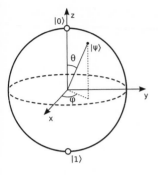

THE QUBIT

in 1935. He publicized it in a famous paper with Boris Podolsky and Nathan Rosen titled "Can Quantum-Mechanical Description of Physical Reality Be Considered Complete?" It was famous in part for provoking Wolfgang Pauli to write to Werner Heisenberg, "Einstein has once again expressed himself publicly on quantum mechanics. . . . As is well known, this is a catastrophe every time it happens." The thought experiment imagined a pair of particles correlated in a special way, as when, for example, a pair of photons are emitted by a single atom. Their polarization is random but identical—now and as long as they last.

Einstein, Podolsky, and Rosen investigated what would happen when the photons are far apart and a measurement is performed on one of them. In the case of entangled particles—the pair of photons, created together and now light-years apart—it seems that the measurement performed on one has an effect on the other. The instant Alice measures the vertical polarization of her photon, Bob's photon will also have a definite polarization state on that axis, whereas its diagonal polarization will be indefinite. The measurement thus creates an influence apparently traveling faster than light. It seemed a paradox, and Einstein abhorred it. "That which really exists in B should not depend on what kind of measurement is carried out in space A," he wrote. The paper concluded sternly, "No reasonable definition of reality could be expected to permit this." He gave it the indelible name *spukhafte Fernwirkung,* "spooky action at a distance."

In 2003 the Israeli physicist Asher Peres proposed one answer to the Einstein-Podolsky-Rosen (EPR) puzzle. The paper was not exactly wrong, he said, but it had been written too soon: before Shannon published his theory of information, "and it took many more years before the latter was included in the physicist's toolbox." Information is physical. It is no use talking about quantum states without considering the *information* about the quantum states.

> Information is not just an abstract notion. It requires a physical carrier, and the latter is (approximately) localized. After all, it was the business of the Bell Telephone Company to transport information from one telephone to another telephone, in a different location.
> . . . When Alice measures her spin, the information she gets is localized at her position, and will remain so until she decides to broadcast it. Absolutely nothing happens at Bob's location. . . . It is only if and when Alice informs Bob of the result she got (by mail, telephone, radio, or by means of any other material carrier, which is naturally restricted to the speed of light) that Bob realizes that his particle has a definite pure state.

For that matter, Christopher Fuchs argues that it is no use talking about quantum states at all. The quantum state is a construct of the observer—from which many troubles spring. Exit states; enter information. "Terminology can say it all: A practitioner in this field, whether she has ever thought an ounce about quantum foundations, is just as likely to say 'quantum information' as 'quantum state' . . . 'What does the quantum teleportation protocol do?' A now completely standard answer would be: 'It transfers quantum information from Alice's site to Bob's.' What we have here is a change of mind-set."

The puzzle of spooky action at a distance has not been altogether resolved. *Nonlocality* has been demonstrated in a variety of clever experiments all descended from the EPR thought experiment. Entanglement turns out to be not only real but ubiquitous. The atom pair in every hydrogen molecule, H_2, is quantumly entangled (*"verschränkt,"* as Schrödinger said). Bennett put entanglement to work in quantum teleportation, presented publicly for the first time in 1993. Teleportation uses an entangled pair to project quantum information from a third particle across an arbitrary distance. Alice cannot measure this third particle directly; rather, she measures something about its relation to one of the entangled particles. Even though Alice herself remains ignorant about the original, because of the uncertainty principle, Bob is able to receive an exact replica. Alice's object is disembodied in the process. Communication is not faster than light, because Alice must also send Bob a classical (nonquantum) message on the side. "The net result of teleportation is completely prosaic: the removal of [the quantum object] from Alice's hands and its appearance in Bob's hands a suitable time later," wrote Bennett and his colleagues. "The only remarkable feature is that in the interim, the information has been cleanly separated into classical and nonclassical parts."

Researchers quickly imagined many applications, such as transfer of volatile information into secure storage, or memory. With or without goulash, teleportation created excitement, because it opened up new possibilities for the very real but still elusive dream of quantum computing.

The idea of a quantum computer is strange. Richard Feynman chose the strangeness as his starting point in 1981, speaking at MIT, when he first explored the possibility of using a quantum system to compute hard quantum problems. He began with a supposedly naughty digression—"Secret! Secret! Close the doors . . ."

> We have always had a great deal of difficulty in understanding the world view that quantum mechanics represents. At least I do, because I'm an old enough man [he was sixty-two] that I haven't got to the point that this stuff is obvious to me. Okay, I still get nervous with it. . . . It has not yet become obvious to me that there is no real problem. I cannot define the real problem, therefore I suspect there's no real problem, but I'm not sure there's no real problem.

He knew very well what the problem was for computation—for simulating quantum physics with a computer. The problem was probability.

Every quantum variable involved probabilities, and that made the difficulty of computation grow exponentially. "The number of information bits is the same as the number of points in space, and therefore you'd have to have something like N^N configurations to be described to get the probability out, and that's too big for our computer to hold. . . . It is therefore impossible, according to the rules stated, to simulate by calculating the probability."

So he proposed fighting fire with fire. "The other way to simulate a probabilistic Nature, which I'll call N for the moment, might still be to simulate the probabilistic Nature by a computer C which itself is probabilistic." A quantum computer would not be a Turing machine, he said. It would be something altogether new.

"Feynman's insight," says Bennett, "was that a quantum system is, in a sense, computing its own future all the time. You may say it's an analog computer of its own dynamics." Researchers quickly realized that if a quantum computer had special powers in cutting through problems in simulating physics, it might be able to solve other types of intractable problems as well.

The power comes from that shimmering, untouchable object the qubit. The probabilities are built in. Embodying a superposition of states gives the qubit more power than the classical bit, always in only one state or the other, zero or one, "a pretty miserable specimen of a two-dimensional vector," as David Mermin says. "When we learned to count on our sticky little classical fingers, we were misled," Rolf Landauer said dryly. "We thought that an integer had to have a particular and unique value." But no—not in the real world, which is to say the quantum world.

In quantum computing, multiple qubits are entangled. Putting qubits at work together does not merely multiply their power; the power increases exponentially. In classical computing, where a bit is either-or, n bits can encode any one of 2^n values. Qubits can encode these Boolean values along with all their possible superpositions. This gives a quantum computer a potential for parallel processing that has no classical equivalent.

So quantum computers—in theory—can solve certain classes of problems that had otherwise been considered computationally infeasible.

An example is finding the prime factors of very large numbers. This happens to be the key to cracking the most widespread cryptographic algorithms in use today, particularly RSA encryption. The world's Internet commerce depends on it. In effect, the very large number is a public key used to encrypt a message; if eavesdroppers can figure out its prime factors (also large), they can decipher the message. But whereas multiplying a pair of large prime numbers is easy, the inverse is exceedingly difficult. The procedure is an informational one-way street. So factoring RSA numbers has been an ongoing challenge for classical computing. In December 2009 a team distributed in Lausanne, Amsterdam, Tokyo, Paris, Bonn, and Redmond, Washington, used many hundreds of machines working almost two years to discover that 1230186684530117 7551304949583849627207728535695953347921973224521517264005072636575187452021997864693899564749427740638459251925573263034537315482685079170261221429134616704292143116022212404792747377940806653514195974598569021434413 is the product of 3347807169 8956898786044169848212690817704794983713768568912431388982883793878002287614711652531743087737814467999489 and 3674604366 67995904282446337996279526322791581643430876426760322838157396665112792333734171433968102700927987363089177. They estimated that the computation used more than 10^{20} operations.

This was one of the smaller RSA numbers, but, had the solution come earlier, the team could have won a $50,000 prize offered by RSA Laboratories. As far as classical computing is concerned, such encryption is considered quite secure. Larger numbers take exponentially longer time, and at some point the time exceeds the age of the universe.

Quantum computing is another matter. The ability of a quantum computer to occupy many states at once opens new vistas. In 1994, before anyone knew how actually to build any sort of quantum computer, a mathematician at Bell Labs figured out how to program one to solve the factoring problem. He was Peter Shor, a problem-solving prodigy who made an early mark in math olympiads and prize competitions. His ingenious algorithm, which broke the field wide open, is known by him simply as the factoring algorithm, and by everyone else as Shor's algorithm. Two years later Lov Grover, also at Bell Labs, came up with a quantum algorithm for searching a vast unsorted database. That is the canonical hard problem for a world of limitless information—needles and haystacks.

"Quantum computers were basically a revolution," Dorit Aharonov of Hebrew University told an audience in 2009. "The revolution was launched into the air by Shor's algorithm. But the *reason* for the revolution—other than the amazing practical implications—is that they redefine what is an *easy* and what is a *hard* problem."

What gives quantum computers their power also makes them exceedingly difficult to work with. Extracting information from a system means observing it, and observing a system means interfering with the quantum magic. Qubits cannot be watched as they do their exponentially many operations in parallel; measuring that shadow-mesh of possibilities reduces it to a classical bit. Quantum information is fragile. The only way to learn the result of a computation is to wait until after the quantum work is done.

Quantum information is like a dream—evanescent, never quite existing as firmly as a word on a printed page. "Many people can read a book and get the same message," Bennett says, "but trying to tell people about your dream changes your memory of it, so that eventually you forget the dream and remember only what you said about it." Quantum erasure, in turn, amounts to a true undoing: "One can fairly say that even God has forgotten."

As for Shannon himself, he was unable to witness this flowering of the seeds he had planted. "If Shannon were around now, I would say he would be very enthusiastic about the entanglement-assisted capacity of a channel," says Bennett. "The same form, a generalization of Shannon's formula, covers both classic and quantum channels in a very elegant way. So it's pretty well established that the quantum generalization of classical information has led to a cleaner and more powerful theory, both of computing and communication." Shannon lived till 2001, his last years dimmed and isolated by the disease of erasure, Alzheimer's. His life had spanned the twentieth century and helped to define it. As much as any one person, he was the progenitor of the information age. Cyberspace is in part his creation; he never knew it, though he told his last interviewer, in 1987, that he was investigating the idea of mirrored rooms: "to work out all the possible mirrored rooms that make sense, in that if you looked everywhere from inside one, space would be divided into a bunch of rooms, and you would be in each room and this would go on to infinity without contradiction." He hoped to build a gallery of mirrors in his house near MIT, but he never did.

It was John Wheeler who left behind an agenda for quantum information science—a modest to-do list for the next generation of physicists and computer scientists together:

"Translate the quantum versions of string theory and of Einstein's geometrodynamics from the language of continuum to the language of bit," he exhorted his heirs.

"Survey one by one with an imaginative eye the powerful tools that mathematics—including mathematical logic—has won . . . and for each such technique work out the transcription into the world of bits."

And, "From the wheels-upon-wheels-upon-wheels evolution of computer programming dig out, systematize and display every feature that illuminates the level-upon-level-upon-level structure of physics."

And, "*Finally*. Deplore? No, celebrate the absence of a clean clear definition of the term 'bit' as elementary unit in the establishment of meaning. . . . If and when we learn how to combine bits in fantastically large numbers to obtain what we call existence, we will know better what we mean both by bit and by existence."

This is the challenge that remains, and not just for scientists: the establishment of meaning.

4 | AFTER THE FLOOD

(A Great Album of Babel)

> *Suppose within every book there is another book, and within every letter on every page another volume constantly unfolding; but these volumes take no space on the desk. Suppose knowledge could be reduced to a quintessence, held within a picture, a sign, held within a place which is no place.*
>
> —Hilary Mantel (2009)

"THE UNIVERSE (which others call the Library) . . ."

Thus Jorge Luis Borges began his 1941 story "The Library of Babel," about the mythical library that contains all books, in all languages, books of apology and prophecy, the gospel and the commentary upon that gospel and the commentary upon the commentary upon the gospel, the minutely detailed history of the future, the interpolations of all books in all other books, the faithful catalogue of the library and the innumerable false catalogues. This library (which others call the universe) enshrines all the information. Yet no knowledge can be discovered there, precisely because all knowledge *is* there, shelved side by side with all falsehood. In the mirrored galleries, on the countless shelves, can be found everything and nothing. There can be no more perfect case of information glut.

We make our own storehouses. The persistence of information, the difficulty of forgetting, so characteristic of our time, accretes confusion. As the free, amateur, collaborative online encyclopedia called Wikipedia began to overtake all the world's printed encyclopedias in volume and comprehensiveness, the editors realized that too many names had multiple identities. They worked out a disambiguation policy, which led to the creation of disambiguation pages—a hundred thousand and more. For example, a user foraging in Wikipedia's labyrinthine galleries for "Babel" finds "Babel (disambiguation)," which leads in turn to the Hebrew name for ancient Babylon, to the Tower of Babel, to an Iraqi newspaper, a book by Patti Smith, a Soviet journalist, an Australian language teachers' journal, a film, a record label, an island in Australia, two different mountains in Canada, and "a neutrally aligned planet in the fictional Star Trek universe." And more. The paths of disambiguation fork again and again. For example, "Tower of Babel (disambiguation)" lists, besides the story in the Old Testament, songs, games, books, a Brueghel painting, an Escher woodcut, and "the tarot card." We have made many towers of Babel.

Long before Wikipedia, Borges also wrote about the encyclopedia "fallaciously called *The Anglo-American Cyclopedia* (New York, 1917)," a warren of fiction mingling with fact, another hall of mirrors and misprints, a compendium of pure and impure information that projects its own world. That world is called Tlön. "It is conjectured that this brave new world is the work of a secret society of astronomers, biologists, engineers, metaphysicians, poets, chemists, algebraists, moralists, painters, geometers. . . ." writes Borges. "This plan is so vast that each writer's contribution is infinitesimal. At first it was believed that Tlön was a mere chaos, an irresponsible license of the imagination; now it is known that it is a cosmos." With good reason, the Argentine master has been taken up as a prophet ("our heresiarch uncle," William Gibson says) by another generation of writers in the age of information.

Long before Borges, the imagination of Charles Babbage had conjured another library of Babel. He found it in the very air: a record, scrambled yet permanent, of every human utterance.

What a strange chaos is this wide atmosphere we breathe! . . . The air itself is one vast library, on whose pages are for ever written all that man has ever said or woman whispered. There, in their mutable but unerring characters, mixed with the earliest, as well as the latest sighs of mortality, stand for ever recorded, vows unredeemed, promises unfulfilled, perpetuating in the united movements of each particle, the testimony of man's changeful will.

Edgar Allan Poe, following Babbage's work eagerly, saw the point. "No thought can perish," he wrote in 1845, in a dialogue between two angels. "Did there not cross your mind some thought of the *physical power of words*? Is not every word an impulse on the air?" Further, every impulse vibrates outward indefinitely, "upward and onward in their influences upon all particles of all matter," until it must, "*in the end*, impress every individual thing that exists *within the universe*." Poe was also reading Newton's champion Pierre-Simon Laplace. "A being of infinite understanding," wrote Poe, "—one to whom the *perfection* of the algebraic analysis lay unfolded" could trace the undulations backward to their source.

Babbage and Poe took an information-theoretic view of the new physics. Laplace had expounded a perfect Newtonian mechanical determinism; he went further than Newton himself, arguing for a clockwork universe in which nothing is left to chance. Since the laws of physics apply equally to the heavenly bodies and the tiniest particles, and since they operate with perfect reliability, then surely (said Laplace) the state of the universe at every instant follows inexorably from the past and must lead just as relentlessly to the future. It was too soon to conceive of quantum uncertainty, chaos theory, or the limits of computability. To dramatize his perfect determinism, Laplace asked us to imagine a being—an "intelligence"—capable of perfect knowledge:

> It would embrace in the same formula the movements of the greatest bodies of the universe and those of the lightest atom; for it, nothing would be uncertain and the future, as the past, would be present to its eyes.

Nothing else Laplace wrote ever became as famous as this thought experiment. It rendered useless not only God's will but Man's. To scientists this extreme Newtonianism seemed cause for optimism. To Babbage, all nature suddenly resembled a vast calculating engine, a grand version of his own deterministic machine: "In turning our views from these simple consequences of the juxtaposition of a few wheels, it is impossible not to perceive the parallel reasoning, as applied to the mighty and far more complex phenomena of nature." Each atom, once disturbed, must communicate its motion to others, and they in turn influence waves of air, and no impulse is ever entirely lost. The track of every canoe remains somewhere in the oceans. Babbage, whose railroad pen recorder traced on a roll of paper the history of a journey, saw information, formerly evanescent, as a series of physical impressions that were, or could be preserved. The phonograph, impressing sound into foil or wax, had yet to be invented, but Babbage could view the atmosphere as an engine of motion with meaning: "every atom impressed with good and with ill . . . which philosophers and sages have imparted to it, mixed and combined in ten thousand ways with all that is worthless and base." Every word ever said, whether heard by a hundred listeners or none, far from having vanished into the air, leaves its indelible mark, the complete record of human utterance being encrypted by the laws of motion and capable, in theory, of being recovered—given enough computing power.

This was overoptimistic. Still, the same year Babbage published his essay, the artist and chemist Louis Daguerre in Paris perfected his means of capturing visual images on silver-coated plates. His English competitor, William Fox Talbot, called this "the art of photogenic drawing, or of forming pictures and images of natural objects by means of solar light." Talbot saw something meme-like. "By means of this contrivance," he wrote, "it is not the artist who makes the picture, but the picture which makes *itself*." Now the images that fly before our eyes could be frozen, impressed upon substance, made permanent.

By painting or drawing, an artist—with skill, training, and long labor—reconstructs what the eye might see. By contrast, a daguerreotype is in some sense the thing itself—the information, stored, in an instant. It was unimaginable, but there it was. The possibilities made the mind reel. Once storage began, where would it stop? An American essayist immediately connected photography to Babbage's atmospheric library of sounds: Babbage said that every word was registered somewhere in the air, so perhaps every image, too, left its permanent mark—somewhere.

In fact, there is a great album of Babel. But what too, if the great business of the sun be to act registrar likewise, and to give out impressions of our looks, and pictures of our actions; and so . . . for all we know to the contrary, other worlds may be peopled and conducted with the images of persons and transactions thrown off from this and from each other; the whole universal nature being nothing more than phonetic and photogenic structures.

The universe, which others called a library or an album, then came to resemble a computer. Alan Turing may have noticed this first: observing that the computer, like the universe, is best seen as a collection of states, and the state of the machine at any instant leads to the state at the next instant, and thus all the future of the machine should be predictable from its initial state and its input signals.

The universe is computing its own destiny.

Turing noticed that Laplace's dream of perfection might be possible in a machine but not in the universe, because of a phenomenon which, a generation later, would be discovered by chaos theorists and named the butterfly effect. Turing described it this way in 1950:

The system of the "universe as a whole" is such that quite small errors in initial conditions can have an overwhelming effect at a later time. The displacement of a single electron by a billionth of a centimetre at one moment might make the difference between a man being killed by an avalanche a year later, or escaping.

If the universe is a computer, we may still struggle to access its memory. If it is a library, it is a library without shelves. When all the world's sounds disperse through the atmosphere, no word is left attached to any particular bunch of atoms. The words are anywhere and everywhere. That was why Babbage called this information store a "chaos." Once again he was ahead of his time.

When the ancients listed the Seven Wonders of the World, they included the Lighthouse of Alexandria, a 400-foot stone tower built to aid sailors, but overlooked the library nearby. The library, amassing hundreds of thousands of papyrus rolls, maintained the greatest collection of knowledge on earth, then and for centuries to come. Beginning in the third century BCE, it served the Ptolemies' ambition to buy, steal, or copy all the writings of the known world. The library enabled Alexandria to surpass Athens as an intellectual center. Its racks and cloisters held the dramas of Sophocles, Aeschylus, and Euripides; the mathematics of Euclid, Archimedes, and Eratosthenes; poetry, medical texts, star charts, mystic writings—"such a blaze of knowledge and discovery," H. G. Wells declared, "as the world was not to see again until the sixteenth century. . . . It is the true beginning of Modern History." The lighthouse loomed large, but the library was the real wonder. And then it burned.

Exactly when and how that happened, no one can ever know. Probably more than once. Vengeful conquerors burn books as if the enemy's souls reside there, too. "The Romans burnt the books of the Jews, of the Christians, and the philosophers," Isaac D'Israeli noted in the nineteenth century; "the Jews burnt the books of the Christians and the Pagans; and the Christians burnt the books of the Pagans and the Jews." The Qin dynasty burned China's books in order to erase previous history. The erasure was effective, the written word being fragile. What we have of Sophocles is not even a tenth of his plays. What we have of Aristotle is mostly second- or thirdhand. For historians peering into the past, the destruction of the Great Library is an event horizon, a boundary across which information does not pass. Not even a partial catalogue survived the flames.

"All the lost plays of the Athenians!" wails Thomasina (a young mathematician who resembles Ada Byron) to her tutor, Septimus, in Tom Stoppard's drama *Arcadia*. "Thousands of poems—Aristotle's own library . . . How can we sleep for grief?"

"By counting our stock," Septimus replies.

> You should no more grieve for the rest than for a buckle lost from your first shoe, or for your lesson book which will be lost when you are old. We shed as we pick up, like travelers who must carry everything in their arms, and what we let fall will be picked up by those behind. The procession is very long and life is very short. We die on the march. But there is nothing outside the march so nothing can be lost to it. The missing plays of Sophocles will turn up piece by piece, or be written again in another language.

Anyway, according to Borges, the missing plays can be found in the Library of Babel.

In honor of the lost library, Wikipedia drew hundreds of its editors to Alexandria in the eighth summer of its existence—people called Shipmaster, Brassratgirl, Notafish, and Jimbo who ordinarily meet only online. More than 7 million such user names had been registered by then; the pilgrims came from forty-five countries, paying their own way, toting laptops, exchanging tradecraft, wearing their fervor on their T-shirts. By then, July 2008, Wikipedia comprised 2.5 million articles in English, more than all the world's paper encyclopedias combined, and a total of 11 million in 264 languages, including Wolof, Twi, and Dutch Low Saxon, but not including Choctaw, closed by community vote after achieving only fifteen articles, or Klingon, found to be a "constructed," if not precisely fictional, language. The Wikipedians consider themselves as the Great Library's heirs, their mission the gathering of all recorded knowledge. They do not, however, collect and preserve existing texts. They attempt to summarize shared knowledge, apart from and outside of the individuals who might have thought it was theirs.

Like the imaginary library of Borges, Wikipedia begins to appear boundless. Several dozen of the non-English Wikipedias have, each, one article on Pokémon, the trading-card game, manga series, and media franchise. The English Wikipedia began with one article and then a jungle grew. There is a page for "Pokémon (disambiguation)," needed, among other reasons, in case anyone is looking for the Zbtb7 oncogene, which was called Pokemon (for POK erythroid myeloid ontogenic factor), until Nintendo's trademark lawyers threatened to sue. There are at least five major articles about the popular-culture Pokémons, and these spawn secondary and side articles, about the Pokémon regions, items, television episodes, game tactics, and all 493 creatures, heroes, protagonists, rivals, companions, and clones, from Bulbasaur to Arceus. All are carefully researched and edited for accuracy, to ensure that they are reliable and true to the Pokémon universe, which does not actually, in some senses of the word, exist. Back in the real world, Wikipedia has, or aspires to have, detailed entries describing the routes, intersections, and histories of every numbered highway and road in the United States. ("Route 273 [New York State, decommissioned in 1980] began at an intersection with U.S. Route 4 in Whitehall. After the intersection, the route passed the Our Lady of Angels Cemetery, where it turned to the southeast. Route 273 ran along the base of Ore Red Hill, outside of Whitehall. Near Ore Red Hill, the highway intersected with a local road, which connected to US 4.") There are pages for every known enzyme and human gene. The *Encyclopaedia Britannica* never aspired to such breadth. How could it, being made of paper?

Alone among the great enterprises of the early Internet, Wikipedia was not a business; made no money, only lost money. It was supported by a nonprofit charity established for the purpose. By the time the encyclopedia had 50 million users daily, the foundation had a payroll of eighteen people, including one in Germany, one in the Netherlands, one in Australia, and one lawyer, and everyone else was a volunteer: the millions of contributors, the thousand or more designated "administrators," and, always a looming presence, the founder and self-described "spiritual leader," Jimmy Wales. Wales did not plan initially the scrappy, chaotic, dilettantish, amateurish, upstart free-for-all that Wikipedia quickly became. The would-be encyclopedia began with a roster of experts, academic credentials, verification, and peer review. But the wiki idea took over, willy-nilly. A "wiki," from a Hawaiian word for "quick," was a web site that could be not just viewed but edited, by anyone. A wiki was therefore self-created, or at least self-sustaining.

Wikipedia first appeared to Internet users with a simple self-description:

HomePage

You can edit this page right now! It's a free, community project

Welcome to Wikipedia! We're writing a complete encyclopedia from scratch, collaboratively. We started work in January 2001. We've got over 3,000 pages already. We want to make over 100,000. So, let's get to work! Write a little (or a lot) about what you know! Read our welcome message here: Welcome, newcomers!

The sparseness of the coverage that first year could be gauged by the list of requested articles. Under the heading of Religion: "Catholicism? —Satan?—Zoroaster?—Mythology?" Under Technology: "internal combustion engine?—dirigible?—liquid crystal display?—bandwidth?" Under Folklore: "(If you want to write about folklore, please come up with a list of folklore topics that are actually recognized as distinct, significant topics in folklore, a subject that you are not likely to know much about if all you've done along these lines is play Dungeons and Dragons, q.v.)." Dungeons and Dragons was already well covered. Wikipedia was not looking for flotsam and jetsam but did not scorn them. Years later, in Alexandria, Jimmy Wales said: "All those people who are obsessively writing about Britney Spears or the Simpsons or Pokémon—it's just not true that we should try to redirect them into writing about obscure concepts in physics. Wiki is not paper, and their time is not owned by us. We can't say, 'Why do we have these employees doing stuff that's so useless?' They're not hurting anything. Let them write it."

"Wiki is not paper" was the unofficial motto. Self-referentially, the phrase has its own encyclopedia page (see also "*Wiki ist kein Papier*" and "*Wikipédia n'est pas sur papier*"). It means there is no physical or economic limit on the number or the length of articles. Bits are free. "Any kind of metaphor around paper or space is dead," as Wales said.

Wikipedia found itself a mainstay of the culture with unexpected speed, in part because of its unplanned synergistic relationship with Google. It became a test case for ideas of crowd intelligence: users endlessly debated the reliability—in theory and in actuality—of articles written in an authoritative tone by people with no credentials, no verifiable identity, and unknown prejudices. Wikipedia was notoriously subject to vandalism. It exposed the difficulties—perhaps the impossibility—of reaching a neutral, consensus view of disputed, tumultuous reality. The process was plagued by so-called edit wars, when battling contributors reversed one another's alterations without surcease. At the end of 2006, people concerned with the "Cat" article could not agree on whether a human with a cat is its "owner," "caregiver," or "human companion." Over a three-week period, the argument extended to the length of a small book. There were edit wars over commas and edit wars over gods, futile wars over spelling and pronunciation and geopolitical disputes. Other edit wars exposed the malleability of words. Was the Conch Republic (Key West, Florida) a "micronation"? Was a particular photograph of a young polar bear "cute"? Experts differed, and everyone was an expert.

After the occasional turmoil, articles tend to settle toward permanence; still, if the project seems to approach a kind of equilibrium, it is nonetheless dynamic and unstable. In the Wikipedia universe, reality cannot be pinned down with finality. That idea was an illusion fostered in part by the solidity of a leather-and-paper encyclopedia. Denis Diderot aimed in the *Encyclopédie*, published in Paris beginning in 1751, "to collect all the knowledge that now lies scattered over the face of the earth, to make known its general structure to the men with whom we live, and to transmit it to those who will come after us." The *Britannica*, first produced in

Edinburgh in 1768 in one hundred weekly installments, sixpence apiece, wears the same halo of authority. It seemed finished—in every edition. It has no equivalent in any other language. Even so, the experts responsible for the third edition ("in Eighteen Volumes, Greatly Improved"), a full century after Isaac Newton's *Principia*, could not bring themselves to endorse his, or any, theory of gravity, or gravitation. "There have been great disputes," the *Britannica* stated.

> Many eminent philosophers, and among the rest Sir Isaac Newton himself, have considered it as the first of all second causes; an incorporeal or spiritual substance, which never can be perceived any other way than by its effects; an universal property of matter, &c. Others have attempted to explain the phenomena of gravitation by the action of a very subtile etherial fluid; and to this explanation Sir Isaac, in the latter part of his life, seems not to have been averse. He hath even given a conjecture concerning the matter in which this fluid might occasion these phenomena. But for a full account of . . . the state of the dispute at present, see the articles, Newtonian Philosophy, Astronomy, Atmosphere, Earth, Electricity, Fire, Light, Attraction, Repulsion, Plenum, Vacuum, &c.

As the *Britannica* was authoritative, Newton's theory of gravitation was not yet knowledge.

Wikipedia disclaims this sort of authority. Academic institutions officially distrust it. Journalists are ordered not to rely upon it. Yet the authority comes. If one wants to know how many American states contain a county named Montgomery, who will disbelieve the tally of eighteen in Wikipedia? Where else could one look for a statistic so obscure— generated by a summing of the knowledge of hundreds or thousands of people, each of whom may know of only one particular Montgomery County? Wikipedia features a popular article called "Errors in the *Encyclopaedia Britannica* that have been corrected in Wikipedia." This article is, of course, always in flux. All Wikipedia is. At any moment the reader is catching a version of truth on the wing.

When Wikipedia states, in the article "Aging,"

> After a period of near perfect renewal (in humans, between 20 and 35 years of age [citation needed]), organismal senescence is characterized by the declining ability to respond to stress, increasing homeostatic imbalance and increased risk of disease. This irreversible series of changes inevitably ends in death,

a reader may trust this; yet for one minute in the early morning of December 20, 2007, the entire article comprised instead a single sentence: "Aging is what you get when you get freakin old old old." Such obvious vandalism lasts hardly any time at all. Detecting it and reversing it are automated vandalbots and legions of human vandal fighters, many of them proud members of the Counter-Vandalism Unit and Task Force. According to a popular saying that originated with a frustrated vandal, "On Wikipedia, there is a giant conspiracy attempting to have articles agree with reality." This is about right. A conspiracy is all the Wikipedians can hope for, and often it is enough.

Lewis Carroll, near the end of the nineteenth century, described in fiction the ultimate map, representing the world on a unitary scale, a mile to a mile: "It has never been spread out, yet. The farmers objected: they said it would cover the whole country, and shut out the sunlight." The point is not lost on Wikipedians. Some are familiar with a debate carried out by the German branch about the screw on the left rear brake pad of Ulrich Fuchs's bicycle. Fuchs, as a Wikipedia editor, proposed the question, Does this item in the universe of objects merit its own Wikipedia entry? The screw was agreed to be small but real and specifiable. "This is an object in space, and I've seen it," said Jimmy Wales. Indeed, an article appeared in the German Meta-Wiki (that is, the Wikipedia *about* Wikipedia) titled "*Die Schraube an der hinteren linken Bremsbacke am Fahrrad von Ulrich Fuchs.*" As Wales noted, the very existence of this article was "a meta-irony." It was written by the very people who were arguing against its suitability. The article was not really about the screw, however. It is about a controversy: whether Wikipedia should strive, in theory or in practice, to describe the whole world in all its detail.

Opposing factions coalesced around the labels "deletionism" and "inclusionism." Inclusionists take the broadest view of what belongs in Wikipedia. Deletionists argue for, and often perform, the removal of trivia: articles too short or poorly written or unreliable, on topics lacking notability. All these criteria are understood to be variable and subjective. Deletionists want to raise the bar of quality. In 2008 they succeeded in removing an entry on the Port Macquarie Presbyterian Church, New South Wales, Australia, on grounds of non-notability. Jimmy Wales himself leaned toward inclusionism. In the late summer of 2007, he visited Cape Town, South Africa, ate lunch at a place called Mzoli's, and created a "stub" with a single sentence: "Mzoli's Meats is a butcher shop and restuarant located in Guguletu township near Cape Town, South Africa." It survived for twenty-two minutes before a nineteen-year-old administrator called ^demon deleted it on grounds of insignificance. An hour later, another user re-created the article and expanded it based on information from a local Cape Town blog and a radio interview transcribed online. Two minutes passed, and yet another user objected on grounds that "this article or section is written like an advertisement." And so on. The word "famous" was inserted and deleted several times. The user ^demon weighed in again, saying, "We are not the white pages and we are not a travel guide." The user EVula retorted, "I think if we give this article a bit more than a couple of hours of existence, we might have something worthwhile." Soon the dispute attracted newspaper coverage in Australia and England. By the next year, the article had not only survived but had grown to include a photograph, an exact latitude and longitude, a list of fourteen references, and separate sections for History, Business, and Tourism. Some hard feelings evidently remained, for in March 2008 an anonymous user replaced the entire article with one sentence: "Mzoli's is an insignificant little restaurant whose article only exists here because Jimmy Wales is a bumbling egomaniac." That lasted less than a minute.

Wikipedia evolves dendritically, sending off new shoots in many directions. (In this it resembles the universe.) So deletionism and inclusionism spawn mergism and incrementalism. They lead to factionalism, and the factions fission into Associations of Deletionist Wikipedians and Inclusionist Wikipedians side by side with the Association of Wikipedians Who Dislike Making Broad Judgments About the Worthiness of a General Category of Article, and Who Are in Favor of the Deletion of Some Particularly Bad Articles, but That Doesn't Mean They Are Deletionists. Wales worried particularly about Biographies of Living Persons. In an ideal world, where Wikipedia could be freed from practical concerns of maintenance and reliability, Wales said he would be happy to see a biography of every human on the planet. It outdoes Borges.

Even then, at the impossible extreme—every person, every bicycle screw—the collection would possess nothing like All Knowledge. For encyclopedias, information tends to come in the form of topics and categories. *Britannica* framed its organization in 1790 as "a plan entirely new." It advertised "the different sciences and arts" arranged as "distinct Treatises or Systems"—

And *full Explanations given* of the
Various Detached Parts of Knowledge, whether relating to Natural and Artificial
Objects, or to Matters Ecclesiastical, Civil, Military, Commercial, &c.

In Wikipedia the detached parts of knowledge tend to keep splitting. The editors analyzed the logical dynamics as Aristotle or Boole might have:

Many topics are based on the relationship of *factor X* to *factor Y*, resulting in one or more full articles. This could refer to, for example, *situation X* in *location Y*, or *version X* of *item Y*. This is perfectly valid when the two variables put together represent some culturally significant phenomenon or some otherwise notable interest. Often, separate articles are needed for a subject within a range of different countries due to its substantial differences across international borders. Articles like Slate industry in Wales and Island Fox are fitting examples. But writing about Oak trees in North

Carolina or a Blue truck would likely constitute a POV fork, original research, or would otherwise be outright silly.

Charles Dickens had earlier considered this very problem. In *The Pickwick Papers*, a man is said to have read up in the *Britannica* on Chinese metaphysics. There was, however, no such article: "He read for metaphysics under the letter M, and for China under the letter C, and combined his information."

In 2008 the novelist Nicholson Baker, calling himself Wageless, got sucked into Wikipedia like so many others, first seeking information and then tentatively supplying some, beginning one Friday evening with the article on bovine somatotropin and, the next day, *Sleepless in Seattle*, periodization, and hydraulic fluid. On Sunday it was pornochanchada (Brazilian sex films), a football player of the 1950s called Earl Blair, and back to hydraulic fluid. On Tuesday he discovered the Article Rescue Squadron, dedicated to finding articles in danger of deletion and saving them by making them better instead. Baker immediately signed up, typing a note: "I want to be a part of this." His descent into obsession is documented in the archives, like everything else that happens on Wikipedia, and he wrote about it a few months later in a print publication, *The New York Review of Books*.

> I began standing with my computer open on the kitchen counter, staring at my growing watchlist, checking, peeking. . . . I stopped hearing what my family was saying to me—for about two weeks I all but disappeared into my screen, trying to salvage brief, sometimes overly promotional but nevertheless worthy biographies by recasting them in neutral language, and by hastily scouring newspaper databases and Google Books for references that would bulk up their notability quotient. I had become an "inclusionist."

He concluded with a "secret hope": that all the flotsam and jetsam could be saved, if not in Wikipedia than in "a Wikimorgue—a bin of broken dreams." He suggested calling it Deletopedia. "It would have much to tell us over time." On the principle that nothing online ever perishes, Deletionpedia was created shortly thereafter, and it has grown by degrees. The Port Macquarie Presbyterian Church lives on there, though it is not, strictly speaking, part of the encyclopedia. Which some call the universe.

Names became a special problem: their disambiguation; their complexity; their collisions. The nearly limitless flow of information had the effect of throwing all the world's items into a single arena, where they seemed to play a frantic game of Bumper Car. Simpler times had allowed simpler naming: "The Lord God formed every beast of the field, and every fowl of the air; and brought them unto Adam to see what he would call them," says Genesis; "and whatsoever Adam called every living creature, that was the name thereof." For each creature one name; for each name one creature. Soon, however, Adam had help.

In his novel *The Infinities*, John Banville imagines the god Hermes saying: "A hamadryad is a wood-nymph, also a poisonous snake in India, and an Abyssinian baboon. It takes a god to know a thing like that." Yet according to Wikipedia, *hamadryad* also names a butterfly, a natural history journal from India, and a Canadian progressive rock band. Are we all now as gods? The rock band and the wood nymph could coexist without friction, but more generally the breaking down of information barriers leads to conflict over names and naming rights. Impossible as it seems, the modern world is running out of names. The roster of possibilities seems infinite, but the demand is even greater.

The major telegraph companies, struggling in 1919 with the growing problem of misdirected messages, established a Central Bureau for Registered Addresses. Its central office in the financial district of New York filled an upstairs room on Broad Street with steel filing cabinets. Customers were invited to register code names for their addresses: single words of five to ten letters, required to be "pronounceable"—that is, "made up of syllables that appear in one of eight European languages."

Many customers complained about the yearly charge—$2.50 per code name—but by 1934 the bureau was managing a list of 28,000, including ILLUMINATE (the New York Edison Company), TOOTSWEETS (the Sweet Company of America), and CHERRYTREE (George Washington Hotel). The financier Bernard M. Baruch managed to get BARUCH all to himself. It was first come, first served, and it was a modest harbinger of things to come.

Cyberspace, of course, changes everything. A South Carolina company called Fox & Hound Realty, Billy Benton owner/broker, registered the domain name BARUCH.COM. A Canadian living in High Prairie, Alberta, registered JRRTOLKIEN.COM and held on to it for a decade, until a panel of the World Intellectual Property Organization in Geneva took it away from him. The name had value; others who claimed an interest in it, as a brand and a trademark, either registered or unregistered, included the late writer's heirs, publisher, and filmmakers, not to mention the several thousand people worldwide who happened to share his surname. The same High Prairie man was basing a business on his possession of famous names: Céline Dion, Albert Einstein, Michael Crichton, Pierce Brosnan, and about 1,500 more. Some of these people fought back. A select few names—the pinnacles and hilltops—have developed a tremendous concentration of economic value. The word *Nike* is thought by economists to be worth $7 billion; *Coca-Cola* is valued at ten times more.

In the study of onomastics it is axiomatic that growing social units lead to growing name systems. For life in tribes and villages, single names like Albin and Ava were enough, but tribes gave way to clans, cities to nations, and people had to do better: surnames and patronyms; names based on geography and occupation. More complex societies demand more complex names. The Internet represents not just a new opportunity for fights over names but a leap in scale causing a phase transition.

An Atlanta music writer known as Bill Wyman received a cease-and-desist letter from lawyers representing the former Rolling Stone bass player also known as Bill Wyman; demanding, that is, that he "cease and desist" using his name. In responding, the first Bill Wyman pointed out that the second Bill Wyman had been born William George Perks. The car company known in Germany as Dr. Ing. h.c. F. Porsche AG fought a series of battles to protect the name Carrera. Another contender was the Swiss village, postal code 7122. "The village Carrera existed prior to the Porsche trademark," Christoph Reuss of Switzerland wrote to Porsche's lawyers. "Porsche's use of that name constitutes a misappropriation of the goodwill and reputation developed by the villagers of Carrera." He added for good measure, "The village emits much less noise and pollution than Porsche Carrera." He did not mention that José Carreras, the opera singer, was embroiled in a name dispute of his own. The car company, meanwhile, also claimed trademark ownership of the numerals 911.

A useful term of art emerged from computer science: *namespace*, a realm within which all names are distinct and unique. The world has long had namespaces based on geography and other namespaces based on economic niche. You could be Bloomingdale's as long as you stayed out of New York; you could be Ford if you did not make automobiles. The world's rock bands constitute a namespace, where Pretty Boy Floyd and Pink Floyd and Pink coexist, along with the 13th Floor Elevators and the 99th Floor Elevators and Hamadryad. Finding new names in this space becomes a challenge. The singer and songwriter long called simply "Prince" was given that name at birth; when he tired of it, he found himself tagged with a meta-name, "the Artist Formerly Known as Prince." The Screen Actors Guild maintains a formal namespace of its own—only one Julia Roberts allowed. Traditional namespaces are overlapping and melting together. And many grow overcrowded.

Pharmaceutical names are a special case: a subindustry has emerged to coin them, research them, and vet them. In the United States, the Food and Drug Administration reviews proposed drug names for possible collisions, and this process is complex and uncertain. Mistakes cause death. Methadone, for opiate dependence, has been administered in place of Metadate, for attention-deficit disorder, and Taxol, a cancer drug, for

Taxotere, a different cancer drug, with fatal results. Doctors fear both look-alike errors and sound-alike errors: Zantac/Xanax; Verelan/Virilon. Linguists devise scientific measures of the "distance" between names. But Lamictal and Lamisil and Ludiomil and Lomotil are all approved drug names.

In the corporate namespace, signs of overcrowding could be seen in the fading away of what might be called simple, meaningful names. No new company could be called anything like General Electric or First National Bank or International Business Machines. Similarly, A.1. Steak Sauce could only refer to a food product with a long history. Millions of company names exist, and vast sums of money go to professional consultants in the business of creating more. It is no coincidence that the spectacular naming triumphs of cyberspace verge on nonsense: Yahoo!, Google, Twitter.

The Internet is not just a churner of namespaces; it is also a namespace of its own. Navigation around the globe's computer networks relies on the special system of domain names, like COCA-COLA.COM. These names are actually addresses, in the modern sense of that word: "a register, location, or a device where information is stored." The text encodes numbers; the numbers point to places in cyberspace, branching down networks, subnetworks, and devices. Although they are code, these brief text fragments also carry the great weight of meaning in the most vast of namespaces. They blend together features of trademarks, vanity license plates, postal codes, radio-station call letters, and graffiti. Like the telegraph code names, anyone could register a domain name, for a small fee, beginning in 1993. It was first come, first served. The demand exceeds the supply.

Too much work for short words. Many entities own "apple" trademarks, but there is only one APPLE.COM; when the domains of music and computing collided, so did the Beatles and the computer company. There is only one MCDONALDS.COM, and a journalist named Joshua Quittner registered it first. Much as the fashion empire of Giorgio Armani wanted ARMANI.COM, so did Anand Ramnath Mani of Vancouver, and he got there first. Naturally a secondary market emerged for trade in domain names. In 2006, one entrepreneur paid another entrepreneur $14 million for SEX.COM. By then nearly every word in every well-known language had been registered; so had uncountable combinations of words and variations of words—more than 100 million. It is a new business for corporate lawyers. A team working for DaimlerChrysler in Stuttgart, Germany, managed to wrest back MERCEDESSHOP.COM, DRIVEAMERCEDES.COM, DODGEVIPER.COM, CRYSLER.COM, CHRISLER .COM, CHRYSTLER.COM, and CHRISTLER.COM.

The legal edifices of intellectual property were rattled. The response was a species of panic—a land grab in trademarks. As recently as 1980, the United States registered about ten thousand a year. Three decades later, the number approached three hundred thousand, jumping every year. The vast majority of trademark applications used to be rejected; now the opposite is true. All the words of the language, in all possible combinations, seem eligible for protection by governments. A typical batch of early twenty-first century United States trademarks: GREEN CIRCLE, DESERT ISLAND, MY STUDENT BODY, ENJOY A PARTY IN EVERY BOWL!, TECHNOLIFT, MEETINGS IDEAS, TAMPER PROOF KEY RINGS, THE BEST FROM THE WEST, AWESOME ACTIVITIES.

The collision of names, the exhaustion of names—it has happened before, if never on this scale. Ancient naturalists knew perhaps five hundred different plants and, of course, gave each a name. Through the fifteenth century, that is as many as anyone knew. Then, in Europe, as printed books began to spread with lists and drawings, an organized, collective knowledge came into being, and with it, as the historian Brian Ogilvie has shown, the discipline called natural history. The first botanists discovered a profusion of names. Caspar Ratzenberger, a student at Wittenberg in the 1550s, assembled a herbarium and tried to keep track: for one species he noted eleven names in Latin and German: *Scandix, Pecten veneris, Herba scanaria, Cerefolium aculeatum, Nadelkrautt, Hechelkam, NadelKoerffel, Venusstrahl, Nadel Moehren, Schnabel Moehren,*

Schnabelkoerffel. In England it would have been called *shepherd's needle* or *shepherd's comb.* Soon enough the profusion of species overtook the profusion of names. Naturalists formed a community; they corresponded, and they traveled. By the end of the century a Swiss botanist had published a catalogue of 6,000 plants. Every naturalist who discovered a new one had the privilege and the responsibility of naming it; a proliferation of adjectives and compounds was inevitable, as were duplication and redundancy. To *shepherd's needle* and *shepherd's comb* were added, in English alone, *shepherd's bag, shepherd's purse, shepherd's beard, shepherd's bedstraw, shepherd's bodkin, shepherd's cress, shepherd's hour-glass, shepherd's rod, shepherd's gourd, shepherd's joy, shepherd's knot, shepherd's myrtle, shepherd's peddler, shepherd's pouche, shepherd's staff, shepherd's teasel, shepherd's scrip,* and *shepherd's delight.*

Carl Linnaeus had yet to invent taxonomy; when he did, in the eighteenth century, he had 7,700 species of plants to name, along with 4,400 animals. Now there are about 300,000, not counting insects, which add millions more. Scientists still try to name them all: there are beetle species named after Barack Obama, Darth Vader, and Roy Orbison. Frank Zappa has lent his name to a spider, a fish, and a jellyfish.

"The name of a man is like his shadow," said the Viennese onomatologist Ernst Pulgram in 1954. "It is not of his substance and not of his soul, but it lives with him and by him. Its presence is not vital, nor its absence fatal." Those were simpler times.

When Claude Shannon took a sheet of paper and penciled his outline of the measures of information in 1949, the scale went from tens of bits to hundreds to thousands, millions, billions, and trillions. The transistor was one year old and Moore's law yet to be conceived. The top of the pyramid was Shannon's estimate for the Library of Congress—one hundred trillion bits, 10^{14}. He was about right, but the pyramid was growing.

After bits came kilobits, naturally enough. After all, engineers had coined the word *kilobuck*—"a scientist's idea of a short way to say 'a thousand dollars,'" *The New York Times* helpfully explained in 1951. The measures of information climbed up an exponential scale, as the realization dawned in the 1960s that everything to do with information would now grow exponentially. That idea was casually expressed by Gordon Moore, who had been an undergraduate studying chemistry when Shannon jotted his note and found his way to electronic engineering and the development of integrated circuits. In 1965, three years before he founded the Intel Corporation, Moore was merely, modestly suggesting that within a decade, by 1975, as many as 65,000 transistors could be combined on a single wafer of silicon. He predicted a doubling every year or two—a doubling of the number of components that could be packed on a chip, but then also, as it turned out, the doubling of all kinds of memory capacity and processing speed, a halving of size and cost, seemingly without end.

Kilobits could be used to express speed of transmission as well as quantity of storage. As of 1972, businesses could lease high-speed lines carrying data as fast as 240 kilobits per second. Following the lead of IBM, whose hardware typically processed information in chunks of eight bits, engineers soon adopted the modern and slightly whimsical unit, the byte. Bits and bytes. A kilobyte, then, represented 8,000 bits; a megabyte (following hard upon), 8 million. In the order of things as worked out by international standards committees, *mega-* led to *giga-, tera-, peta-*, and *exa-*, drawn from Greek, though with less and less linguistic fidelity. That was enough, for everything measured, until 1991, when the need was seen for the zettabyte (1,000,000,000,000,000,000,000) and the inadvertently comic sounding yottabyte (1,000,000,000,000, 000,000,000,000). In this climb up the exponential ladder information left other gauges behind. Money, for example, is scarce by comparison. After kilobucks, there were megabucks and gigabucks, and people can joke about inflation leading to terabucks, but all the money in the world, all the wealth amassed by all the generations of humanity, does not amount to a petabuck.

The 1970s were the decade of megabytes. In the summer of 1970, IBM introduced two new computer models with more memory than ever before: the Model 155, with 768,000 bytes of memory, and the larger Model 165, with a full megabyte, in a large cabinet. One of these room-filling mainframes could be purchased for $4,674,160. By 1982 Prime Computer was marketing a megabyte of memory on a single circuit board, for $36,000. When the publishers of the *Oxford English Dictionary* began digitizing its contents in 1987 (120 typists; an IBM mainframe), they estimated its size at a gigabyte. A gigabyte also encompasses the entire human genome. A thousand of those would fill a terabyte. A terabyte was the amount of disk storage Larry Page and Sergey Brin managed to patch together with the help of $15,000 spread across their personal credit cards in 1998, when they were Stanford graduate students building a search-engine prototype, which they first called BackRub and then renamed Google. A terabyte is how much data a typical analog television station broadcasts daily, and it was the size of the United States government's database of patent and trademark records when it went online in 1998. By 2010, one could buy a terabyte disc drive for a hundred dollars and hold it in the palm of one hand. The books in the Library of Congress represent about 10 terabytes (as Shannon guessed), and the number is many times more when images and recording music are counted. The library now archives web sites; by February 2010 it had collected 160 terabytes' worth.

As the train hurtled onward, its passengers sometimes felt the pace foreshortening their sense of their own history. Moore's law had looked simple on paper, but its consequences left people struggling to find metaphors with which to understand their experience. The computer scientist Jaron Lanier describes the feeling this way: "It's as if you kneel to plant the seed of a tree and it grows so fast that it swallows your whole town before you can even rise to your feet."

A more familiar metaphor is the cloud. All that information—all that information capacity—looms over us, not quite visible, not quite tangible, but awfully real; amorphous, spectral; hovering nearby, yet not situated in any one place. Heaven must once have felt this way to the faithful. People talk about shifting their lives to the cloud—their informational lives, at least. You may store photographs in the cloud; Google will manage your business in the cloud; Google is putting all the world's books into the cloud; e-mail passes to and from the cloud and never really leaves the cloud. All traditional ideas of privacy, based on doors and locks, physical remoteness and invisibility, are upended in the cloud.

Money lives in the cloud; the old forms are vestigial tokens of knowledge about who owns what, who owes what. To the twenty-first century these will be seen as anachronisms, quaint or even absurd: bullion carried from shore to shore in fragile ships, subject to the tariffs of pirates and the god Poseidon; metal coins tossed from moving cars into baskets at highway tollgates and thereafter trucked about (now the history of your automobile is in the cloud); paper checks torn from pads and signed in ink; tickets for trains, performances, air travel, or anything at all, printed on weighty perforated paper with watermarks, holograms, or fluorescent fibers; and, soon enough, all forms of cash. The economy of the world is transacted in the cloud.

Its physical aspect could not be less cloudlike. Server farms proliferate in unmarked brick buildings and steel complexes, with smoked windows or no windows, miles of hollow floors, diesel generators, cooling towers, seven-foot intake fans, and aluminum chimney stacks. This hidden infrastructure grows in a symbiotic relationship with the electrical infrastructure it increasingly resembles. There are information switchers, control centers, and substations. They are clustered and distributed. These are the wheel-works; the cloud is their avatar.

The information produced and consumed by humankind used to vanish—that was the norm, the default. The sights, the sounds, the songs, the spoken word just melted away. Marks on stone, parchment, and paper were the special case. It did not occur to Sophocles' audiences that it would be sad for his plays to be lost; they enjoyed the show. Now expectations have inverted. Everything may be recorded and preserved, at least potentially: every musical performance; every crime in a shop, elevator, or city street; every volcano or tsunami on the remotest shore; every card played or piece moved in an online game; every rugby scrum and cricket match. Having a camera at hand is normal, not exceptional; something like 500 billion images were captured in 2010. YouTube was streaming more than a billion videos a day. Most of this is haphazard and unorganized, but there are extreme cases. The computer pioneer Gordon Bell, at Microsoft Research in his seventies, began recording every moment of his day, every conversation, message, document, a megabyte per hour or a gigabyte per month, wearing around his neck what he called a "SenseCam" to create what he called a "LifeLog." Where does it end? Not with the Library of Congress.

It is finally natural—even inevitable—to ask how much information is in the universe. It is the consequence of Charles Babbage and Edgar Allan Poe saying, "No thought can perish." Seth Lloyd does the math. He is a moon-faced, bespectacled quantum engineer at MIT, a theorist and designer of quantum computers. The universe, by existing, registers information, he says. By evolving in time, it processes information. How much? To figure that out, Lloyd takes into account how fast this "computer" works and how long it has been working. Considering the fundamental limit on speed, $2E/\pi\hbar$ operations per second ("where E is the system's average energy above the ground state and $\hbar = 1.0545 \times 10^{-34}$ joule-sec is Planck's reduced constant"), and on memory space, limited by entropy to $S/k_B \ln 2$ ("where S is the system's thermodynamic entropy and $k_B = 1.38 \times 10^{-23}$ joules/K is Boltzmann's constant"), along with the speed of light and the age of the universe since the Big Bang, Lloyd calculates that the universe can have performed something on the order of 10^{120} "ops" in its entire history. Considering "every degree of freedom of every particle in the universe," it could now hold something like 10^{90} bits. And counting.

15 | NEW NEWS EVERY DAY

(And Such Like)

> *Sorry for all the ups and downs of the web site in recent days. The way I understand it, freakish accumulations of ice weigh down the branches of the Internet and trucks carrying packets of information skid all over the place.*
>
> —Andrew Tobias (2007)

AS THE PRINTING PRESS, the telegraph, the typewriter, the telephone, the radio, the computer, and the Internet prospered, each in its turn, people said, as if for the first time, that a burden had been placed on human communication: new complexity, new detachment, and a frightening new excess. In 1962 the president of the American Historical Association, Carl Bridenbaugh, warned his colleagues that human existence was undergoing a "Great Mutation"—so sudden and so radical "that we are now suffering something like historical amnesia." He lamented the decline of reading; the distancing from nature (which he blamed in part on "ugly yellow Kodak boxes" and "the transistor radio everywhere"); and the loss of shared culture. Most of all, for the preservers and recorders of the past, he worried about the new tools and techniques available to scholars: "that Bitch-goddess, Quantification"; "the data processing machines"; as well as "those frightening projected scanning devices, which we are told will read documents and books for us." *More* was not *better*, he declared:

Notwithstanding the incessant chatter about communication that we hear daily, it has not improved; actually it has become more difficult.

These remarks became well known in several iterations: first, the oral address, heard by about a thousand people in the ballroom of Conrad Hilton's hotel in Chicago on the last Saturday evening on 1962; next, the printed version in the society's journal in 1963; and then, a generation later, an online version, with its far greater reach and perhaps greater durability as well.

Elizabeth Eisenstein encountered the printed version in 1963, when she was teaching history as a part-time adjunct lecturer at American University in Washington (the best job she could get, as a woman with a Harvard Ph.D.). Later she identified that moment as the starting point of fifteen years of research that culminated in her landmark of scholarship, two volumes titled *The Printing Press as an Agent of Change*. Before Eisenstein's work appeared in 1979, no one had attempted a comprehensive study of printing as the communications revolution essential to the transition from medieval times to modernity. Textbooks, as she noted, tended to slot the printing press somewhere between the Black Death and the discovery of America. She placed Gutenberg's invention at center stage: the shift from script to print; the rise of printing shops in the cities of fifteenth-century Europe; the transformation in "data collection, storage and retrieval systems and communications networks." She emphasized modestly that she would treat printing only as *an* agent of change, but she left readers convinced of its indispensable part in the transformations of early modern Europe: the Renaissance, the Protestant Reformation, and the birth of science. It was "a decisive point of no return in human history." It shaped the modern mind.

It shaped the minds of historians, too; she was interested in the unconscious mental habits of her profession. As she embarked on her project, she began to believe that scholars were too often blinded to the effects of the very medium in which they swam. She gave credit to Marshall McLuhan, whose *Gutenberg Galaxy* had appeared in 1962, for forcing them to refocus their gaze. In the age of scribes, the culture had only primitive reckonings of chronology: muddled timelines counted the generations from Adam, or Noah, or Romulus and Remus. "Attitudes toward historical change," she wrote, "will be found only occasionally in writings ostensibly devoted to 'history' and often have to be read into such writings. They must also be read into sagas and epics, sacred scriptures, funerary inscriptions, glyphs and ciphers, vast stone monuments, documents locked in chests in muniment rooms, and marginal notations on manuscript." The sense of *when* we are—the ability to see the past spread out before one; the internalization of mental time charts; the appreciation of anachronism—came with the shift to print.

As a duplicating machine, the printing press not only made texts cheaper and more accessible; its real power was to make them stable. "Scribal culture," Eisenstein wrote, was "constantly enfeebled by erosion, corruption, and loss." Print was trustworthy, reliable, and permanent. When Tycho Brahe spent his countless hours poring over planetary and star tables, he could count on others checking the same tables, now and in the future. When Kepler computed his own far more accurate catalogue, he was leveraging the tables of logarithms published by Napier. Meanwhile, print shops were not only spreading Martin Luther's theses but, more important, the Bible itself. The revolution of Protestantism hinged more on Bible reading than on any point of doctrine—print overcoming script; the codex supplanting the scroll; and the vernacular replacing the ancient languages. Before print, scripture was not truly fixed. All forms of knowledge achieved stability and permanence, not because paper was more durable than papyrus but simply because there were many copies.

In 1963, reading the warnings of the president of the American Historical Association, Eisenstein found herself agreeing that the profession faced a crisis, of sorts. But she felt Bridenbaugh had it exactly backward.

He thought the problem was forgetfulness: "As I see it," he said dramatically, "mankind is faced with nothing short of the loss of its memory, and this memory is history." Eisenstein, looking at the same new information technologies that so troubled older historians, drew the opposite lesson. The past is not receding from view but, on the contrary, becoming *more* accessible and *more* visible. "In an age that has seen the deciphering of Linear B and the discovery of the Dead Sea Scrolls," she wrote, "there appears to be little reason to be concerned about 'the loss of mankind's memory.' There are good reasons for being concerned about the overloading of its circuits." As for the amnesia lamented by Bridenbaugh and so many of his colleagues:

> This is a misreading of the predicament confronting historians today. It is not the onset of amnesia that accounts for present difficulties but a more complete recall than any prior generation has ever experienced. Steady recovery, not obliteration, accumulation, rather than loss, have led to the present impasse.

From her point of view, a five-centuries-old communications revolution was still gathering momentum. How could they not see this?

"Overloading of circuits" was a fairly new metaphor to express a sensation—*too much information*—that felt new. It had always felt new. One hungers for books; rereads a cherished few; begs or borrows more; waits at the library door, and perhaps, in the blink of an eye, finds oneself in a state of surfeit: *too much to read*. In 1621 the Oxford scholar Robert Burton (who amassed one of the world's largest private libraries, 1,700 books, but never a thesaurus) gave voice to the feeling:

> I hear new news every day, and those ordinary rumours of war, plagues, fires, inundations, thefts, murders, massacres, meteors, comets, spectrums,

prodigies, apparitions, of towns taken, cities besieged in France, Germany, Turkey, Persia, Poland, &c. daily musters and preparations, and such like, which these tempestuous times afford, battles fought, so many men slain, monomachies, shipwrecks, piracies, and sea-fights, peace, leagues, stratagems, and fresh alarms. A vast confusion of vows, wishes, actions, edicts, petitions, lawsuits, pleas, laws, proclamations, complaints, grievances are daily brought to our ears. New books every day, pamphlets, currantoes, stories, whole catalogues of volumes of all sorts, new paradoxes, opinions, schisms, heresies, controversies in philosophy, religion, &c. Now come tidings of weddings, maskings, mummeries, entertainments, jubilees, embassies, tilts and tournaments, trophies, triumphs, revels, sports, plays: then again, as in a new shifted scene, treasons, cheating tricks, robberies, enormous villanies in all kinds, funerals, burials, deaths of Princes, new discoveries, expeditions; now comical then tragical matters. To-day we hear of new Lords and officers created, to-morrow of some great men deposed, and then again of fresh honours conferred; one is let loose, another imprisoned; one purchaseth, another breaketh: he thrives, his neighbour turns bankrupt; now plenty, then again dearth and famine; one runs, another rides, wrangles, laughs, weeps &c. Thus I daily hear, and such like.

He thought information glut was new then. He was not complaining; just amazed. Protests followed soon enough, however. Leibniz feared a return to barbarism—"to which result that horrible mass of books which keeps on growing might contribute very much. For in the end the disorder will become nearly insurmountable." Alexander Pope wrote satirically of "those days, when (after Providence had permitted the invention of Printing as a scourge for the sins of the learned) Paper also became so cheap, and printers so numerous, that a deluge of Authors covered the land."

Deluge became a common metaphor for people describing information surfeit. There is a sensation of drowning: information as a rising, churning flood. Or it calls to mind bombardment, data impinging in a series of blows, from all sides, too fast. Fear of the cacophony of voices can have a religious motivation, a worry about secular noise overwhelming the truth. T. S. Eliot expressed that in 1934:

> Knowledge of speech, but not of silence;
> Knowledge of words, and ignorance of the Word.
> All our knowledge brings us nearer to our ignorance,
> All our ignorance brings us nearer to death,
> But nearness to death no nearer to GOD.

Or one may dread the breaching of walls that stand before what is unfamiliar, or horrible, or terrifying. Or one may lose the ability to impose order on the chaos of sensations. The truth seems harder to find amid the multitude of plausible fictions.

After "information theory" came to be, so did "information overload," "information glut," "information anxiety," and "information fatigue," the last recognized by the *OED* in 2009 as a timely syndrome: "Apathy, indifference, or mental exhaustion arising from exposure to too much information, esp. (in later use) stress induced by the attempt to assimilate excessive amounts of information from the media, the Internet, or at work." Sometimes information anxiety can coexist with boredom, a particularly confusing combination. David Foster Wallace had a more ominous name for this modern condition: Total Noise. "The tsunami of available fact, context, and perspective"—that, he wrote in 2007, constitutes Total Noise. He talked about the sensation of drowning and also of a loss of autonomy, of personal responsibility for being *informed*. To keep up with all the information we need proxies and subcontractors.

Another way to speak of the anxiety is in terms of the gap between information and knowledge. A barrage of data so often fails to tell us what we need to know. Knowledge, in turn, does not guarantee enlightenment or wisdom. (Eliot said that, too: "Where is the wisdom we have lost in knowledge? / Where is the knowledge we have lost in information?") It is an ancient observation, but one that seemed to bear restating when information became plentiful—particularly in a world where all bits are created equal and information is divorced from meaning. The humanist and philosopher of technology Lewis Mumford, for example, restated it in 1970: "Unfortunately, 'information retrieving,' however swift, is no substitute for discovering by direct personal inspection knowledge whose very existence one had possibly never been aware of, and following it at one's own pace through the further ramification of relevant literature." He begged for a return to "moral self-discipline." There is a whiff of nostalgia in this sort of warning, along with an undeniable truth: that in the pursuit of knowledge, slower can be better. Exploring the crowded stacks of musty libraries has its own rewards. Reading—even browsing—an old book can yield sustenance denied by a database search. Patience is a virtue, gluttony a sin.

Even in 1970, however, Mumford was not thinking about databases or any of the electronic technologies that loomed. He complained about "the multiplication of microfilms." He also complained about too many books. Without "self-imposed restraints," he warned, "the overproduction of books will bring about a state of intellectual enervation and depletion hardly to be distinguished from massive ignorance." Restraints were not imposed. Titles continue to multiply. Books about information glut join the cornucopia; no irony is intended when the online bookseller Amazon.com transmits messages like "Start reading *Data Smog* on your Kindle in *under a minute*" and "Surprise me! See a random page in this book."

The electronic communication technologies arrived so quickly, almost without warning. The word *e-mail* appeared in print (so far as the *OED* can determine) in 1982, in *Computerworld* magazine, which had barely heard reports: "ADR/Email is reportedly easy to use and features simple, English verbs and prompt screens." Next year, the journal *Infosystems* declared, "Email promotes movement of information through space." And the year after that—still a full decade before most people heard the word—a Swedish computer scientist named Jacob Palme at the QZ Computer Center in Stockholm issued a prescient warning—as simple, accurate, and thorough as any that followed in the next decades. Palme began:

Electronic mail system can, if used by many people, cause severe information overload problems. The cause of this problem is that it is so easy to send a message to a large number of people, and that systems are often designed to give the sender too much control of the communication process, and the receiver too little control. . . .

People get too many messages, which they do not have time to read. This also means that the really important messages are difficult to find in a large flow of less important messages.

In the future, when we get larger and larger message systems, and these systems get more and more interconnected, this will be a problem for almost all users of these systems.

He had statistics from his local network: the average message took 2 minutes, 36 seconds to write and just 28 seconds to read. Which would have been fine, except that people could so easily send many copies of the same message.

When psychologists or sociologists try to study information overload with the methods of their disciplines, they get mixed results. As early as 1963, a pair of psychologists set out to quantify the effect of extra information on the process of clinical diagnosis. As they expected, they found that "too much information"—not easy to define, they admitted—often contaminated judgment. They titled their paper "Does One Sometimes Know Too Much?" and somewhat gleefully listed alternative titles, as a bonus: "Never Have So Many Done So Little"; "Are You Getting More Now But Predicting It Less?"; and "Too Much Information Is a Dangerous Thing." Others tried to measure the effects of information load on blood pressure, heart rhythms, and respiration rates.

One worker in the area was Siegfried Streufert, who reported in a series of papers in the 1960s that the relation between information load and information handling typically looked like an "inverted U": more information was helpful at first, then not so helpful, and then actually harmful. One of his studies took 185 university students (all male) and had them pretend to be commanders making decisions in a tactical game. They were told:

The information you are receiving is prepared for you in the same way it would be prepared for real commanders by a staff of intelligence officers. . . . You may instruct these intelligence officers to increase or decrease the amount of information they present to you. . . . Please check your preference: I would prefer to:

> receive much more information
> receive a little more information
> receive about the same amount of information
> receive a little less information
> receive much less information.

No matter what they chose, their preferences were ignored. The experimenter, not the subjects, predetermined the amount of information. Streufert concluded from the data that "superoptimal" information loads caused poor performance, "yet it should be noted that even at highly superoptimal information loads (i.e., 25 messages per 30-minute period), the subjects are still asking for increased information levels." Later, he used similar methodology to study the effects of drinking too much coffee.

By the 1980s, researchers were speaking confidently about the "information-load paradigm." This was a paradigm based on a truism: that people can only "absorb" or "process" a limited amount of information. Various investigators found surfeits causing not only confusion and frustration, but also blurred vision and dishonesty. Experiments themselves had a broad menu of information to process: measurements of memory span; ideas of channel capacity drawn from Shannon; and variations on the theme of signal-to-noise ratio. A common, if dubious, approach to research was direct introspection. One small project in 1998 took as a "community or folk group" graduate students in library and information science at the University of Illinois; all agreed, when asked, that they suffered from information overload, due to "e-mail, meetings, listservs, and in-basket paper piles." Most felt that a surfeit of information tainted their leisure time as well as their work time. Some reported headaches. The tentative conclusion: information overload is real; also, it is both a "code phrase" and a myth. The research can only press onward.

Having to think of information as a burden is confusing, as Charles Bennett says. "We pay to have newspapers delivered, not taken away." But the thermodynamics of computation shows that yesterday's newspaper takes up space that Maxwell's demon needs for today's work, and modern experience teaches the same. Forgetting used to be a failing, a waste, a sign of senility. Now it takes effort. It may be as important as remembering.

Facts were once dear; now they are cheap. Once, people would turn to the pages of *Whitaker's Almanack*, published yearly in Britain, or the *World Almanac*, in the United States, to find the names and dates of monarchs and presidents, tables of holidays and high water, sizes and populations of faraway places, or the ships and chief officers of the navy. Lacking the almanac, or seeking an even more obscure fact, they might call on a man or woman of experience behind a desk at a public library. When George Bernard Shaw needed the whereabouts of the nearest crematorium—his wife was dying—he opened the almanac and was aggrieved. "I have just found an astonishing omission in Whitaker," he wrote to the editor. "As the desired information is just what one goes to your invaluable almanack for, may I suggest that a list of the 58 crematoria now working in the country, and instructions what to do, would be a very desirable addition." His letter is poignant. He does not mention his wife—only "a case of serious illness"—and refers to himself as "the bereaved enquirer." Shaw had a telegraph address and a telephone but took it for granted that facts were to be found in print.

For many, the telephone had already begun to extend the reach of the inquisitive. Twentieth-century people realized that they could know instantly the scores of sporting events they had not witnessed; so many came up with the idea of telephoning the newspaper that *The New York Times* felt compelled to print a front-page notice in 1929 begging readers to desist: "Don't Ask by Telephone for World's Series Scores." Now the information, in "real time," is considered a birthright.

What do you do when you have everything at last? Daniel Dennett imagined—in 1990, just before the Internet made this dream possible—that electronic networks could upend the economics of publishing poetry. Instead of slim books, elegant specialty items marketed to connoisseurs, what if poets could publish online, instantly reaching not hundreds but millions of readers, not for tens of dollars but for fractions of pennies? That same year, Sir Charles Chadwyck-Healey, a publisher, conceived of the English Poetry Full-Text Database as he walked one day through the British Library, and four years later he had produced it—not the present or future of poetry, but the past, and not, at first, online but in four compact discs, 165,000 poems by 1,250 poets spanning thirteen centuries, priced at $51,000. Readers and critics had to figure out what to make of this. Not *read* it, surely, the way they would read a book. Read *in* it, perhaps. Search it, for a word or an epigraph or a fragment half remembered.

Anthony Lane, reviewing the database for *The New Yorker*, found himself swinging from elation to dismay and back. "You hunch like a pianist over the keys," he wrote, "knowing what awaits you, thinking, Ah, the untold wealth of English literature! What hidden jewels I shall excavate from the deepest mines of human fancy!" Then come the macaronics, the clunkers, the flood of bombast and mediocrity. The sheer unordered mass begins to wear you down. Not that Lane sounds at all weary. "What a steaming *heap*," he cries, and he revels in it. "Never have I beheld such a magnificent tribute to the powers of human incompetence—and also, by the same token, to the blessings of human forgetfulness." Where else would he have found the utterly forgotten Thomas Freeman (not in Wikipedia) and this lovely self-referential couplet:

Whoop, whoop, me thinkes I heare my Reader cry,
Here is rime doggrell: I confesse it I.

The CD-ROMs are already obsolete. All English poetry is in the network now—or if not all, some approximation thereof, and if not now, then soon.

The past folds accordion-like into the present. Different media have different event horizons—for the written word, three millennia; for recorded sound, a century and a half—and within their time frames the old becomes as accessible as the new. Yellowed newspapers come back to life. Under headings of *50 Years Ago* and *100 Years Ago*, veteran publications recycle their archives: recipes, card-play techniques, science, gossip, once out of print and now ready for use. Record companies rummage through their attics to release, or re-release, every scrap of music, rarities, B-sides, and bootlegs. For a certain time, collectors, scholars, or fans *possessed* their books and their records. There was a line between what they had and what they did not. For some, the music they owned (or the books, or the videos) became part of who they were. That line fades away. Most of Sophocles' plays are lost, but those that survive are available at the touch of a button. Most of Bach's music was unknown to Beethoven; we have it all—partitas, cantatas, and ringtones. It comes to us instantly, or at light speed. It is a symptom of omniscience. It is what the critic Alex Ross calls the Infinite Playlist, and he sees how mixed is the blessing: "anxiety in place of fulfillment, an addictive cycle of craving and malaise. No sooner has one experience begun than the thought of what else is out there intrudes." The embarrassment of riches. Another reminder that information is not knowledge, and knowledge is not wisdom.

Strategies emerge for coping. There are many, but in essence they all boil down to two: filter and search. The harassed consumer of information turns to filters to separate the metal from the dross; filters include blogs and aggregators—the choice raises issues of trust and taste. The need for filters intrudes on any thought experiment about the wonders of abundant information. When Dennett imagined his Complete Poetry Network, he saw the problem. "The obvious counterhypothesis arises from population memetics," he said. "If such a network were established, no poetry lover would be willing to wade through thousands of electronic files filled with doggerel, looking for good poems." Filters would be needed—editors and critics. "They flourish because of the short supply and limited capacity of minds, whatever the transmission media between minds." When information is cheap, attention becomes expensive.

For the same reason, mechanisms of search—*engines*, in cyberspace— find needles in haystacks. By now we've learned that it is not enough for information to *exist*. A "file" was originally—in sixteenth-century England—a wire on which slips and bills and notes and letters could be strung for preservation and reference. Then came file folders, file drawers, and file cabinets; then the electronic namesakes of all these; and the inevitable irony. Once a piece of information is *filed*, it is statistically unlikely ever to be seen again by human eyes. Even in 1847, Augustus De Morgan, Babbage's friend, knew this. For any random book, he said, a library was no better than a wastepaper warehouse. "Take the library of the British Museum, for instance, valuable and useful and accessible as it is: what chance has a work of being known to be there, merely because it is there? If it be wanted, it can be asked for; but to be wanted it must be known. Nobody can rummage the library."

Too much information, and so much of it lost. An unindexed Internet site is in the same limbo as a misshelved library book. This is why the successful and powerful business enterprises of the information economy are built on filtering and searching. Even Wikipedia is a combination of the two: powerful search, mainly driven by Google, and a vast, collaborative filter, striving to gather the true facts and screen out the false ones. Searching and filtering are all that stand between this world and the Library of Babel.

In their computer-driven incarnations these strategies seem new. But they are not. In fact, a considerable part of the gear and tackle of print media—now taken for granted, invisible as old wallpaper—evolved in direct response to the sense of information surfeit. They are mechanisms of selection and sorting: alphabetical indexes, book reviews, library shelving schemes and card catalogues, encyclopedias, anthologies and digests, books of quotation and concordances and gazetteers. When Robert Burton held forth on all his "new news every day," his "new paradoxes, opinions, schisms, heresies, controversies in philosophy, religion, &c," it was by way of justifying his life's great project, *The Anatomy of Melancholy*, a rambling compendium of all previous knowledge. Four centuries earlier, the Dominican monk Vincent of Beauvais tried to set down his own version of everything that was known, creating one of the first medieval encyclopedias, *Speculum Maius*, "The Great Mirror"—his manuscripts organized into eighty books, 9,885 chapters. His justification: "The multitude of books, the shortness of time and the slipperiness of memory do not allow all things which are written to be equally retained in the mind." Ann Blair, a Harvard historian of early modern Europe, puts it simply: "The perception of an overabundance of books fueled the production of many more books." In their own way, too, the natural sciences such as botany arose in answer to information overload. The explosion of recognized species (and names) in the sixteenth century demanded new routines of standardized description. Botanical encyclopedias appeared, with glossaries and indexes. Brian Ogilvie sees the story of Renaissance botanists as "driven by the need to master the information overload that they had unwittingly produced." They created a "*confusio rerum*," he says, "accompanied by a *confusio verborum*." Confused mass of new things; confusion of words. Natural history was born to channel information.

When new information technologies alter the existing landscape, they bring disruption: new channels and new dams rerouting the flow of irrigation and transport. The balance between creators and consumers is upset: writers and readers, speakers and listeners. Market forces are confused; information can seem too cheap and too expensive at the same time. The old ways of organizing knowledge no longer work. Who will search; who will filter? The disruption breeds hope mixed with fear. In the first days of radio Bertolt Brecht, hopeful, fearful, and quite obsessed, expressed this feeling aphoristically: "A man who has something to say and finds no listeners is bad off. Even worse off are listeners who can't find anyone with something to say to them." The calculus always changes. Ask bloggers and tweeters: Which is worse, too many mouths or too many ears?

EPILOGUE

(The Return of Meaning)

It was inevitable that meaning would force its way back in.
—Jean-Pierre Dupuy (2000)

THE EXHAUSTION, the surfeit, the pressure of information have all been seen before. Credit Marshall McLuhan for this insight—his most essential—in 1962:

> We are today as far into the electric age as the Elizabethans had advanced into the typographical and mechanical age. And we are experiencing the same confusions and indecisions which they had felt when living simultaneously in two contrasted forms of society and experience.

But as much as it is the same, this time it is different. We are a half century further along now and can begin to see how vast the scale and how strong the effects of connectedness.

Once again, as in the first days of the telegraph, we speak of the annihilation of space and time. For McLuhan this was prerequisite to the creation of global consciousness—global *knowing*. "Today," he wrote, "we have extended our central nervous systems in a global embrace, abolishing both space and time as far as our planet is concerned. Rapidly, we approach the final phase of the extensions of man—the technological simulation of consciousness, when the creative process of knowing will be collectively and corporately extended to the whole of human society." Walt Whitman had said it better a century before:

What whispers are these O lands, running ahead of you, passing under the seas?

Are all nations communing? is there going to be but one heart to the globe?

The wiring of the world, followed hard upon by the spread of wireless communication, gave rise to romantic speculation about the birth of a new global organism. Even in the nineteenth century mystics and theologians began speaking of a shared mind or collective consciousness, formed through the collaboration of millions of people placed in communication with one another.

Some went so far as to view this new creature as a natural product of continuing evolution—a way for humans to fulfill their special destiny, after their egos had been bruised by Darwinism. "It becomes absolutely necessary," wrote the French philosopher Édouard Le Roy in 1928, "to place [man] above the lower plane of nature, in a position which enables him to dominate it." How? By creating the "noosphere"—the sphere of mind—a climactic "mutation" in evolutionary history. His friend the Jesuit philosopher Pierre Teilhard de Chardin did even more to promote the noosphere, which he called a "new skin" on the earth:

Does it not seem as though a great body is in the process of being born—with its limbs, its nervous system, its centers of perception, its memory—the very body of that great something to come which was to fulfill the aspirations that had been aroused in the reflective being by the freshly acquired consciousness of its interdependence with and responsibility for a whole in evolution?

That was a mouthful even in French, and less mystical spirits considered it bunkum ("nonsense, tricked out with a variety of tedious metaphysical conceits," judged Peter Medawar), but many people were testing the same idea, not least among them the writers of science fiction. Internet pioneers a half century later liked it, too.

H. G. Wells was known for his science fiction, but it was as a purposeful social critic that he published a little book in 1938, late in his life, with the title *World Brain*. There was nothing fanciful about what he wanted to promote: an improved educational system throughout the whole "body" of humanity. Out with the hodgepodge of local fiefdoms: "our multitude of unco-ordinated ganglia, our powerless miscellany of universities, research institutions, literatures with a purpose." In with "a reconditioned and more powerful Public Opinion." His World Brain would rule the globe. "We do not want dictators, we do not want oligarchic parties or class rule, we want a widespread world intelligence conscious of itself." Wells believed that a new technology was poised to revolutionize the production and distribution of information: microfilm. Tiny pictures of printed materials could be made for less than a penny per page, and librarians from Europe and the United States met to discuss the possibilities in Paris in 1937 for a World Congress of Universal Documentation. New ways of indexing the literature would be needed, they realized. The British Museum embarked on a program of microfilming four thousand of its oldest books. Wells made this prediction: "In a few score years there will be thousands of workers at this business of ordering and digesting knowledge where now you have one." He admitted that he meant to be controversial and provocative. Attending the congress himself on behalf of England, he foresaw a "sort of cerebrum for humanity, a cerebral cortex which will constitute a memory and a perception of current reality for the whole human race." Yet he was imagining something mundane, as well as utopian: an encyclopedia. It would be a successor to the great national encyclopedias—the French encyclopedia of Diderot, the *Britannica*, the German *Konversations-Lexikon* (he did not mention China's *Four Great Books of Song*)—which had stabilized and equipped "the general intelligence."

This new world encyclopedia would transcend the static form of the book, printed in volumes, said Wells. Under the direction of a wise professional staff ("very important and distinguished men in the new world"), it would be in a state of constant change—"a sort of mental clearinghouse for the mind, a depot where knowledge and ideas are received, sorted, summarized, digested, clarified and compared." Who knows whether Wells would recognize his vision in Wikipedia? The hurly-burly of competing ideas did not enter into it. His world brain was to be authoritative, but not centralized.

> It need not be vulnerable as a human head or a human heart is vulnerable. It can be reproduced exactly and fully, in Peru, China, Iceland, Central Africa. . . . It can have at once the concentration of a craniate animal and the diffused vitality of an amoeba.

For that matter, he said, "It might have the form of a network."

It is not the amount of knowledge that makes a brain. It is not even the distribution of knowledge. It is the interconnectedness. When Wells used the word *network*—a word he liked very much—it retained its original, physical meaning for him, as it would for anyone in his time. He visualized threads or wires interlacing: "A network of marvellously gnarled and twisted stems bearing little leaves and blossoms"; "an intricate network of wires and cables." For us that sense is almost lost; a network is an abstract object, and its domain is information.

The birth of information theory came with its ruthless sacrifice of meaning—the very quality that gives information its value and its purpose. Introducing *The Mathematical Theory of Communication*, Shannon had to be blunt. He simply declared meaning to be "irrelevant to the engineering problem." Forget human psychology; abandon subjectivity.

He knew there would be resistance. He could hardly deny that messages can have meaning, "that is, they refer to or are correlated according to some system with certain physical or conceptual entities." (Presumably a "system with certain physical or conceptual entities" would be the world and its inhabitants, the kingdom and the power and the glory, amen.) For some, this was just too cold. There was Heinz von Foerster at one of the early cybernetics conferences, complaining that information theory was merely about "beep beeps," saying that only when understanding begins, in the human brain, "*then* information is born—it's not in the beeps." Others dreamed of extending information theory with a semantic counterpart. Meaning, as ever, remained hard to pin down. "I know an uncouth region," wrote Borges of the Library of Babel, "whose librarians repudiate the vain and superstitious custom of finding a meaning in books and equate it with that of finding a meaning in dreams or in the chaotic lines of one's palm."

Epistemologists cared about knowledge, not beeps and signals. No one would have bothered to make a philosophy of dots and dashes or puffs of smoke or electrical impulses. It takes a human—or, let's say, a "cognitive agent"—to take a signal and turn it into information. "Beauty is in the eye of the beholder, and information is in the head of the receiver," says Fred Dretske. At any rate that is a common view, in epistemology— that "we *invest* stimuli with meaning, and apart from such investment, they are informationally barren." But Dretske argues that distinguishing information and meaning can set a philosopher free. The engineers have provided an opportunity and a challenge: to understand how meaning can evolve; how life, handling and coding information, progresses to interpretation, belief, and knowledge.

Still, who could love a theory that gives false statements as much value as true statements (at least, in terms of quantity of information)? It was mechanistic. It was desiccated. A pessimist, looking backward, might call it a harbinger of a soulless Internet at its worst. "The more we 'communicate' the way we do, the more we create a *hellish* world," wrote the Parisian philosopher—also a historian of cybernetics—Jean-Pierre Dupuy.

> I take "hell" in its theological sense, i.e., a place which is void of *grace*— the undeserved, unnecessary, surprising, unforeseen. A paradox is at work

here: ours is a world about which we pretend to have more and more *information* but which seems to us increasingly devoid of meaning.

That hellish world, devoid of grace—has it arrived? A world of information glut and gluttony; of bent mirrors and counterfeit texts; scurrilous blogs, anonymous bigotry, banal messaging. Incessant chatter. The false driving out the true.

That is not the world I see.

It was once thought that a perfect language should have an exact one-to-one correspondence between words and their meanings. There should be no ambiguity, no vagueness, no confusion. Our earthly Babel is a falling off from the lost speech of Eden: a catastrophe and a punishment. "I imagine," writes the novelist Dexter Palmer, "that the entries of the dictionary that lies on the desk in God's study must have one-to-one correspondences between the words and their definitions, so that when God sends directives to his angels, they are completely free from ambiguity. Each sentence that He speaks or writes must be perfect, and therefore a miracle." We know better now. With or without God, there is no perfect language.

Leibniz thought that if natural language could not be perfect, at least the calculus could: a language of symbols rigorously assigned. "All human thoughts might be entirely resolvable into a small number of thoughts considered as primitive." These could then be combined and dissected mechanically, as it were. "Once this had been done, whoever uses such characters would either never make an error, or, at least, would have the possibility of immediately recognizing his mistakes, by using the simplest of tests." Gödel ended that dream.

On the contrary, the idea of perfection is contrary to the nature of language. Information theory has helped us understand that—or, if you are a pessimist, forced us to understand it. "We are forced to see," Palmer continues, that words are not themselves ideas, but merely strings of ink marks; we see that sounds are nothing more than waves. In a modern age without an Author looking down on us from heaven, language is not a thing of definite certainty, but infinite possibility; without the comforting illusion of meaningful order we have no choice but to stare into the face of meaningless disorder; without the feeling that meaning can be certain, we find ourselves overwhelmed by all the things that words *might* mean.

Infinite possibility is good, not bad. Meaningless disorder is to be challenged, not feared. Language maps a boundless world of objects and sensations and combinations onto a finite space. The world changes, always mixing the static with the ephemeral, and we know that language changes, not just from edition to edition of the *Oxford English Dictionary* but from one moment to the next, and from one person to the next. Everyone's language is different. We can be overwhelmed or we can be emboldened.

More and more, the lexicon is in the network now—preserved, even as it changes; accessible and searchable. Likewise, human knowledge soaks into the network, into the cloud. The web sites, the blogs, the search engines and encyclopedias, the analysts of urban legends and the debunkers of the analysts. Everywhere, the true rubs shoulders with the false. No form of digital communication has earned more mockery than the service known as Twitter—banality shrink-wrapped, enforcing triviality by limiting all messages to 140 characters. The cartoonist Garry Trudeau twittered satirically in the guise of an imaginary newsman who could hardly look up from his twittering to gather any news. But then, eyewitness Twitter messages provided emergency information and comfort during terrorist attacks in Mumbai in 2008, and it was Twitter feeds from Tehran that made the Iranian protests visible to the world in 2009. The aphorism is a form with an honorable history. I barely twitter myself, but even this odd medium, microblogging so quirky and confined, has its uses and its enchantment. By 2010 Margaret

Atwood, a master of a longer form, said she had been "sucked into the Twittersphere like Alice down the rabbit hole."

> Is it signaling, like telegraphs? Is it Zen poetry? Is it jokes scribbled on the washroom wall? Is it John Hearts Mary carved on a tree? Let's just say it's communication, and communication is something human beings like to do.

Shortly thereafter, the Library of Congress, having been founded to collect every book, decided to preserve every tweet, too. Possibly undignified, and probably redundant, but you never know. It is human communication.

And the network has learned a few things that no individual could ever know.

It identifies CDs of recorded music by looking at the lengths of their individual tracks and consulting a vast database, formed by accretion over years, by the shared contributions of millions of anonymous users. In 2007 this database revealed something that had eluded distinguished critics and listeners: that more than one hundred recordings released by the late English pianist Joyce Hatto—music by Chopin, Beethoven, Mozart, Liszt, and others—were actually stolen performances by other pianists. MIT established a Center for Collective Intelligence, devoted to finding group wisdom and "harnessing" it. It remains difficult to know when and how much to trust the *wisdom of crowds*—the title of a 2004 book by James Surowiecki, to be distinguished from the *madness of crowds* as chronicled in 1841 by Charles Mackay, who declared that people "go mad in herds, while they recover their senses slowly, and one by one." Crowds turn all too quickly into mobs, with their time-honored manifestations: manias, bubbles, lynch mobs, flash mobs, crusades, mass hysteria, herd mentality, goose-stepping, conformity, groupthink—all potentially magnified by network effects and studied under the rubric of information cascades. Collective judgment has appealing possibilities;

collective self-deception and collective evil have already left a cataclysmic record. But knowledge in the network is different from group decision making based on copying and parroting. It seems to develop by accretion; it can give full weight to quirks and exceptions; the challenge is to recognize it and gain access to it. In 2008, Google created an early warning system for regional flu trends based on data no firmer than the incidence of Web searches for the word *flu;* the system apparently discovered outbreaks a week sooner than the Centers for Disease Control and Prevention. This was Google's way: it approached classic hard problems of artificial intelligence—machine translation and voice recognition—not with human experts, not with dictionaries and linguists, but with its voracious data mining of trillions of words in more than three hundred languages. For that matter, its initial approach to searching the Internet relied on the harnessing of collective knowledge.

Here is how the state of search looked in 1994. Nicholson Baker—in a later decade a Wikipedia obsessive; back then the world's leading advocate for the preservation of card catalogues, old newspapers, and other apparently obsolete paper—sat at a terminal in a University of California library and typed, BROWSE SU[BJECT] CENSORSHIP. He received an error message,

> LONG SEARCH: Your search consists of one or more very common words, which will retrieve over 800 headings and take a long time to complete,

and a knuckle rapping:

> Long searches slow the system down for everyone on the catalog and often do not produce useful results. Please type HELP or see a reference librarian for assistance.

All too typical. Baker mastered the syntax needed for Boolean searches with complexes of ANDs and ORs and NOTs, to little avail. He cited research on screen fatigue and search failure and information overload and admired a theory that electronic catalogues were "in effect, conducting a program of 'aversive operant conditioning'" against online search.

Here is how the state of search looked two years later, in 1996. The volume of Internet traffic had grown by a factor of ten each year, from 20 terabytes a month worldwide in 1994 to 200 terabytes a month in 1995, to 2 petabytes in 1996. Software engineers at the Digital Equipment Corporation's research laboratory in Palo Alto, California, had just opened to the public a new kind of search engine, named AltaVista, continually building and revising an index to every page it could find on the Internet—at that point, tens of millions of them. A search for the phrase *truth universally acknowledged* and the name *Darcy* produced four thousand matches. Among them:

- The complete if not reliable text of *Pride and Prejudice*, in several versions, stored on computers in Japan, Sweden, and elsewhere, downloadable free or, in one case, for a fee of $2.25.
- More than one hundred answers to the question, "Why did the chicken cross the road?" including "Jane Austen: Because it is a truth universally acknowledged that a single chicken, being possessed of a good fortune and presented with a good road, must be desirous of crossing."
- The statement of purpose of the *Princeton Pacific Asia Review:* "The strategic importance of the Asia Pacific is a truth universally acknowledged . . ."
- An article about barbecue from the Vegetarian Society UK: "It is a truth universally acknowledged among meat-eaters that . . ."
- The home page of Kevin Darcy, Ireland. The home page of Darcy Cremer, Wisconsin. The home page and boating pictures of Darcy Morse. The vital statistics of Tim Darcy, Australian footballer. The résumé of Darcy Hughes, a fourteen-year-old yard worker and babysitter in British Columbia.

Trivia did not daunt the compilers of this ever-evolving index. They were acutely aware of the difference between making a library catalogue—its target fixed, known, and finite—and searching a world of information without boundaries or limits. They thought they were onto something grand. "We have a lexicon of the current language of the world," said the project manager, Allan Jennings.

Then came Google. Brin and Page moved their fledgling company from their Stanford dorm rooms into offices in 1998. Their idea was that cyberspace possessed a form of self-knowledge, inherent in the links from one page to another, and that a search engine could exploit this knowledge. As other scientists had done before, they visualized the Internet as a graph, with nodes and links: by early 1998, 150 million nodes joined by almost 2 billion links. They considered each link as an expression of value—a recommendation. And they recognized that all links are not equal. They invented a recursive way of reckoning value: the rank of a page depends on the value of its incoming links; the value of a link depends on the rank of its containing page. Not only did they invent it, they published it. Letting the Internet know how Google worked did not hurt Google's ability to leverage the Internet's knowledge.

At the same time, the rise of this network of all networks was inspiring new theoretical work on the topology of interconnectedness in very large systems. The science of networks had many origins and evolved along many paths, from pure mathematics to sociology, but it crystallized in the summer of 1998, with the publication of a letter to *Nature* from Duncan Watts and Steven Strogatz. The letter had three things that combined to make it a sensation: a vivid catchphrase, a nice result, and a surprising assortment of applications. It helped that one of the applications was All the World's People. The catchphrase was *small world*. When two strangers discover that they have a mutual friend—an unexpected connection—they may say, "It's a small world," and it was in this sense that Watts and Strogatz talked about small-world networks.

The defining quality of a small-world network is the one unforgettably captured by John Guare in his 1990 play, *Six Degrees of Separation*. The canonical explanation is this:

> I read somewhere that everybody on this planet is separated by only six other people. Six degrees of separation. Between us and everyone else on this planet. The President of the United States. A gondolier in Venice. Fill in the names.

The idea can be traced back to a 1967 social-networking experiment by the Harvard psychologist Stanley Milgram and, even further, to a 1929 short story by a Hungarian writer, Frigyes Karinthy, titled "*Láncszemek*"—*Chains*. Watts and Strogatz took it seriously: it seems to be true, and it is counterintuitive, because in the kinds of networks they studied, nodes tended to be highly clustered. They are cliquish. You may know many people, but they tend to be your neighbors—in a social space, if not literally—and they tend to know mostly the same people. In the real world, clustering is ubiquitous in complex networks: neurons in the brain, epidemics of infectious disease, electric power grids, fractures and channels in oil-bearing rock. Clustering alone means fragmentation: the oil does not flow, the epidemics sputter out. Faraway strangers remain estranged.

But some nodes may have distant links, and some nodes may have an exceptional degree of connectivity. What Watts and Strogatz discovered in their mathematical models is that it takes astonishingly few of these exceptions—just a few distant links, even in a tightly clustered network—to collapse the average separation to almost nothing and create a small world. One of their test cases was a global epidemic: "Infectious diseases are predicted to spread much more easily and quickly in a small world; the alarming and less obvious point is how few short cuts are needed to make the world small." A few sexually active flight attendants might be enough.

In cyberspace, almost everything lies in the shadows. Almost everything is connected, too, and the connectedness comes from a relatively few nodes, especially well linked or especially well trusted. However, it is one thing to prove that every node is close to every other node; that does not provide a way of finding the path between them. If the gondolier in Venice cannot find his way to the president of the United States, the mathematical existence of their connection may be small comfort. John Guare understood this, too; the next part of his *Six Degrees of Separation* explanation is less often quoted:

> I find that A) tremendously comforting that we're so close, and B) like Chinese water torture that we're so close. Because you have to find the right six people to make the connection.

There is not necessarily an algorithm for that.

The network has a structure, and that structure stands upon a paradox. Everything is close, and everything is far, at the same time. This is why cyberspace can feel not just crowded but lonely. You can drop a stone into a well and never hear a splash.

No deus ex machina waits in the wings; no man behind the curtain. We have no Maxwell's demon to help us filter and search. "We want the Demon, you see," wrote Stanislaw Lem, "to extract from the dance of atoms only information that is genuine, like mathematical theorems, fashion magazines, blueprints, historical chronicles, or a recipe for ion crumpets, or how to clean and iron a suit of asbestos, and poetry too, and scientific advice, and almanacs, and calendars, and secret documents, and everything that ever appeared in any newspaper in the Universe, and telephone books of the future." As ever, it is the choice that *informs* us (in the original sense of that word). Selecting the genuine takes work; then forgetting takes even more work. This is the curse of omniscience: the answer to any question may arrive at the fingertips—via Google or Wikipedia or IMDb or YouTube or Epicurious or the National DNA

Database or any of their natural heirs and successors—and still we wonder what we know.

We are all patrons of the Library of Babel now, and we are the librarians, too. We veer from elation to dismay and back. "When it was proclaimed that the Library contained all books," Borges tells us, "the first impression was one of extravagant happiness. All men felt themselves to be the masters of an intact and secret treasure. There was no personal or world problem whose eloquent solution did not exist in some hexagon. The universe was justified." Then come the lamentations. What good are the precious books that cannot be found? What good is complete knowledge, in its immobile perfection? Borges worries: "The certitude that everything has been written negates us or turns us into phantoms." To which, John Donne had replied long before, "He that desires to print a book, should much more desire, to be a book."

The library will endure; it is the universe. As for us, everything has not been written; we are not turning into phantoms. We walk the corridors, searching the shelves and rearranging them, looking for lines of meaning amid leagues of cacophony and incoherence, reading the history of the past and of the future, collecting our thoughts and collecting the thoughts of others, and every so often glimpsing mirrors, in which we may recognize creatures of the information.

Acknowledgments

I am indebted and grateful to Charles H. Bennett, Gregory J. Chaitin, Neil J. A. Sloane, Susanna Cuyler, Betty Shannon, Norma Barzman, John Simpson, Peter Gilliver, Jimmy Wales, Joseph Straus, Craig Townsend, Janna Levin, Katherine Bouton, Dan Menaker, Esther Schor, Siobhan Roberts, Douglas Hofstadter, Martin Seligman, Christopher Fuchs, the late John Archibald Wheeler, Carol Hutchins, and Betty Alexandra Toole; also my agent, Michael Carlisle, and, as always, for his brilliance and his patience, my editor, Dan Frank.

Notes

PROLOGUE

3 *MY MIND WANDERS AROUND:* Robert Price, "A Conversation with Claude Shannon: One Man's Approach to Problem Solving," *IEEE Communications Magazine* 22 (1984): 126.

3 *TRANSISTOR . . . BIT:* The committee got *transistor* from John R. Pierce; Shannon got *bit* from John W. Tukey.

4 SHANNON SUPPOSEDLY BELONGED: Interview, Mary Elizabeth Shannon, 25 July 2006.

5 BY 1948 MORE THAN 125 MILLION: *Statistical Abstract of the United States 1950.* More exactly: 3,186 radio and television broadcasting stations, 15,000 newspapers and periodicals, 500 million books and pamphlets, and 40 billion pieces of mail.

6 CAMPBELL'S SOLUTION: George A. Campbell, "On Loaded Lines in Telephonic Transmission," *Philosophical Magazine* 5 (1903): 313.

6 "THEORIES PERMIT CONSCIOUSNESS TO 'JUMP OVER ITS OWN SHADOW'": Hermann Weyl, "The Current Epistemological Situation in Mathematics" (1925), quoted in John L. Bell, "Hermann Weyl on Intuition and the Continuum," *Philosophia Mathematica* 8, no. 3 (2000): 261.

6 "SHANNON WANTS TO FEED NOT JUST *DATA*": Andrew Hodges, *Alan Turing: The Enigma* (London: Vintage, 1992), 251.

7 "OFF AND ON . . . I HAVE BEEN WORKING": Letter, Shannon to Vannevar Bush, 16 February 1939, in Claude Elwood Shannon, *Collected Papers*, ed. N. J. A. Sloane and Aaron D. Wyner (New York: IEEE Press, 1993), 455.

7 "NOWE USED FOR AN ELEGANT WORDE": Thomas Elyot, *The Boke Named The Governour* (1531), III: xxiv.

8 "MAN THE FOOD-GATHERER REAPPEARS": Marshall McLuhan, *Understanding Media: The Extensions of Man* (New York: McGraw-Hill, 1965), 302.

8 "WHAT LIES AT THE HEART OF EVERY LIVING THING": Richard Dawkins, *The Blind Watchmaker* (New York: Norton, 1986), 112.

9 "THE INFORMATION CIRCLE BECOMES THE UNIT OF LIFE": Werner R. Loewenstein, *The Touchstone of Life: Molecular Information, Cell Communication, and the Foundations of Life* (New York: Oxford University Press, 1999), xvi.

10 "EVERY IT—EVERY PARTICLE, EVERY FIELD OF FORCE": John Archibald Wheeler, "It from Bit," in *At Home in the Universe* (New York: American Institute of Physics, 1994), 296.

11 "THE BIT COUNT OF THE COSMOS": John Archibald Wheeler, "The Search for Links," in Anthony J. G. Hey, ed., *Feynman and Computation* (Boulder, Colo.: Westview Press, 2002), 321.

11 "NO MORE THAN 10^{120} OPS": Seth Lloyd, "Computational Capacity of the Universe," *Physical Review Letters* 88, no. 23 (2002).

11 "TOMORROW . . . WE WILL HAVE LEARNED TO UNDERSTAND": John Archibald Wheeler, "It from Bit," 298.

11 "IT IS HARD TO PICTURE THE WORLD BEFORE SHANNON": John R. Pierce, "The Early Days of Information Theory," *IEEE Transactions on Information Theory* 19, no. 1 (1973): 4.

11 "NUMBERS TOO, CHIEFEST OF SCIENCES": Aeschylus, *Prometheus Bound*, trans. H. Smyth, 460–61.

12 "THE INVENTION OF PRINTING, THOUGH INGENIOUS": Thomas Hobbes, *Leviathan* (London: Andrew Crooke, 1660), ch. 4.

1. DRUMS THAT TALK

13 "ACROSS THE DARK CONTINENT SOUND": Irma Wassall, "Black Drums," *Phylon Quarterly* 4 (1943): 38.

13 "MAKE YOUR FEET COME BACK": Walter J. Ong, *Interfaces of the Word* (Ithaca, N.Y.: Cornell University Press, 1977), 105.

14 IN 1730 FRANCIS MOORE SAILED EASTWARD: Francis Moore, *Travels into the Inland Parts of Africa* (London: J. Knox, 1767).

14 "SUDDENLY HE BECAME TOTALLY ABSTRACTED": William Allen and Thomas R. H. Thompson, *A Narrative of the Expedition to the River Niger in 1841*, vol. 2 (London: Richard Bentley, 1848), 393.

15 A MISSIONARY, ROGER T. CLARKE: Roger T. Clarke, "The Drum Language of the Tumba People," *American Journal of Sociology* 40, no. 1 (1934): 34–48.

16 "VERY OFTEN ARRIVING BEFORE THE MESSENGERS": G. Suetonius Tranquillus, *The Lives of the Caesars*, trans. John C. Rolfe (Cambridge, Mass.: Harvard University Press, 1998), 87.

16 "YET WHO SO SWIFT COULD SPEED THE MESSAGE": Aeschylus, *Agamemnon*, trans. Charles W. Eliot, 335.

16 A GERMAN HISTORIAN, RICHARD HENNIG: Gerard J. Holzmann and Björn Pehrson, *The Early History of Data Networks* (Washington, D.C.: IEEE Computer Society, 1995), 17.

17 A "CONCEIT . . . WHISPERED THOROW THE WORLD": Thomas Browne, *Pseudoxia Epidemica: Or, Enquiries Into Very Many Received Tenents, and Commonly Presumed Truths*, 3rd ed. (London: Nath. Ekins, 1658), 59.

17 IN ITALY A MAN TRIED TO SELL GALILEO: Galileo Galilei, *Dialogue Concerning the Two Chief World Systems: Ptolemaic and Copernican*, trans. Stillman Drake (Berkeley, Calif.: University of California Press, 1967), 95.

19 "A SYSTEM OF SIGNS FOR LETTERS": *Samuel F. B. Morse: His Letters and Journals*, vol. 2, ed. Edward Lind Morse (Boston: Houghton Mifflin, 1914), 12.

19 "THE DICTIONARY OR VOCABULARY CONSISTS OF WORDS": U. S. Patent 1647, 20 June 1840, 6.

19 "THE SUPERIORITY OF THE ALPHABETIC MODE": Samuel F. B. Morse, letter to Leonard D. Gale, in *Samuel F. B. Morse: His Letters and Journals*, vol. 2, 65.

20 "WHEN THE CIRCUIT WAS CLOSED A LONGER TIME": Ibid., 64.

20 "THE CLERKS WHO ATTEND AT THE RECORDING INSTRUMENT": "The Atlantic Telegraph," *The New York Times*, 7 August 1858.

21 IN SEARCH OF DATA ON THE LETTERS' RELATIVE FREQUENCIES: Morse claimed that this was he, and their partisans differ. Cf. *Samuel F. B. Morse: His Letters and Journals*, vol. 2, 68; George P. Oslin, *The Story of Telecommunications* (Macon, Ga.: Mercer University Press, 1992), 24; Franklin Leonard Pope, "The American Inventors of the Telegraph," *Century Illustrated Magazine* (April 1888): 934; Kenneth Silverman, *Lightning Man: The Accursed Life of Samuel F. B. Morse* (New York: Knopf, 2003), 167.

21 LONG AFTERWARD, INFORMATION THEORISTS CALCULATED: John R. Pierce, *An Introduction to Information Theory: Symbols, Signals, and Noise*, 2nd ed. (New York: Dover, 1980), 25.

21 "ONLY A FEW DAYS AGO I READ IN THE *TIMES*": Robert Sutherland Rattray, "The Drum Language of West Africa: Part II," *Journal of the Royal African Society* 22, no. 88 (1923): 302.

22 "HE IS NOT REALLY A EUROPEAN": John F. Carrington, *La Voix des tambours: comment comprendre le langage tambouriné d'Afrique* (Kinshasa: Protestant d'Édition et de Diffusion, 1974), 66, quoted in Walter J. Ong, *Interfaces of the Word*, 95.

23 "I MUST HAVE BEEN GUILTY MANY A TIME": John F. Carrington, *The Talking Drums of Africa* (London: Carey Kingsgate, 1949), 19.

24 EVEN THE LIMITED DICTIONARY OF THE MISSIONARIES: Ibid., 33.

25 "AMONG PEOPLES WHO KNOW NOTHING OF WRITING": Robert Sutherland Rattray, "The Drum Language of West Africa: Part I," *Journal of the Royal African Society* 22, no. 87 (1923): 235.

25 FOR THE YAUNDE, THE ELEPHANT: Theodore Stern, "Drum and Whistle 'Languages': An Analysis of Speech Surrogates," *American Anthropologist* 59 (1957): 489.

26 "THIS COUNTERSPELL MAY SAVE YOUR SOUL": James Merrill, "Eight Bits," in *The Inner Room* (New York: Knopf, 1988), 48.

26 A PAPER BY A BELL LABS TELEPHONE ENGINEER: Ralph V. L. Hartley, "Transmission of Information," *Bell System Technical Journal* 7 (1928): 535–63.

27 HE SAW LOKELE YOUTH PRACTICING THE DRUMS LESS AND LESS: John F. Carrington, *The Talking Drums of Africa*, 83.

27 A VISITOR FROM THE UNITED STATES FOUND HIM: Israel Shenker, "Boomlay," *Time*, 22 November 1954.

2. THE PERSISTENCE OF THE WORD

28 "ODYSSEUS WEPT": Ward Just, *An Unfinished Season* (New York: Houghton Mifflin, 2004), 153.

28 "TRY TO IMAGINE": Walter J. Ong, *Orality and Literacy: The Technologizing of the Word* (London: Methuen, 1982), 31.

28 THE PASTNESS OF THE PAST: Jack Goody and Ian Watt, "The Consequences of Literacy," *Comparative Studies in Society and History* 5, no. 3 (1963): 304–45.

29 "THE OTHER EMINENT CATHOLIC-ELECTRONIC PROPHET": Frank Kermode, "Free Fall," *New York Review of Books* 10, no. 5 (14 March 1968).

29 "HORSES AS AUTOMOBILES WITHOUT WHEELS": Walter J. Ong, *Orality and Literacy*, 12.

30 "LANGUAGE IN FACT BEARS THE SAME RELATIONSHIP": Jonathan Miller, *Marshall McLuhan* (New York: Viking, 1971), 100.

30 "FOR THIS INVENTION WILL PRODUCE FORGETFULNESS": Plato, *Phaedrus*, trans. Benjamin Jowett (Fairfield, Iowa: First World Library, 2008), 275a.

31 "TWO THOUSAND YEARS OF MANUSCRIPT CULTURE": Marshall McLuhan, "Culture Without Literacy," in Eric McLuhan and Frank Zingrone, eds., *Essential McLuhan* (New York: Basic Books, 1996), 305.

31 "THIS MIRACULOUS REBOUNDING OF THE VOICE": Pliny the Elder, *The Historie of the World*, vol. 2, trans. Philemon Holland (London: 1601), 581.

31 "THE WRITTEN SYMBOL EXTENDS INFINITELY": Samuel Butler, *Essays on Life, Art, and Science* (Port Washington, N.Y.: Kennikat Press, 1970), 198.

34 "THERE NEVER WAS A MAN": David Diringer and Reinhold Regensburger, *The Alphabet: A Key to the History of Mankind*, 3rd ed., vol. 1 (New York: Funk & Wagnalls, 1968), 166.

34 "IT WAS SOMETHING LIKE A THUNDER-CLAP": "The Alphabetization of Homer," in Eric Alfred Havelock and Jackson P. Hershbell, *Communication Arts in the Ancient World* (New York: Hastings House, 1978), 3.

35 "HAPPENS, UP TO THE PRESENT DAY": Aristotle, *Poetics*, trans. William Hamilton Fyfe (Cambridge, Mass.: Harvard University Press, 1953), 1447b.

35 HAVELOCK DESCRIBED IT AS CULTURAL WARFARE: Eric A. Havelock, *Preface to Plato* (Cambridge, Mass.: Harvard University Press, 1963), 300–301.

36 "A *BEGINNING* IS THAT WHICH ITSELF DOES NOT FOLLOW": Aristotle, *Poetics*, 1450b.

36 "THE MULTITUDE CANNOT ACCEPT": *Republic*, 6.493e. Cf. in Eric A. Havelock, *Preface to Plato*, 282.

36 "LOSE THEMSELVES AND WANDER": *Republic*, 6.484b.

37 "TRYING FOR THE FIRST TIME IN HISTORY": Eric A. Havelock, *Preface to Plato*, 282.

37 LOGIC DESCENDED FROM THE WRITTEN WORD: Not everyone agrees with all this. A counterargument: John Halverson, "Goody and the Implosion of the Literacy Thesis," *Man* 27, no. 2 (1992): 301–17.

38 *IF IT IS POSSIBLE FOR NO MAN TO BE A HORSE*: Aristotle, *Prior Analytics*, trans. A. J. Jenkinson, 1:3.

38 "WE KNOW THAT FORMAL LOGIC": Walter J. Ong, *Orality and Literacy*, 49.

38 FIELDWORK OF THE RUSSIAN PSYCHOLOGIST: A. R. Luria, *Cognitive Development, Its Cultural and Social Foundations* (Cambridge, Mass.: Harvard University Press, 1976), 86.

39 "BASICALLY THE PEASANT WAS RIGHT": Walter J. Ong, *Orality and Literacy*, 53.

39 "IN THE INFANCY OF LOGIC": Benjamin Jowett, introduction to Plato's *Theaetetus* (Teddington, U.K.: Echo Library, 2006), 7.

40 "WHEN A WHITE HORSE IS NOT A HORSE": Gongsun Long, "When a White Horse Is Not a Horse," trans. by A. C. Graham, in P. J. Ivanhoe et al., *Readings in Classical Chinese Philosophy*, 2nd ed. (Indianapolis, Ind.: Hackett Publishing, 2005), 363–66. Also A. C. Graham, *Studies in Chinese Philosophy and Philosophical Literature*, SUNY Series in Chinese Philosophy and Culture (Albany: State University of New York Press, 1990), 178.

42 "WRITING, LIKE A THEATER CURTAIN GOING UP": Julian Jaynes, *The Origin of Consciousness in the Breakdown of the Bicameral Mind* (Boston: Houghton Mifflin, 1977), 177.

42 "TO THE ASSYRIANS, THE CHALDEANS, AND EGYPTIANS": Thomas Sprat, *The History of the Royal Society of London, for the Improving of Natural Knowledge*, 3rd ed. (London: 1722), 5.

43 "THIS PROCESS OF CONQUEST AND INFLUENCE": Julian Jaynes, *The Origin of Consciousness in the Breakdown of the Bicameral Mind*, 198.

43 TO FORM LARGE NUMBERS, THE BABYLONIANS: Donald E. Knuth, "Ancient Babylonian Algorithms," *Communications of the Association for Computing Machinery* 15, no. 7 (1972): 671–77.

44 "IT WAS ASSUMED THAT THE BABYLONIANS": Asger Aaboe, *Episodes from the Early History of Mathematics* (New York: L. W. Singer, 1963), 5.

45 "OUR TASK CAN THEREFORE PROPERLY BE COMPARED": Otto Neugebauer, *The Exact Sciences in Antiquity*, 2nd ed. (Providence, R.I.: Brown University Press, 1957), 30 and 40–46.

45 "A CISTERN. THE HEIGHT IS 3,20": Donald E. Knuth, "Ancient Babylonian Algorithms," 672.

46 "FUNDAMENTALLY LETTERS ARE SHAPES": John of Salisbury, *Metalogicon*, I:13,

quoted and translated by M. T. Clanchy, *From Memory to Written Record, England, 1066-1307* (Cambridge, Mass.: Harvard University Press, 1979), 202.

47 "OH! ALL YE WHO SHALL HAVE HEARD": Ibid.

47 "I CANNOT HELP FEELING": *Phaedrus*, trans. Benjamin Jowett, 275d.

48 "WE ARE IN OUR CENTURY 'WINDING THE TAPE BACKWARD'": Marshall McLuhan, "Media and Cultural Change," in *Essential McLuhan*, 92.

48 "THE LARGER THE NUMBER OF SENSES INVOLVED": Jonathan Miller, *Marshall McLuhan*, 3.

49 "ACOUSTIC SPACE IS ORGANIC": *Playboy* interview, March 1969, in *Essential McLuhan*, 240.

49 "MEN LIVED UPON GROSS EXPERIENCE": Thomas Hobbes, *Leviathan, or The Matter, Forme and Power of a Commonwealth, Ecclesiasticall, and Civill*, (1651; repr., London: George Routledge and Sons, 1886), 299.

50 "MOST LITERATE PERSONS, WHEN YOU SAY": Walter J. Ong, "This Side of Oral Culture and of Print," *Lincoln Lecture* (1973), 2.

50 "IT IS DEMORALIZING TO REMIND ONESELF": Walter J. Ong, *Orality and Literacy*, 14.

3. TWO WORDBOOKS

51 "IN SUCH BUSIE, AND ACTIVE TIMES": Thomas Sprat, *The History of the Royal Society of London, for the Improving of Natural Knowledge*, 3rd ed. (London: 1722), 42.

51 A BOOK IN 1604 WITH A RAMBLING TITLE: Robert Cawdrey, *A Table Alphabeticall* (London: Edmund Weaver, 1604) may be found in the Bodleian Library; in a facsimile edition, Robert A. Peters, ed. (Gainesville, Fla.: Scholars' Facsimiles & Reprints, 1966); online via the University of Toronto Library; and, most satisfyingly, reprinted as John Simpson, ed., *The First English Dictionary, 1604: Robert Cawdrey's A Table Alphabeticall* (Oxford: Bodleian Library, 2007).

53 A SINGLE 1591 PAMPHLET: Robert Greene, *A Notable Discovery of Coosnage* (1591; repr., Gloucester, U.K.: Dodo Press, 2008); Albert C. Baugh, *A History of the English Language*, 2nd ed. (New York: Appleton-Century-Crofts, 1957), 252.

54 "IT WERE A THING VERIE PRAISEWORTHIE": Richard Mulcaster, *The First Part of the Elementarie Which Entreateth Chefelie of the Right Writing of Our English Tung* (London: Thomas Vautroullier, 1582).

55 "SOME MEN SEEK SO FAR FOR OUTLANDISH ENGLISH": John Simpson, ed., *The First English Dictionary*, 41.

55 "NOT CONFORMING HIMSELF": John Strype, *Historical Collections of the Life and Acts of the Right Reverend Father in God, John Aylmer* (London: 1701), 129, quoted in John Simpson, ed., *The First English Dictionary*, 10.

57 HE COPIED THE REMARKS ABOUT INKHORN TERMS: Gertrude E. Noyes, "The First English Dictionary, Cawdrey's *Table Alphabeticall*," *Modern Language Notes* 58, no. 8 (1943): 600.

57 "SO MORE KNOWLEDGE WILL BE BROUGHT INTO THIS LAND": Edmund Coote, *The English Schoole-maister* (London: Ralph Jackson & Robert Dexter, 1596), 2.

58 "FOR EXAMPLE I INTEND TO DISCUSS *AMO*": Lloyd W. Daly, *Contributions to a History of Alphabeticization in Antiquity and the Middle Ages* (Brussels: Latomus, 1967), 73.

58 NOT UNTIL 1613 WAS THE FIRST ALPHABETICAL CATALOGUE: William Dunn Macray, *Annals of the Bodleian Library, Oxford, 1598–1867* (London: Rivingtons, 1868), 39.

59 "LET ME MENTION THAT THE WORDS OR NAMES": Gottfried Leibniz, *Unvorgreifliche Gedanken*, quoted and translated by Werner Hüllen, *English Dictionaries 800–1700: The Topical Tradition* (Oxford: Clarendon Press, 1999), 16*n*.

60 "SAYWHAT, CORRUPTLY CALLED A DEFINITION": Ralph Lever, *The Arte of Reason* (London: H. Bynneman, 1573).

60 "DEFINITION . . . BEING NOTHING BUT MAKING ANOTHER UNDERSTAND": John Locke, *An Essay Concerning Human Understanding*, ch. 3, sect. 10.

62 "SO LONG AS MEN WERE IN FACT OBLIGED": Galileo, letter to Mark Welser, 4 May 1612, trans. Stillman Drake, in *Discoveries and Opinions of Galileo*, 92.

62 "I DO NOT DEFINE TIME, SPACE, PLACE, AND MOTION": Isaac Newton, *Philosophiae Naturalis Principia Mathematica*, trans. Andrew Motte (Scholium) 6.

63 JOHN BULLOKAR, OTHERWISE LEFT AS FAINT A MARK: Jonathon Green, *Chasing the Sun: Dictionary Makers and the Dictionaries They Made* (New York: Holt, 1996), 181.

65 "WE REALLY DON'T LIKE BEING PUSHED": Interview, John Simpson, 13 September 2006.

66 "DICTIONARY, A MALEVOLENT LITERARY DEVICE": Ambrose Bierce, *The Devil's Dictionary* (New York: Dover, 1993), 25.

66 "IN GIVING EXPLANATIONS I ALREADY HAVE TO USE LANGUAGE": Ludwig Wittgenstein, *Philosophical Investigations*, trans. G. E. M. Anscombe (New York: Macmillan, 1953), 47.

67 "THE ENGLISH DICTIONARY, LIKE THE ENGLISH CONSTITUTION": James A. H. Murray, "The Evolution of English Lexicography," Romanes Lecture (1900).

69 W. H. AUDEN DECLARED: Peter Gilliver et al., *The Ring of Words: Tolkien and the Oxford English Dictionary* (Oxford: Oxford University Press, 2006), 82.

69 ANTHONY BURGESS WHINGED: Anthony Burgess, "OED +," in *But Do Blondes Prefer Gentlemen? Homage to Qwert Yuiop and Other Writings* (New York: McGraw-Hill, 1986), 139. He could not let go, either. In a later essay, "Ameringlish," he complained again.

69 "*EVERY* FORM IN WHICH A WORD": "Writing the *OED:* Spellings," Oxford English Dictionary, http://www.oed.com/about/writing/spellings.html (accessed 6 April 2007).

71 "WHICH, WHILE IT WAS EMPLOYED IN THE CULTIVATION": Samuel Johnson, preface to *A Dictionary of the English Language* (1755).

74 WE POSSESS NOW A MORE COMPLETE DICTIONARY: John Simpson, ed., *The First English Dictionary,* 24.

75 "WHAT I SHALL HEREAFTER CALL MONDEGREENS": "The Death of Lady Mondegreen," *Harper's Magazine*, November 1954, 48.

76 "THE INTERESTING THING ABOUT MONDEGREENS": Steven Pinker, *The Language Instinct: How the Mind Creates Language* (New York: William Morrow, 1994), 183.

4. TO THROW THE POWERS OF THOUGHT INTO WHEEL-WORK

The original writings of Charles Babbage and, to a lesser extent, Ada Lovelace are increasingly accessible. The comprehensive, thousand-dollar, eleven-volume edition, *The Works of Charles Babbage*, edited by Martin Campbell-Kelly, was published in 1989. Online, the full texts of Babbage's *Passages from the Life of a Philosopher* (1864), *On the Economy of Machinery and Manufactures* (1832), and *The Ninth Bridgewater Treatise* (1838) can now be found in editions scanned from libraries by Google's book program. Not yet available there (as of 2010), but also useful, is his son's volume, *Babbage's Calculating Engines: Being a Collection of Papers Relating to Them* (1889). As interest grew during the era of computing, much of the useful material in these books was reprinted in collections; most valuable are *Charles Babbage and His Calculating Engines*, edited by Philip Morrison and Emily Morrison (1961); and Anthony Hyman's *Science and Reform: Selected Works of Charles Babbage* (1989). Other manuscripts were published in J. M. Dubbey, *The Mathematical Work of Charles Babbage* (1978). The notes that follow refer to one or more of these sources, depending on what seems most useful for the reader. The translation and astounding "notes" on L. F. Menabrea's "Sketch of the Analytical Engine" by Ada Augusta, Countess of Lovelace, have been made available online at http://www.fourmilab.ch/babbage/sketch.html thanks to John Walker; they are also reproduced in the Morrisons' collection. As for the Lovelace letters and papers, they are in the British Library, the Bodleian, and elsewhere, but many have been published by Betty Alexandra Toole in *Ada: The Enchantress of Numbers* (1992 and 1998); where possible I try to cite the published versions.

78 "LIGHT ALMOST SOLAR HAS BEEN EXTRACTED": Charles Babbage, *On the Economy of Machinery and Manufactures* (1832), 300; reprinted in *Science and Reform: Selected Works of Charles Babbage*, ed. Anthony Hyman (Cambridge: Cambridge University Press, 1989), 200.

78 THE *TIMES* OBITUARIST: "The Late Mr. Charles Babbage, F.R.S.," *The Times* (London), 23 October 1871. Babbage's crusade against organ-grinders and hurdy-gurdies was not in vain; a new law against street music in 1864 was known as Babbage's Act. Cf. Stephanie Pain, "Mr. Babbage and the Buskers," *New Scientist* 179, no. 2408 (2003): 42.

78 "HE SHOWED A GREAT DESIRE TO INQUIRE": N. S. Dodge, "Charles Babbage," *Smithsonian Annual Report of 1873*, 162–97, reprinted in *Annals of the History of Computing* 22, no. 4 (October–December 2000), 20.

79 NOT "THE MANUAL LABOR OF ROWING": Charles Babbage, *Passages from the Life of a Philosopher* (London: Longman, Green, Longman, Roberts, & Green, 1864), 37.

79 " 'THE TALL GENTLEMAN IN THE CORNER' ": Ibid., 385–86.

79 "THOSE WHO ENJOY LEISURE": Charles Babbage, *On the Economy of Machinery and Manufactures*, 4th ed. (London: Charles Knight, 1835), v.

80 HE COMPUTED THE COST OF EACH PHASE: Ibid., 146.

80 "AT THE EXPENSE OF THE NATION": Henry Prevost Babbage, ed., *Babbage's Calculating Engines: Being a Collection of Papers Relating to Them; Their History and Construction* (London: E. & F. N. Spon, 1889), 52.

81 "ON TWO OCCASIONS I HAVE BEEN ASKED": Charles Babbage, *Passages from the Life of a Philosopher*, 67.

81 TABLE OF CONSTANTS OF THE CLASS MAMMALIA: *Charles Babbage and His Calculating Engines: Selected Writings*, ed. Philip Morrison and Emily Morrison (New York: Dover Publications, 1961), xxiii.

82 "LO! THE RAPTURED ARITHMETICIAN!": Élie de Joncourt, *De Natura Et Praeclaro Usu Simplicissimae Speciei Numerorum Trigonalium* (Hagae Comitum: Husson, 1762), quoted in Charles Babbage, *Passages from the Life of a Philosopher*, 54.

83 "TO ASTROLOGERS, LAND-MEASURERS, MEASURERS OF TAPESTRY": Quoted in Elizabeth L. Eisenstein, *The Printing Press as an Agent of Change: Communications and Cultural Transformations in Early-Modern Europe* (Cambridge: Cambridge University Press, 1979), 468.

84 THIRTY-FOUR MEN AND ONE WOMAN: Mary Croarken, "Mary Edwards: Computing for a Living in 18th-Century England," *IEEE Annals of the History of Computing* 25, no. 4 (2003): 9–15; and—with fascinating detective work— Mary Croarken, "Tabulating the Heavens: Computing the Nautical Almanac in 18th-Century England," *IEEE Annals of the History of Computing* 25, no. 3 (2003): 48–61.

84 "LOGARITHMES ARE NUMBERS INVENTED": Henry Briggs, *Logarithmicall Arithmetike: Or Tables of Logarithmes for Absolute Numbers from an Unite to 100000* (London: George Miller, 1631), 1.

85 "TAKE AWAY ALL THE DIFFICULTIE": John Napier, "Dedicatorie," in *A Description of the Admirable Table of Logarithmes*, trans. Edward Wright (London: Nicholas Okes, 1616), 3.

85 "NAPER, LORD OF MARKINSTON, HATH SET": Henry Briggs to James Ussher, 10 March 1615, quoted by Graham Jagger in Martin Campbell-Kelly et al., eds., *The History of Mathematical Tables: From Sumer to Spreadsheets* (Oxford: Oxford University Press, 2003), 56.

85 A QUARTER HOUR OF SILENCE: "SPENT, EACH BEHOLDING OTHER": William Lilly, *Mr. William Lilly's History of His Life and Times, from the Year 1602 to 1681* (London: Charles Baldwyn, 1715), 236.

87 *POLE STARRE, GIRDLE OF ANDROMEDA, WHALES BELLIE*: Henry Briggs, *Logarithmicall Arithmetike*, 52.

87 "IT MAY BE HERE ALSO NOTED THAT THE USE OF A 100 POUND": Ibid., 11.

87 "A SCOTTISH BARON HAS APPEARED ON THE SCENE": Ole I. Franksen, "Introducing 'Mr. Babbage's Secret,'" *APL Quote Quad* 15, no. 1 (1984): 14.

87 THE MAJORITY OF HUMAN COMPUTATION: Michael Williams, *A History of Computing Technology* (Washington, D.C.: IEEE Computer Society, 1997), 105.

87 "IT IS NOT FITTING FOR A PROFESSOR": Michael Mästlin, quoted in Ole I. Franksen, "Introducing 'Mr. Babbage's Secret,'" 14.

88 "THIS LADY ATTITUDINIZED": Charles Babbage, *Passages from the Life of a Philosopher*, 17.

88 INSTALLED IT ON A PEDESTAL: Simon Schaffer, "Babbage's Dancer," in Francis Spufford and Jenny Uglow, eds., *Cultural Babbage: Technology, Time and Invention* (London: Faber and Faber, 1996), 58.

89 FROM A SPECIALTY BOOKSELLER: Charles Babbage, *Passages from the Life of a Philosopher*, 26–27.

89 "A SIN AGAINST THE MEMORY OF NEWTON": W. W. Rouse Ball, *A History of the Study of Mathematics at Cambridge* (Cambridge: Cambridge University Press, 1889), 117.

89 "THE DOTS OF NEWTON, THE *D*'S OF LEIBNITZ": *Charles Babbage and His Calculating Engines*, 23.

89 "TO THINK AND REASON IN A NEW LANGUAGE": Ibid., 31.

90 "A NEW KIND OF AN INSTRUMENT INCREASING THE POWERS OF REASON": C. Gerhardt, ed., *Die Philosophischen Schriften von Gottfried Wilhelm Leibniz*, vol. 7 (Berlin: Olms, 1890), 12, quoted by Kurt Gödel in "Russell's Mathematical Logic" (1944), in *Kurt Gödel: Collected Works*, vol. 2, ed. Solomon Feferman (New York: Oxford University Press, 1986), 140.

90 "BY THE APPARENT IMPOSSIBILITY OF ARRANGING SIGNS": Charles Babbage, *Passages from the Life of a Philosopher*, 25.

90 "THE DOT-AGE OF THE UNIVERSITY": *Charles Babbage and His Calculating Engines*, 25.

90 "WE HAVE NOW TO RE-IMPORT THE EXOTIC": Charles Babbage, *Memoirs of the Analytical Society*, preface (1813), in Anthony Hyman, ed., *Science and Reform: Selected Works of Charles Babbage* (Cambridge: Cambridge University Press, 1989), 15–16.

91 "THE BROWS OF MANY A CAMBRIDGE MODERATOR": Agnes M. Clerke, *The Herschels and Modern Astronomy* (New York: Macmillan, 1895), 144.

91 "EVERY MEMBER SHALL COMMUNICATE HIS ADDRESS": Charles Babbage, *Passages from the Life of a Philosopher*, 34.

92 "I AM THINKING THAT ALL THESE TABLES": Ibid., 42.

93 "WHETHER, WHEN THE NUMBERS": Ibid., 41.

93 "WE MAY GIVE FINAL PRAISE": "*Machina arithmetica in qua non additio tantum et subtractio sed et multipicatio nullo, divisio vero paene nullo animi labore peragantur,*" trans. M. Kormes, 1685, in D. E. Smith, *A Source Book in Mathematics* (New York: McGraw-Hill, 1929), 173.

93 "INTOLERABLE LABOUR AND FATIGUING MONOTONY": Charles Babbage, *A Letter to Sir Humphry Davy on the Application of Machinery to the Purpose of Calculating and Printing Mathematical Tables* (London: J. Booth & Baldwin, Cradock & Joy, 1822), 1.

93 "I WILL YET VENTURE TO PREDICT": Babbage to David Brewster, 6 November 1822, in Martin Campbell-Kelly, ed., *The Works of Charles Babbage* (New York: New York University Press, 1989) 2:43.

94 "CONFUSION IS WORSE CONFOUNDED": Dionysius Lardner, "Babbage's Calculating Engine," *Edinburgh Review* 59, no. 120 (1834), 282; and Edward Everett, "The Uses of Astronomy," in *Orations and Speeches on Various Occasions* (Boston: Little, Brown, 1870), 447.

94 250 SETS OF LOGARITHMIC TABLES: Martin Campbell-Kelly, "Charles Babbage's Table of Logarithms (1827)," *Annals of the History of Computing* 10 (1988): 159–69.

94 "WOULD AFFORD A CURIOUS SUBJECT OF METAPHYSICAL SPECULATION": Dionysius Lardner, "Babbage's Calculating Engines," 282.

95 "IF PAPA FAIL TO INFORM HIM": Charles Babbage, *Passages from the Life of a Philosopher*, 52.

97 "IF THIS COULD BE ACCOMPLISHED": Ibid., 60–62.

98 "IT IS SCARCELY TOO MUCH TO ASSERT": Babbage to John Herschel, 10 August 1814, quoted in Anthony Hyman, *Charles Babbage: Pioneer of the Computer* (Princeton, N.J.: Princeton University Press, 1982), 31.

99 "IT IS WITH NO INCONSIDERABLE DEGREE OF RELUCTANCE": David Brewster to Charles Babbage, 3 July 1821, quoted in J. M. Dubbey, *The Mathematical Work of Charles Babbage* (Cambridge: Cambridge University Press, 1978), 94.

99 "LOGARITHMIC TABLES AS CHEAP AS POTATOES": Babbage to John Herschel, 27 June 1823, quoted in Anthony Hyman, *Charles Babbage*, 53.

99 "PROPOSITION TO REDUCE ARITHMETIC TO THE DOMINION OF MECHANISM": Dionysius Lardner, "Babbage's Calculating Engines," 264.

99 "THE QUESTION IS SET TO THE INSTRUMENT": "Address of Presenting the Gold Medal of the Astronomical Society to Charles Babbage," in *Charles Babbage and His Calculating Engines*, 219.

101 LARDNER'S OWN EXPLANATION OF "CARRYING": Dionysius Lardner, "Babbage's Calculating Engines," 288–300.

102 IN 1826 HE PROUDLY REPORTED TO THE ROYAL SOCIETY: Charles Babbage, "On a Method of Expressing by Signs the Action of Machinery," *Philosophical Transactions of the Royal Society of London* 116, no. 3 (1826): 250–65.

103 "I NEED HARDLY POINT OUT TO YOU THAT THIS CALCULATION": Quoted in *Charles Babbage and His Calculating Engines*, xxiii. The Morrisons point out that Tennyson apparently did change "minute" to "moment" in editions after 1850.

104 "THE PROS AND CONS IN PARALLEL COLUMNS": Harriet Martineau, *Autobiography* (1877), quoted in Anthony Hyman, *Charles Babbage,* 129.

104 "IF YOU SPEAK TO HIM OF A MACHINE FOR PEELING A POTATO": Quoted in Doron Swade, *The Difference Engine: Charles Babbage and the Quest to Build the First Computer* (New York: Viking, 2001), 132.

105 "I THINK IT LIKELY HE LIVES IN A SORT OF DREAM": Quoted in ibid., 38.

105 FOR A GUINEA, SHE COULD SIT: Advertisement in *The Builder,* 31 December 1842, http://www.victorianlondon.org/photography/adverts.htm (accessed 7 March 2006).

105 "THE CHILD OF LOVE, . . . —THOUGH BORN IN BITTERNESS": Lord Byron, "Childe Harold's Pilgrimage," canto 3, 118.

106 "IS THE GIRL IMAGINATIVE?": Byron to Augusta Leigh, 12 October 1823, in Leslie A. Marchand, ed., *Byron's Letters and Journals,* vol. 9 (London: John Murray, 1973–94), 47.

106 "I AM GOING TO BEGIN MY PAPER WINGS": Ada to Lady Byron, 3 February 1828, in Betty Alexandra Toole, *Ada, the Enchantress of Numbers: Prophet of the Computer Age* (Mill Valley, Calif.: Strawberry Press, 1998), 25.

106 "MISS STAMP DESIRES ME TO SAY": Ada to Lady Byron, 2 April 1828, ibid., 27.

106 "WHEN I AM WEAK": Ada to Mary Somerville, 20 February 1835, ibid., 55.

107 AN "OLD MONKEY": Ibid., 33.

107 "WHILE OTHER VISITORS GAZED": Sophia Elizabeth De Morgan, *Memoir of Augustus De Morgan* (London: Longmans, Green, 1882), 89.

107 "I DO NOT CONSIDER THAT I KNOW": Ada to Dr. William King, 24 March 1834, in Betty Alexandra Toole, *Ada, the Enchantress of Numbers,* 45.

107 "GEM OF ALL MECHANISM": Ada to Mary Somerville, 8 July 1834, ibid., 46.

109 "PUNCHES HOLES IN A SET OF PASTEBOARD CARDS": "Of the Analytical Engine," in *Charles Babbage and His Calculating Engines,* 55.

109 "HOW THE MACHINE COULD PERFORM THE ACT OF JUDGMENT": Ibid., 65.

110 "I AM AT PRESENT A CONDEMNED SLAVE": Ada to Mary Somerville, 22 June 1837, in Betty Alexandra Toole, *Ada, the Enchantress of Numbers,* 70.

110 "THE ONLY OTHER PERSON WAS A MIDDLE-AGED GENTLEMAN": Ada to Lady Byron, 26 June 1838, ibid., 78.

111 "I HAVE A PECULIAR *WAY* OF *LEARNING*": Ada to Babbage, November 1839, ibid., 82.

111 "you know i am by nature a bit of a philosopher": Ada to Babbage, 16 February 1840, ibid., 83.

111 "an original mathematical investigator": Augustus De Morgan to Lady Byron, quoted in Betty Alexandra Toole, "Ada Byron, Lady Lovelace, an Analyst and Metaphysician," *IEEE Annals of the History of Computing* 18, no. 3 (1996), 7.

112 "i *have* done it by trying": Ada to Babbage, 16 February 1840, in Betty Alexandra Toole, *Ada, the Enchantress of Numbers*, 83.

112 "of certain sprites & fairies": Ada to Augustus De Morgan, 3 February 1841, ibid., 99.

112 "we talk *much* of imagination": Untitled essay, 5 January 1841, ibid., 94.

113 "i have on my mind most strongly": Ada to Woronzow Greig, 15 January 1841, ibid., 98.

113 "*what* a mountain i have to climb": Ada to Lady Byron, 6 February 1841, ibid., 101.

114 "it will enable our clerks to plunder us": *Charles Babbage and His Calculating Engines*, 113. He added: "possibly we might send lightning to outstrip the culprit . . ."

115 "the discovery of the analytical engine": Quoted in Anthony Hyman, *Charles Babbage*, 185.

115 "*notions sur la machine analytique*": *Bibliothèque Universelle de Genève*, no. 82 (October 1842).

115 not to "*proclaim* who has written it": Ada to Babbage, 4 July 1843, in Betty Alexandra Toole, *Ada, the Enchantress of Numbers*, 145.

116 "any process which alters the mutual relation": Note A (by the translator, Ada Lovelace) to L. F. Menabrea, "Sketch of the Analytical Engine Invented by Charles Babbage," in *Charles Babbage and His Calculating Engines*, 247.

116 "the analytical engine does not occupy common ground": Ibid., 252.

118 "the engine eating its own tail": H. Babbage, "The Analytical Engine," paper read at Bath, 12 September 1888, in *Charles Babbage and His Calculating Engines*, 331.

118 "we easily perceive that since every successive function": Note D (by the translator, Ada Lovelace) to L. F. Menabrea, "Sketch of the Analytical Engine Invented by Charles Babbage."

118 "that *brain* of mine": Ada to Babbage, 5 July 1843, in Betty Alexandra Toole, *Ada, the Enchantress of Numbers*, 147.

119 "how multifarious and how mutually complicated": Note D (by the translator, Ada Lovelace) to L. F. Menabrea, "Sketch of the Analytical Engine Invented by Charles Babbage."

119 "i am in much dismay": Ada to Babbage, 13 July 1843, in Betty Alexandra Toole, *Ada, the Enchantress of Numbers*, 149.

119 "I FIND THAT MY PLANS & IDEAS": Ada to Babbage, 22 July 1843, ibid., 150.

119 "I DO NOT THINK YOU POSSESS HALF *MY* FORETHOUGHT": Ada to Babbage, 30 July 1843, ibid., 157.

119 "IT WOULD BE LIKE USING THE STEAM HAMMER": H. P. Babbage, "The Analytical Engine," 333.

120 "WHAT SHALL WE THINK OF THE CALCULATING MACHINE": "Maelzel's Chess-Player," in *The Prose Tales of Edgar Allan Poe: Third Series* (New York: A. C. Armstrong & Son, 1889), 230.

120 "STEAM IS AN APT SCHOLAR": Ralph Waldo Emerson, *Society and Solitude* (Boston: Fields, Osgood, 1870), 143.

120 "WHAT A SATIRE IS THAT MACHINE": Oliver Wendell Holmes, *The Autocrat of the Breakfast-Table* (New York: Houghton Mifflin, 1893), 11.

121 "ONE OF THE MOST FASCINATING OF ARTS": Charles Babbage, *Passages from the Life of a Philosopher*, 235.

121 "EVERY SHOWER THAT FALLS": "On the Age of Strata, as Inferred from the Rings of Trees Embedded in Them," from Charles Babbage, *The Ninth Bridgewater Treatise: A Fragment* (London: John Murray, 1837), in *Charles Babbage and His Calculating Engines*, 368.

121 "ADMITTING IT TO BE POSSIBLE BETWEEN LONDON AND LIVERPOOL": Charles Babbage, *On the Economy of Machinery*, 10.

122 "ENCLOSED IN SMALL CYLINDERS ALONG WIRES": Charles Babbage, *Passages from the Life of a Philosopher*, 447.

122 "A COACH AND APPARATUS": Charles Babbage, *On the Economy of Machinery*, 273.

122 "ZENITH-LIGHT SIGNALS": Charles Babbage, *Passages from the Life of a Philosopher*, 460.

123 "THIS LED TO A NEW THEORY OF STORMS": Ibid., 301.

123 "A DIFFERENT SENSE OF ANACHRONISM": Jenny Uglow, "Possibility," in Francis Spufford and Jenny Uglow, *Cultural Babbage*, 20.

123 "IF, UNWARNED BY MY EXAMPLE": Charles Babbage, *Passages from the Life of a Philosopher*, 450.

124 "THEY SAY THAT '*COMING EVENTS*'": Ada to Lady Byron, 10 August 1851, in Betty Alexandra Toole, *Ada, the Enchantress of Numbers*, 287.

124 "MY BEING *IN TIME* AN *AUTOCRAT*": Ada to Lady Byron, 29 October 1851, ibid., 291.

5. A NERVOUS SYSTEM FOR THE EARTH

125 "IS IT A FACT—OR HAVE I DREAMT IT": Nathaniel Hawthorne, *The House of the Seven Gables* (Boston: Ticknor, Reed, & Fields, 1851), 283.

125 THREE CLERKS IN A SMALL ROOM: They managed the traffic "easily, and not

very continuously." "Central Telegraph Stations," *Journal of the Society of Telegraph Engineers* 4 (1875): 106.

126 "WHO WOULD THINK THAT BEHIND THIS NARROW FOREHEAD": Andrew Wynter, "The Electric Telegraph," *Quarterly Review* 95 (1854): 118–64.

126 HE WAS NEITHER THE FIRST NOR THE LAST: Iwan Rhys Morus, "'The Nervous System of Britain': Space, Time and the Electric Telegraph in the Victorian Age," *British Journal of the History of Science* 33 (2000): 455–75.

126 ALFRED SMEE: Quoted in Iwan Rhys Morus, "'The Nervous System of Britain,'" 471.

126 "THE DOCTOR CAME AND LOOKED": "Edison's Baby," *The New York Times*, 27 October 1878, 5.

126 "THE TIME IS CLOSE AT HAND": "The Future of the Telephone," *Scientific American*, 10 January 1880.

127 "ELECTRICITY IS THE POETRY OF SCIENCE": Alexander Jones, *Historical Sketch of the Electric Telegraph: Including Its Rise and Progress in the United States* (New York: Putnam, 1852), v.

127 "AN INVISIBLE, INTANGIBLE, IMPONDERABLE AGENT": William Robert Grove, quoted in Iwan Rhys Morus, "'The Nervous System of Britain,'" 463.

127 "THE WORLD OF SCIENCE IS NOT AGREED": Dionysus Lardner, *The Electric Telegraph*, revised and rewritten by Edward B. Bright (London: James Walton, 1867), 6.

128 "WE ARE NOT TO CONCEIVE OF THE ELECTRICITY": "The Telegraph," *Harper's New Monthly Magazine*, 47 (August 1873), 337.

128 "BOTH OF THEM ARE POWERFUL": "The Electric Telegraph," *The New York Times*, 11 November 1852.

128 "CANST THOU SEND LIGHTNINGS": Job 38:35; Dionysus Lardner, *The Electric Telegraph*.

129 COUNT MIOT DE MELITO CLAIMED: *Memoirs of Count Miot de Melito*, vol. 1, trans. Cashel Hoey and John Lillie (London: Sampson Low, 1881), 44n.

130 MEANWHILE THE CHAPPES MANAGED: Gerard J. Holzmann and Björn Pehrson, *The Early History of Data Networks* (Washington, D.C.: IEEE Computer Society, 1995), 52 ff.

131 "THE DAY WILL COME": *"Lettre sur une nouveau télégraphe,"* quoted in Jacques Attali and Yves Stourdze, "The Birth of the Telephone and the Economic Crisis: The Slow Death of Monologue in French Society," in Ithiel de Sola Poolin, ed., *The Social Impact of the Telephone* (Cambridge, Mass.: MIT Press, 1977), 97.

131 "CITIZEN CHAPPE OFFERS AN INGENIOUS METHOD": Gerard J. Holzmann and Björn Pehrson, *The Early History of Data Networks*, 59.

132 ONE DEPUTY NAMED A PANTHEON: Bertrand Barère de Vieuzac, 17 August 1794, quoted in ibid., 64.

132 CHAPPE ONCE CLAIMED: Taliaferro P. Shaffner, *The Telegraph Manual: A Complete History and Description of the Semaphoric, Electric and Magnetic Telegraphs*

of Europe, Asia, Africa, and America, Ancient and Modern (New York: Pudney & Russell, 1859), 42.

134 "THEY HAVE PROBABLY NEVER PERFORMED EXPERIMENTS": Gerard J. Holzmann and Björn Pehrson, *The Early History of Data Networks*, 81.

134 "IF YOU'LL ONLY JUST PROMISE": Charles Dibdin, "The Telegraph," in *The Songs of Charles Dibdin, Chronologically Arranged*, vol. 2 (London: G. H. Davidson, 1863), 69.

135 "THESE STATIONS ARE NOW SILENT": Taliaferro P. Shaffner, *The Telegraph Manual*, 31.

135 "ANYTHING THAT COULD BE THE SUBJECT": Gerard J. Holzmann and Björn Pehrson, *The Early History of Data Networks*, 56.

136 "ANYONE PERFORMING UNAUTHORIZED TRANSMISSIONS": Ibid., 91.

136 "WHAT CAN ONE EXPECT": Ibid., 93.

137 "OTHER BODIES THAT CAN BE AS EASILY ATTRACTED": J. J. Fahie, *A History of Electric Telegraphy to the Year 1837* (London: E. & F. N. Spon, 1884), 90.

137 "THIS SECONDARY OBJECT, THE ALARUM": E. A. Marland, *Early Electrical Communication* (London: Abelard-Schuman, 1964), 37.

137 HARRISON GRAY DYER TRIED SENDING SIGNALS: "An attempt made by Dyer to introduce his telegraph to general use encountered intense prejudice, and, becoming frightened at some of the manifestations of this feeling, he left the country." Chauncey M. Depew, *One Hundred Years of American Commerce* (New York: D. O. Haynes, 1895), 126.

139 "IT MUST BE EVIDENT TO THE MOST COMMON OBSERVER": John Pickering, *Lecture on Telegraphic Language* (Boston: Hilliard, Gray, 1833), 11.

139 "TELEGRAPHY IS AN ELEMENT OF POWER AND ORDER": Quoted in Daniel R. Headrick, *When Information Came of Age: Technologies of Knowledge in the Age of Reason and Revolution, 1700–1850* (Oxford: Oxford University Press, 2000), 200.

139 "IF THERE ARE NOW ESSENTIAL ADVANTAGES": John Pickering, *Lecture on Telegraphic Language*, 26.

140 "A SINGLE LETTER MAY BE INDICATED": Davy manuscript, quoted in J. J. Fahie, *A History of Electric Telegraphy to the Year 1837*, 351.

141 "I WORKED OUT EVERY POSSIBLE PERMUTATION": William Fothergill Cooke, *The Electric Telegraph: Was it Invented By Professor Wheatstone?* (London: W. H. Smith & Son, 1857), 27.

142 "SUPPOSE THE MESSAGE TO BE SENT": Alfred Vail, *The American Electro Magnetic Telegraph: With the Reports of Congress, and a Description of All Telegraphs Known, Employing Electricity Or Galvanism* (Philadelphia: Lea & Blanchard, 1847), 178.

142 "THE WORDY BATTLES WAGED": *Samuel F. B. Morse: His Letters and Journals*, vol. 2 (Boston: Houghton Mifflin, 1914), 21.

142 "THE MAILS IN OUR COUNTRY ARE TOO SLOW": Recalled by R. W. Habersham, *Samuel F. B. Morse: His Letters and Journals*.

142 "IT WOULD NOT BE DIFFICULT": Alfred Vail, *The American Electro Magnetic Telegraph*, 70.

144 "SEND A MESSENGER TO MR HARRIS": Andrew Wynter, "The Electric Telegraph," 128.

144 AT THE STROKE OF THE NEW YEAR: Laurence Turnbull, *The Electro-Magnetic Telegraph, With an Historical Account of Its Rise, Progress, and Present Condition* (Philadelphia: A. Hart, 1853), 87.

144 "IN THE GARB OF A KWAKER": "The Trial of John Tawell for the Murder of Sarah Hart by Poison, at the Aylesbury Spring Assizes, before Mr. Baron Parks, on March 12th 1845," in William Otter Woodall, *A Collection of Reports of Celebrated Trials* (London: Shaw & Sons, 1873).

144 "IN CONVEYING THE MOVES, THE ELECTRICITY TRAVELLED": John Timbs, *Stories of Inventors and Discoverers in Science and the Useful Arts* (London: Kent, 1860), 335.

145 "WHEN YOU CONSIDER THAT BUSINESS IS EXTREMELY DULL": Quoted in Tom Standage, *The Victorian Internet: The Remarkable Story of the Telegraph and the Nineteenth Century's On-Line Pioneers* (New York: Berkley, 1998), 55.

145 ALEXANDER JONES SENT HIS FIRST STORY: Alexander Jones, *Historical Sketch of the Electric Telegraph*, 121.

145 "THE FIRST INSTALMENT OF THE INTELLIGENCE": Charles Maybury Archer, ed., *The London Anecdotes: The Electric Telegraph*, vol. 1 (London: David Bogue, 1848), 85.

145 "THE RAPID AND INDISPENSABLE CARRIER": *Littell's Living Age* 6, no. 63 (26 July 1845): 194.

146 "SWIFTER THAN A ROCKET COULD FLY": Andrew Wynter, "The Electric Telegraph," 138.

146 "ALL IDEA OF CONNECTING EUROPE WITH AMERICA": Alexander Jones, *Historical Sketch of the Electric Telegraph*, 6.

146 "A RESULT SO PRACTICAL, YET SO INCONCEIVABLE": "The Atlantic Telegraph," *The New York Times*, 6 August 1858, 1.

147 DERBY, VERY DULL: Charles Maybury Archer, *The London Anecdotes*, 51.

147 "THE PHENOMENA OF THE ATMOSPHERE": Ibid., 73.

148 "ENABLES US TO SEND COMMUNICATIONS": George B. Prescott, *History, Theory, and Practice of the Electric Telegraph* (Boston: Ticknor and Fields, 1860), 5.

148 "FOR ALL PRACTICAL PURPOSES": *The New York Times*, 7 August 1858, 1.

148 "DISTANCE AND TIME HAVE BEEN SO CHANGED": Quoted in Iwan Rhys Morus, "'The Nervous System of Britain,'" 463.

149 LIEUTENANT CHARLES WILKES: Charles Wilkes to S. F. B. Morse, 13 June 1844, in Alfred Vail, *The American Electro Magnetic Telegraph*, 60.

149 "PROFESSOR MORSE'S TELEGRAPH IS NOT ONLY AN ERA": Quoted in Adam Frank, "Valdemar's Tongue, Poe's Telegraphy," *ELH* 72 (2005): 637.

149 "WHAT MIGHT NOT BE GATHERED SOME DAY": Andrew Wynter, "The Electric Telegraph," 133.

150 "MUCH IMPORTANT INFORMATION . . . CONSISTING OF MESSAGES": Alfred Vail, *The American Electro Magnetic Telegraph*, viii.

150 THE GIVING, PRINTING, STAMPING, OR OTHERWISE TRANSMITTING: Agreement between Cooke and Wheatstone, 1843, in William Fothergill Cooke, *The Electric Telegraph*, 46.

151 "THE DIFFICULTY OF FORMING A CLEAR CONCEPTION": "The Telegraph," *Harper's New Monthly Magazine*, 336.

151 "TELEGRAPHIC COMPANIES ARE RUNNING A RACE": Andrew Wynter, *Subtle Brains and Lissom Fingers: Being Some of the Chisel-Marks of Our Industrial and Scientific Progress* (London: Robert Hardwicke, 1863), 363.

151 "THEY STRING AN INSTRUMENT AGAINST THE SKY": Robert Frost, "The Line-Gang," 1920.

151 "A NET-WORK OF NERVES OF IRON WIRE": *Littell's Living Age* 6, no. 63 (26 July 1845): 194.

152 "THE WHOLE NET-WORK OF WIRES": "The Telegraph," *Harper's New Monthly Magazine*, 333.

152 "THE TIME IS NOT DISTANT": Andrew Wynter, *Subtle Brains and Lissom Fingers*, 371.

152 "THE TELEGRAPHIC STYLE BANISHES": Andrew Wynter, "The Electric Telegraph," 132.

152 "WE EARLY INVENTED A SHORT-HAND": Alexander Jones, *Historical Sketch of the Electric Telegraph*, 123.

153 "THE GREAT ADVANTAGE": Alfred Vail, *The American Electro Magnetic Telegraph*, 46.

154 *THE SECRET CORRESPONDING VOCABULARY*: Francis O. J. Smith, *The Secret Corresponding Vocabulary; Adapted for Use to Morse's Electro-Magnetic Telegraph: And Also in Conducting Written Correspondence, Transmitted by the Mails, or Otherwise* (Portland, Maine: Thurston, Ilsley, 1845).

155 *THE A B C UNIVERSAL COMMERCIAL ELECTRIC TELEGRAPH CODE*: Examples from William Clauson-Thue, *The A B C Universal Commercial Electric Telegraph Code*, 4th ed. (London: Eden Fisher, 1880).

157 "IT HAS BEEN BROUGHT TO THE AUTHOR'S KNOWLEDGE": Ibid., iv.

158 "TO GUARD AGAINST MISTAKES OR DELAYS": *Primrose v. Western Union Tel. Co.*, 154 U.S. 1 (1894); "Not Liable for Errors in Ciphers," *The New York Times*, 27 May 1894, 1.

159 AN ANONYMOUS LITTLE BOOK: Later reprinted, with the author identified, as John Wilkins, *Mercury: Or the Secret and Swift Messenger. Shewing, How a Man May With Privacy and Speed Communicate His Thoughts to a Friend At Any Distance*, 3rd ed. (London: John Nicholson, 1708).

159 "HE WAS A VERY INGENIOUS MAN": John Aubrey, *Brief Lives*, ed. Richard Barber (Woodbridge, Suffolk: Boydell Press, 1982), 324.

160 "HOW A MAN MAY WITH THE GREATEST SWIFTNESS": John Wilkins, *Mercury: Or the Secret and Swift Messenger*, 62.

161 "*WHATEVER IS CAPABLE OF A COMPETENT DIFFERENCE*": Ibid., 69.

161 THE CONTRIBUTION OF THE DILETTANTES: David Kahn, *The Codebreakers: The Story of Secret Writing* (London: Weidenfeld & Nicolson, 1968), 189.

162 "WE CAN SCARCELY IMAGINE A TIME": "A Few Words on Secret Writing," *Graham's Magazine*, July 1841; Edgar Allan Poe, *Essays and Reviews* (New York: Library of America, 1984), 1277.

162 "THE SOUL IS A CYPHER": *The Literati of New York* (1846), in Edgar Allan Poe, *Essays and Reviews*, 1172.

162 A BRIDGE BETWEEN SCIENCE AND THE OCCULT: Cf. William F. Friedman, "Edgar Allan Poe, Cryptographer," *American Literature* 8, no. 3 (1936): 266–80; Joseph Wood Krutch, *Edgar Allan Poe: A Study in Genius* (New York: Knopf, 1926).

162 A "KEY-ALPHABET" AND A "MESSAGE-ALPHABET": Lewis Carroll, "The Telegraph-Cipher," printed card 8 x 12 cm., Berol Collection, New York University Library.

162 "ONE OF THE MOST SINGULAR CHARACTERISTICS": Charles Babbage, *Passages from the Life of a Philosopher* (London: Longman, Green, Longman, Roberts, & Green, 1864), 235.

163 POLYALPHABETIC CIPHER KNOWN AS THE VIGENÈRE: Simon Singh, *The Code Book: The Secret History of Codes and Code-breaking* (London: Fourth Estate, 1999), 63 ff.

163 "THE VARIOUS PARTS OF THE MACHINERY": Dionysius Lardner, "Babbage's Calculating Engines," *Edinburgh Review* 59, no. 120 (1834): 315–17.

164 "NAME OF EVERYTHING WHICH IS *BOTH X* AND *Y*": De Morgan to Boole, 28 November 1847, in G. C. Smith, ed., *The Boole–De Morgan Correspondence 1842–1864* (Oxford: Clarendon Press, 1982), 25.

164 "NOW SOME *Z*S ARE NOT *X*S": De Morgan to Boole, draft, not sent, ibid., 27.

164 "IT IS SIMPLY A FACT": quoted by Samuel Neil, "The Late George Boole, LL.D., D.C.L." (1865), in James Gasser, ed., *A Boole Anthology: Recent and Classical Studies in the Logic of George Boole* (Dordrecht, Netherlands: Kluwer Academic, 2000), 16.

164 "THE RESPECTIVE INTERPRETATION OF THE SYMBOLS 0 AND 1": George Boole, *An Investigation of the Laws of Thought, on Which Are Founded the Mathematical Theories of Logic and Probabilities* (London: Walton & Maberly, 1854), 34.

164 "THAT LANGUAGE IS AN INSTRUMENT OF HUMAN REASON": Ibid., 24–25.

165 "UNCLEAN BEASTS ARE ALL": Ibid., 69.

166 "A WORD IS A TOOL FOR THINKING": "The Telegraph," *Harper's New Monthly Magazine*, 359.

166 "BABIES ARE ILLOGICAL": Lewis Carroll, *Symbolic Logic: Part I, Elementary* (London: Macmillan, 1896), 112 and 131. And cf. Steve Martin, *Born Standing Up: A Comic's Life* (New York: Simon & Schuster, 2007), 74.

166 "PURE MATHEMATICS WAS DISCOVERED BY BOOLE": Bertrand Russell, *Mysticism and Logic* (1918; reprinted Mineola, N.Y.: Dover, 2004), 57.

168 "THE PERFECT SYMMETRY OF THE WHOLE APPARATUS": James Clerk Maxwell, "The Telephone," Rede Lecture, Cambridge 1878, "illustrated with the aid of Mr. Gower's telephonic harp," in W. D. Niven, ed., *The Scientific Papers of James Clerk Maxwell*, vol. 2 (Cambridge: Cambridge University Press, 1890; repr. New York: Dover, 1965), 750.

169 GAYLORD AMOUNTED TO LITTLE MORE: "Small enough that if you walked a couple of blocks, you'd be in the countryside." Shannon interview with Anthony Liversidge, *Omni* (August 1987), in Claude Elwood Shannon, *Collected Papers*, ed. N. J. A. Sloane and Aaron D. Wyner (New York: IEEE Press, 1993), xx.

169 "THERE CAN BE NO DOUBT": "In the World of Electricity," *The New York Times*, 14 July 1895, 28.

170 THE MONTANA EAST LINE TELEPHONE ASSOCIATION: David B. Sicilia, "How the West Was Wired," *Inc.*, 15 June 1997.

171 "THE GOLD-BUG": 1843; *Complete Stories and Poems of Edgar Allan Poe* (New York: Doubleday, 1966), 71.

171 "CIRCUMSTANCES, AND A CERTAIN BIAS OF MIND": Ibid., 90.

171 " 'THINKING MACHINE' DOES HIGHER MATHEMATICS": *The New York Times*, 21 October 1927.

172 "A MATHEMATICIAN IS NOT A MAN": Vannevar Bush, "As We May Think," *The Atlantic* (July 1945).

172 UTTERLY CAPTIVATED BY THIS "COMPUTER": Shannon to Rudolf E. Kalman, 12 June 1987, Manuscript Division, Library of Congress.

175 "AUTOMATICALLY ADD TWO NUMBERS": Claude Shannon, "A Symbolic Analysis of Relay and Switching Circuits," *Transactions of the American Institute of Electrical Engineers* 57 (1938): 38–50.

175 HIS "QUEER ALGEBRA": Vannevar Bush to Barbara Burks, 5 January 1938, Manuscript Division, Library of Congress.

175 "AN ALGEBRA FOR THEORETICAL GENETICS": Claude Shannon, *Collected Papers*, 892.

176 EVALUATION FORTY YEARS LATER: Ibid., 921.

176 "OFF AND ON I HAVE BEEN WORKING ON AN ANALYSIS": Claude Shannon to Vannevar Bush, 16 February 1939, in Claude Shannon, *Collected Papers*, 455.

178 "A CERTAIN SCRIPT OF LANGUAGE": Leibniz to Jean Galloys, December 1678, in Martin Davis, *The Universal Computer: The Road from Leibniz to Turing* (New York: Norton, 2000), 16.

178 "HIGHLY ABSTRACT PROCESSES AND IDEAS": Alfred North Whitehead and Bertrand Russell, *Principia Mathematica*, vol. 1 (Cambridge: Cambridge University Press, 1910), 2.

179 "EPIMENIDES THE CRETAN SAID": Bertrand Russell, "Mathematical Logic Based on the Theory of Types," *American Journal of Mathematics* 30, no. 3 (July 1908): 222.

179 "IT WAS IN THE AIR": Douglas R. Hofstadter, *I Am a Strange Loop* (New York: Basic Books, 2007), 109.

180 "HENCE THE NAMES OF SOME INTEGERS": Alfred North Whitehead and Bertrand Russell, *Principia Mathematica*, vol. 1, 61.

180 DOES THE BARBER SHAVE HIMSELF: "The Philosophy of Logical Atomism" (1910), in Bertrand Russell, *Logic and Knowledge: Essays, 1901–1950* (London: Routledge, 1956), 261.

181 "LOOKED AT FROM THE OUTSIDE": Kurt Gödel, "On Formally Undecidable Propositions of *Principia Mathematica* and Related Systems I" (1931), in *Kurt Gödel: Collected Works*, vol. 1, ed. Solomon Feferman (New York: Oxford University Press, 1986), 146.

181 "A SCIENCE PRIOR TO ALL OTHERS": Kurt Gödel, "Russell's Mathematical Logic" (1944), in *Kurt Gödel: Collected Works*, vol. 2, 119.

182 "ONE CAN PROVE ANY THEOREM": Kurt Gödel, "On Formally Undecidable Propositions of *Principia Mathematica* and Related Systems I" (1931), 145.

184 "CONTRARY TO APPEARANCES, SUCH A PROPOSITION": Ibid., 151 n15.

184 "AMAZING FACT"—"THAT OUR LOGICAL INTUITIONS": Kurt Gödel, "Russell's Mathematical Logic" (1944), 124.

184 "A SUDDEN THUNDERBOLT FROM THE BLUEST OF SKIES": Douglas R. Hofstadter, *I Am a Strange Loop*, 166.

185 "THE IMPORTANT POINT": John von Neumann, "Tribute to Dr. Gödel" (1951), quoted in Steve J. Heims, *John von Neumann and Norbert Weiner* (Cambridge, Mass.: MIT Press, 1980), 133.

185 "IT MADE ME GLAD": Russell to Leon Henkin, 1 April 1963.

186 "MATHEMATICS CANNOT BE INCOMPLETE": Ludwig Wittgenstein, *Remarks on the Foundations of Mathematics* (Cambridge, Mass.: MIT Press, 1967), 158.

186 "RUSSELL EVIDENTLY MISINTERPRETS MY RESULT": Gödel to Abraham Robinson, 2 July 1973, in *Kurt Gödel: Collected Works*, vol. 5, 201.

186 HIS NAME WAS RECODED BY THE TELEPHONE COMPANY: Rebecca Goldstein, *Incompleteness: The Proof and Paradox of Kurt Gödel* (New York: Atlas, 2005), 207.

186 "YOUR BIO-MATHEMATICAL PROBLEMS": Hermann Weyl to Claude Shannon, 11 April 1940, Manuscript Division, Library of Congress.

187 "PROJECT 7": David A. Mindell, *Between Human and Machine: Feedback, Control, and Computing Before Cybernetics* (Baltimore: Johns Hopkins University Press, 2002), 289.

187 "APPLYING CORRECTIONS TO THE GUN CONTROL": Vannevar Bush, "Report of the National Defense Research Committee for the First Year of Operation, June 27, 1940, to June 28, 1941," Franklin D. Roosevelt Presidential Library and Museum, 19.

188 "THERE IS AN OBVIOUS ANALOGY": R. B. Blackman, H. W. Bode, and Claude E. Shannon, "Data Smoothing and Prediction in Fire-Control Systems," Summary Technical Report of Division 7, National Defense Research Committee,

vol. 1, *Gunfire Control* (Washington D.C.: 1946), 71–159 and 166–67; David A. Mindell, "Automation's Finest Hour: Bell Labs and Automatic Control in World War II," *IEEE Control Systems* 15 (December 1995): 72–80.

188 "BELL SEEMS TO BE SPENDING ALL HIS ENERGIES": Elisha Gray to A. L. Hayes, October 1875, quoted in Michael E. Gorman, *Transforming Nature: Ethics, Invention and Discovery* (Boston: Kluwer Academic, 1998), 165.

189 "I CAN SCARCE BELIEVE THAT A MAN": Albert Bigelow Paine, *In One Man's Life: Being Chapters from the Personal & Business Career of Theodore N. Vail* (New York: Harper & Brothers, 1921), 114.

189 "I FANCY THE DESCRIPTIONS WE GET": Marion May Dilts, *The Telephone in a Changing World* (New York: Longmans, Green, 1941), 11.

190 "NO MATTER TO WHAT EXTENT A MAN": "The Telephone Unmasked," *The New York Times*, 13 October 1877, 4.

190 "THE SPEAKER TALKS TO THE TRANSMITTER": *The Scientific Papers of James Clerk Maxwell*, ed. W. D. Niven, vol. 2 (Cambridge: Cambridge University Press, 1890; repr. New York: Dover, 1965), 744.

191 "WHAT THE TELEGRAPH ACCOMPLISHED IN YEARS": *Scientific American*, 10 January 1880.

192 "INSTANTANEOUS COMMUNICATION ACROSS SPACE": *Telephones: 1907*, Special Reports, Bureau of the Census, 74.

192 "IT MAY SOUND RIDICULOUS TO SAY THAT BELL": Quoted in Ithiel de Sola Pool, ed., *The Social Impact of the Telephone* (Cambridge, Mass.: MIT Press, 1977), 140.

193 "AFFECTATIONS OF THE SAME SUBSTANCE": J. Clerk Maxwell, "A Dynamical Theory of the Electromagnetic Field," *Philosophical Transactions of the Royal Society* 155 (1865): 459.

194 THE FIRST TELEPHONE OPERATORS: Michèle Martin, *"Hello, Central?": Gender, Technology, and Culture in the Formation of Telephone Systems* (Montreal: McGill–Queen's University Press, 1991), 55.

195 "THEY ARE STEADIER, DO NOT DRINK BEER": Proceedings of the National Telephone Exchange Association, 1881, in Frederick Leland Rhodes, *Beginnings of Telephony* (New York: Harper & Brothers, 1929), 154.

195 "THE ACTION OF STRETCHING HER ARMS": Quoted in Peter Young, *Person to Person: The International Impact of the Telephone* (Cambridge: Granta, 1991), 65.

195 "THE TELEPHONE REMAINS THE ACME": Herbert N. Casson, *The History of the Telephone* (Chicago: A. C. McClurg, 1910), 296.

196 "ANY TWO OF THAT LARGE NUMBER": John Vaughn, "The Thirtieth Anniversary of a Great Invention," *Scribner's* 40 (1906): 371.

196 A MONSTER OF 2 MILLION SOLDERED PARTS: G. E. Schindler, Jr., ed., *A History of Engineering and Science in the Bell System: Switching Technology 1925–1975* (Bell Telephone Laboratories, 1982).

196 "FOR THE MATHEMATICIAN, AN ARGUMENT": T. C. Fry, "Industrial Mathematics," *Bell System Technical Journal* 20 (July 1941): 255.

197 "there was sputtering and bubbling": Bell Canada Archives, quoted in Michèle Martin, *"Hello, Central?"* 23.

199 "speed of transmission of intelligence": H. Nyquist, "Certain Factors Affecting Telegraph Speed," *Bell System Technical Journal* 3 (April 1924): 332.

200 "information is a very elastic term": R. V. L. Hartley, "Transmission of Information," *Bell System Technical Journal* 7 (July 1928): 536.

201 "for example, in the sentence, 'apples are red'": Ibid.

202 "by the speed of transmission of intelligence is meant": H. Nyquist, "Certain Factors Affecting Telegraph Speed," 333.

203 "the capacity of a system to transmit": R. V. L. Hartley, "Transmission of Information," 537.

7. INFORMATION THEORY

204 "perhaps coming up with a theory": Jon Barwise, "Information and Circumstance," *Notre Dame Journal of Formal Logic* 27, no. 3 (1986): 324.

204 said nothing to each other about their work: Shannon interview with Robert Price: "A Conversation with Claude Shannon: One Man's Approach to Problem Solving," *IEEE Communications Magazine* 22 (1984): 125; cf. Alan Turing to Claude Shannon, 3 June 1953, Manuscript Division, Library of Congress.

205 "no, i'm not interested in developing a *powerful* brain": Andrew Hodges, *Alan Turing: The Enigma* (London: Vintage, 1992), 251.

207 "a confirmed solitary": Max H. A. Newman to Alonzo Church, 31 May 1936, quoted in Andrew Hodges, *Alan Turing*, 113.

207 "the justification . . . lies in the fact": Alan M. Turing, "On Computable Numbers, with an Application to the *Entscheidungsproblem*," *Proceedings of the London Mathematical Society* 42 (1936): 230–65.

207 "it was only by turing's work": Kurt Gödel to Ernest Nagel, 1957, in *Kurt Gödel: Collected Works*, vol. 5, ed. Solomon Feferman (New York: Oxford University Press, 1986), 147.

208 "you see . . . the funny little rounds": letter from Alan Turing to his mother and father, summer 1923, AMT/K/1/3, Turing Digital Archive, http://www.turingarchive.org.

208 "in elementary arithmetic the two-dimensional character": Alan M. Turing, "On Computable Numbers," 230–65.

212 "the thing hinges on getting this halting inspector": "On the Seeming Paradox of Mechanizing Creativity," in Douglas R. Hofstadter, *Metamagical Themas: Questing for the Essence of Mind and Pattern* (New York: Basic Books, 1985), 535.

212 "it used to be supposed in science": "The Nature of Spirit," unpublished essay, 1932, in Andrew Hodges, *Alan Turing*, 63.

212 "ONE CAN PICTURE AN INDUSTRIOUS AND DILIGENT CLERK": Herbert B. Enderton, "Elements of Recursion Theory," in Jon Barwise, *Handbook of Mathematical Logic* (Amsterdam: North Holland, 1977), 529.

213 "A LOT OF PARTICULAR AND INTERESTING CODES": Alan Turing to Sara Turing, 14 October 1936, quoted in Andrew Hodges, *Alan Turing*, 120.

215 "THE ENEMY KNOWS THE SYSTEM BEING USED": "Communication Theory of Secrecy Systems" (1948), in Claude Elwood Shannon, *Collected Papers*, ed. N. J. A. Sloane and Aaron D. Wyner (New York: IEEE Press, 1993), 90.

216 "FROM THE POINT OF VIEW OF THE CRYPTANALYST": Ibid., 113.

216 "THE MERE SOUNDS OF SPEECH": Edward Sapir, *Language: An Introduction to the Study of Speech* (New York: Harcourt, Brace, 1921), 21.

216 "*D* MEASURES, IN A SENSE, HOW MUCH A TEXT": "Communication Theory of Secrecy Systems," in Claude Shannon, *Collected Papers*, 85.

218 "THE ENEMY IS NO BETTER OFF": Ibid., 97.

219 "THE 'MEANING' OF A MESSAGE IS GENERALLY IRRELEVANT": "Communication Theory—Exposition of Fundamentals," *IRE Transactions on Information Theory*, no. 1 (February 1950), in Claude Shannon, *Collected Papers*, 173.

221 "WHAT GIBBS DID FOR PHYSICAL CHEMISTRY": Warren Weaver letter to Claude Shannon, 27 January 1949, Manuscript Division, Library of Congress.

221 "SOMETHING OF A DELAYED ACTION BOMB": John R. Pierce, "The Early Days of Information Theory," *IEEE Transactions on Information Theory* 19, no. 1 (1973): 4.

221 "THE FUNDAMENTAL PROBLEM OF COMMUNICATION": Claude Elwood Shannon and Warren Weaver, *The Mathematical Theory of Communication* (Urbana: University of Illinois Press, 1949), 31.

225 "THIS IS ALREADY DONE TO A LIMITED EXTENT": Ibid., 11.

225 LANDMARK 1943 PAPER: "Stochastic Problems in Physics and Astronomy," *Reviews of Modern Physics* 15, no. 1 (January 1943), 1.

226 BOOK NEWLY PUBLISHED FOR SUCH PURPOSES: M. G. Kendall and B. Babbington Smith, *Table of Random Sampling Numbers* (Cambridge: Cambridge University Press, 1939). Kendall and Smith used a "randomizing machine"— a rotating disc with the ten digits illuminated at irregular intervals by a neon light. An earlier effort, by L. H. C. Tippett in 1927, drew 41,000 digits from population census reports, also noting only the last digit of any number. A slightly naïve article in the *Mathematical Gazette* argued in 1944 that machines were unnecessary: "In a modern community, there is, it seems, no need to construct a randomising machine, for so many features of sociological life exhibit randomness. . . . Thus a set of random numbers serviceable for all ordinary purposes can be constructed by reading the registration numbers of cars as they pass us in the street, for cars though numbered serially move about the streets in non-serial fashion, obvious errors, such as those of reading the numbers seen every morning on the way to the station along one's own road when Mr. Smith's

car is always standing outside No. 49 being, of course, avoided." Frank Sandon, "Random Sampling Numbers," *The Mathematical Gazette* 28 (December 1944): 216.

227 TABLES CONSTRUCTED FOR USE BY CODE BREAKERS: Fletcher Pratt, *Secret and Urgent: The Story of Codes and Ciphers* (Garden City, N.Y.: Blue Ribbon, 1939).

228 "HOW MUCH 'CHOICE' IS INVOLVED": Claude Elwood Shannon and Warren Weaver, *The Mathematical Theory of Communication*, 18.

229 "BINARY DIGITS, OR MORE BRIEFLY, *BITS*": "A word suggested by J. W. Tukey," he added. John Tukey, the statistician, had been a roommate of Richard Feynman's at Princeton and spent some time working at Bell Labs after the war.

229 "MORE ERRATIC AND UNCERTAIN": Claude Shannon, "Prediction and Entropy of Printed English," *Bell System Technical Journal* 30 (1951): 50, in Claude Shannon, *Collected Papers*, 94.

230 "TO MAKE THE CHANCE OF ERROR": quoted in M. Mitchell Waldrop, "Reluctant Father of the Digital Age," *Technology Review* (July–August 2001): 64–71.

231 "IT'S A SOLID-STATE AMPLIFIER": Shannon interview with Anthony Liversidge, *Omni* (August 1987), in Claude Shannon, *Collected Papers*, xxiii.

231 "BITS STORAGE CAPACITY": Handwritten note, 12 July 1949, Manuscript Division, Library of Congress.

8. THE INFORMATIONAL TURN

233 "IT IS PROBABLY DANGEROUS TO USE THIS THEORY": Heinz von Foerster, ed., *Cybernetics: Circular Causal and Feedback Mechanisms in Biological and Social Systems: Transactions of the Seventh Conference, March 23–24, 1950* (New York: Josiah Macy, Jr. Foundation, 1951), 155.

233 "AND IT IS NOT ALWAYS CLEAR": J. J. Doob, review (untitled), *Mathematical Reviews* 10 (February 1949): 133.

233 "AT FIRST GLANCE, IT MIGHT APPEAR": A. Chapanis, review (untitled), *Quarterly Review of Biology* 26, no. 3 (September 1951): 321.

233 "SHANNON DEVELOPS A CONCEPT OF *INFORMATION*": Arthur W. Burks, review (untitled), *Philosophical Review* 60, no. 3 (July 1951): 398.

234 SHORT REVIEW OF WIENER'S BOOK: *Proceedings of the Institute of Radio Engineers* 37 (1949), in Claude Elwood Shannon, *Collected Papers*, ed. N. J. A. Sloane and Aaron D. Wyner (New York: IEEE Press, 1993), 872.

235 "WIENER'S HEAD WAS FULL": John R. Pierce, "The Early Days of Information Theory," *IEEE Transactions on Information Theory* 19, no. 1 (1973): 5.

235 THE WORD HE TOOK FROM THE GREEK: André-Marie Ampère had used the word, *cybernétics*, in 1834 (*Essai sur la philosophie des sciences*).

235 "A LAD WHO HAS BEEN PROUDLY TERMED": "Boy of 14 College Graduate," *The New York Times*, 9 May 1909, 1.

236 "AN INFANT PRODIGY NAMED WIENER": Bertrand Russell to Lucy Donnelly, 19 October 1913, quoted in Steve J. Heims, *John von Neumann and Norbert Wiener* (Cambridge, Mass.: MIT Press, 1980), 18.

236 "HE IS AN ICEBERG": Norbert Wiener to Leo Wiener, 15 October 1913, quoted in Flo Conway and Jim Siegelman, *Dark Hero of the Information Age: In Search of Norbert Weiner, the Father of Cybernetics* (New York: Basic Books, 2005), 30.

237 "WE ARE SWIMMING UPSTREAM AGAINST A GREAT TORRENT": Norbert Wiener, *I Am a Mathematician: The Later Life of a Prodigy* (Cambridge, Mass.: MIT Press, 1964), 324.

238 "A NEW INTERPRETATION OF MAN": Ibid., 375.

238 "ANY CHANGE OF AN ENTITY": Arturo Rosenblueth et al., "Behavior, Purpose and Teleology," *Philosophy of Science* 10 (1943): 18.

238 "THAT IT WAS NOT SOME PARTICULAR PHYSICAL THING": Quoted in Warren S. McCulloch, "Recollections of the Many Sources of Cybernetics," *ASC Forum* 6, no. 2 (1974).

239 "THEY ARE GROWING WITH FEARFUL SPEED": "In Man's Image," *Time*, 27 December 1948.

240 "THE ALGEBRA OF LOGIC *PAR EXCELLENCE*": Norbert Wiener, *Cybernetics: Or Control and Communication in the Animal and the Machine*, 2nd ed. (Cambridge, Mass.: MIT Press, 1961), 118.

241 "TRAFFIC PROBLEMS AND OVERLOADING": Ibid., 132.

241 "FOR THE FIRST TIME IN THE HISTORY OF SCIENCE": Warren S. McCulloch, "Through the Den of the Metaphysician," *British Journal for the Philosophy of Science* 5, no. 17 (1954): 18.

242 A NOAH'S ARK RULE: Warren S. McCulloch, "Recollections of the Many Sources of Cybernetics," 11.

242 WIENER TOLD THEM THAT ALL THESE SCIENCES: Steve J. Heims, *The Cybernetics Group* (Cambridge, Mass.: MIT Press, 1991), 22.

243 "THE SUBJECT AND THE GROUP": Heinz von Foerster, ed., *Transactions of the Seventh Conference*, 11.

243 "TO SAY, AS THE PUBLIC PRESS SAYS": Ibid., 12.

244 "I HAVE NOT BEEN ABLE TO PREVENT THESE REPORTS": Ibid., 18.

244 IT WAS, AT BOTTOM, A PERFECTLY ORDINARY SITUATION: Jean-Pierre Dupuy, *The Mechanization of the Mind: On the Origins of Cognitive Science*, trans. M. B. DeBevoise (Princeton, N.J.: Princeton University Press, 2000), 89.

244 COULD PROPERLY BE DESCRIBED AS ANALOG OR DIGITAL: Heinz von Foerster, ed., *Transactions of the Seventh Conference*, 13.

244 "THE STATE OF THE NERVE CELL WITH NO MESSAGE IN IT": Ibid., 20.

245 "IN THIS WORLD IT SEEMS BEST": Warren S. McCulloch and John Pfeiffer, "Of Digital Computers Called Brains," *Scientific Monthly* 69, no. 6 (1949): 368.

245 HE WAS WORKING ON AN IDEA FOR QUANTIZING SPEECH: J. C. R. Licklider, interview by William Aspray and Arthur Norberg, 28 October 1988, Charles Babbage Institute, University of Minnesota, http://special.lib.umn.edu/cbi/oh/pdf.phtml?id=180 (accessed 6 June 2010).

245 "MATHEMATICIANS ARE ALWAYS DOING THAT": Heinz von Foerster, ed., *Transactions of the Seventh Conference*, 66.

246 "YES!" INTERRUPTED WIENER: Ibid., 92.

246 "IF YOU TALK ABOUT ANOTHER KIND OF INFORMATION": Ibid., 100.

246 "IT MIGHT, FOR EXAMPLE, BE A RANDOM SEQUENCE": Ibid., 123.

247 "I WOULDN'T CALL THAT RANDOM, WOULD YOU?": Ibid., 135.

248 "I WANTED TO CALL THE WHOLE": quoted in Flo Conway and Jim Siegelman, *Dark Hero of the Information Age*, 189.

248 "I'M THINKING OF THE OLD MAYA TEXTS": Heinz von Foerster, ed., *Transactions of the Seventh Conference*, 143.

248 "INFORMATION CAN BE CONSIDERED AS ORDER": Heinz von Foerster, ed., *Cybernetics: Circular Causal and Feedback Mechanisms in Biological and Social Systems: Transactions of the Eighth Conference, March 15–16, 1951* (New York: Josiah Macy, Jr. Foundation, 1952), xiii.

249 HIS NEIGHBOR SAID: Heinz von Foerster, ed., *Transactions of the Seventh Conference*, 151.

249 "WHEN THE MACHINE WAS TURNED OFF": Heinz von Foerster, ed., *Transactions of the Eighth Conference*, 173.

250 "IT BUILDS UP A COMPLETE PATTERN OF INFORMATION": "Computers and Automata," in Claude Shannon, *Collected Papers*, 706.

250 "WHEN IT ARRIVES AT A, IT REMEMBERS": Heinz von Foerster, ed. *Transactions of the Eighth Conference*, 175.

251 "LIKE A MAN WHO KNOWS THE TOWN": Ibid., 180.

252 "IN REALITY IT IS THE MAZE WHICH REMEMBERS": Quoted in Roberto Cordeschi, *The Discovery of the Artificial: Behavior, Mind, and Machines Before and Beyond Cybernetics* (Dordrecht, Netherlands: Springer, 2002), 163.

252 FOUND RESEARCHERS TO BE "WELL-INFORMED": Norbert Wiener, *Cybernetics*, 23.

253 "ABOUT FIFTEEN PEOPLE WHO HAD WIENER'S IDEAS": John Bates to Grey Walter, quoted in Owen Holland, "The First Biologically Inspired Robots," *Robotica* 21 (2003): 354.

253 HALF PRONOUNCED IT RAY-SHE-OH: Philip Husbands and Owen Holland, "The Ratio Club: A Hub of British Cybernetics," in *The Mechanical Mind in History* (Cambridge, Mass.: MIT Press, 2008), 103.

253 "A BRAIN CONSISTING OF RANDOMLY CONNECTED IMPRESSIONAL SYNAPSES": Ibid., 110.

253 "THINK OF THE BRAIN AS A TELEGRAPHIC RELAY": "Brain and Behavior," *Comparative Psychology Monograph*, Series 103 (1950), in Warren S. McCulloch, *Embodiments of Mind* (Cambridge, Mass.: MIT Press, 1965), 307.

253 "I PROPOSE TO CONSIDER THE QUESTION": Alan M. Turing, "Computing Machinery and Intelligence," *Minds and Machines* 59, no. 236 (1950): 433–60.

254 "THE PRESENT INTEREST IN 'THINKING MACHINES'": Ibid., 436.

255 "SINCE BABBAGE'S MACHINE WAS NOT ELECTRICAL": Ibid., 439.

255 "IN THE CASE THAT THE FORMULA IS NEITHER PROVABLE NOR DISPROVABLE":

Alan M. Turing, "Intelligent Machinery, A Heretical Theory," unpublished lecture, c. 1951, in Stuart M. Shieber, ed., *The Turing Test: Verbal Behavior as the Hallmark of Intelligence* (Cambridge, Mass.: MIT Press, 2004), 105.

256 THE ORIGINAL QUESTION, "CAN MACHINES THINK?": Alan M. Turing, "Computing Machinery and Intelligence," 442.

256 "THE IDEA OF A MACHINE THINKING": Claude Shannon to C. Jones, 16 June 1952, Manuscript Div., Library of Congress, by permission of Mary E. Shannon.

256 "*PSYCHOLOGIE* IS A DOCTRINE WHICH SEARCHES OUT": Translated in William Harvey, *Anatomical Exercises Concerning the Motion of the Heart and Blood* (London, 1653), quoted in "psychology, *n*," draft revision Dec. 2009, *OED Online*, Oxford University Press, http://dictionary.oed.com/cgi/entry/50191636.

257 "THE SCIENCE OF MIND, IF IT CAN BE CALLED A SCIENCE": *North British Review* 22 (November 1854), 181.

257 "A LOATHSOME, DISTENDED, TUMEFIED, BLOATED, DROPSICAL MASS": William James to Henry Holt, 9 May 1890, quoted in Robert D. Richardson, *William James: In the Maelstrom of American Modernism* (New York: Houghton Mifflin, 2006), 298.

258 "YOU TALK ABOUT MEMORY": George Miller, dialogue with Jonathan Miller, in Jonathan Miller, *States of Mind* (New York: Pantheon, 1983), 22.

259 "NEW CONCEPTS OF THE NATURE AND MEASURE": Homer Jacobson, "The Informational Capacity of the Human Ear," *Science* 112 (4 August 1950): 143–44; "The Informational Capacity of the Human Eye," *Science* 113 (16 March 1951): 292–93.

259 A GROUP IN 1951 TESTED THE LIKELIHOOD: G. A. Miller, G. A. Heise, and W. Lichten, "The Intelligibility of Speech as a Function of the Context of the Test Materials," *Journal of Experimental Psychology* 41 (1951): 329–35.

260 "THE DIFFERENCE BETWEEN A DESCRIPTION": Donald E. Broadbent, *Perception and Communication* (Oxford: Pergamon Press, 1958), 31.

260 "THE MAGICAL NUMBER SEVEN": *Psychological Review* 63 (1956): 81–97.

262 "THOSE WHO TAKE THE INFORMATIONAL TURN": Frederick Adams, "The Informational Turn in Philosophy," *Minds and Machines* 13 (2003): 495.

262 THE MIND CAME IN ON THE BACK: Jonathan Miller, *States of Mind*, 26.

262 "I THINK THAT THIS PRESENT CENTURY": Claude Shannon, "The Transfer of Information," talk presented at the 75th anniversary of the University of Pennsylvania Graduate School of Arts and Sciences, Manuscript Division, Library of Congress. Reprinted by permission of Mary E. Shannon.

263 "OUR FELLOW SCIENTISTS IN MANY DIFFERENT FIELDS": "The Bandwagon," in Claude Shannon, *Collected Papers*, 462.

263 "OUR CONSENSUS HAS NEVER BEEN UNANIMOUS": quoted in Steve J. Heims, *The Cybernetics Group*, 277.

264 THIS WAS CHANGED FOR PUBLICATION: Notes by Neil J. A. Sloane and Aaron D. Wyner in Claude Shannon, *Collected Papers*, 882.

264 "of course, is of no importance": Claude E. Shannon, "Programming a Computer for Playing Chess," first presented at National IRE Convention, 9 March 1949, in Claude Shannon, *Collected Papers*, 637; and "A Chess-Playing Machine," *Scientific American* (February 1950), in Claude Shannon, *Collected Papers*, 657.

265 visited the american champion: Edward Lasker to Claude Shannon, 7 February 1949, Manuscript Division, Library of Congress.

265 "learning chess player": Claude Shannon to C. J. S. Purdy, 28 August 1952, Manuscript Div., Library of Congress, by permission of Mary E. Shannon.

266 scientific aspects of juggling: Unpublished, in Claude Shannon, *Collected Papers*, 861. The actual lines, from Cummings's poem "voices to voices, lip to lip," are: "who cares if some oneeyed son of a bitch / invents an instrument to measure Spring with?"

266 a machine that would repair itself: Claude Shannon to Irene Angus, 8 August 1952, Manuscript Division, Library of Congress.

266 "what happens if you switch on one of these mechanical computers": Robert McCraken, "The Sinister Machines," *Wyoming Tribune*, March 1954.

267 "information theory, photosynthesis, and religion": Peter Elias, "Two Famous Papers," *IRE Transactions on Information Theory* 4, no. 3 (1958): 99.

267 "we have heard of 'entropies'": E. Colin Cherry, *On Human Communication* (Cambridge, Mass.: MIT Press, 1957), 214.

9. ENTROPY AND ITS DEMONS

269 "thought interferes with the probability of events": David L. Watson, "Entropy and Organization," *Science* 72 (1930): 222.

269 the rumor at bell labs: Robert Price, "A Conversation with Claude Shannon: One Man's Approach to Problem Solving," *IEEE Communications Magazine* 22 (1984): 124.

269 "the theoretical study of the steam engine": For example, J. Johnstone, "Entropy and Evolution," *Philosophy* 7 (July 1932): 287.

270 maxwell turned about-face: James Clerk Maxwell, *Theory of Heat*, 2nd ed. (London: Longmans, Green, 1872), 186; 8th edition (London: Longmans, Green, 1891), 189 n.

271 "you can't win": Peter Nicholls and David Langford, eds., *The Science in Science Fiction* (New York: Knopf, 1983), 86.

271 "although mechanical energy is *indestructible*": Lord Kelvin (William Thomson), "Physical Considerations Regarding the Possible Age of the Sun's Heat," lecture at the Meeting of the British Association at Manchester, September 1861, in *Philosophical Magazine* 152 (February 1862): 158.

271 "IN CONSIDERING THE CONVERSION OF PSYCHICAL ENERGY": Sigmund Freud, "From the History of an Infantile Neurosis," 1918*b*, 116, in *The Standard Edition of the Complete Psychological Works of Sigmund Freud* (London: Hogarth Press, 1955).

272 "CONFUSION, LIKE THE CORRELATIVE TERM ORDER": James Clerk Maxwell, "Diffusion," written for the ninth edition of *Encyclopaedia Britannica*, in *The Scientific Papers of James Clerk Maxwell*, ed. W. D. Niven, vol. 2 (Cambridge: Cambridge University Press, 1890; repr. New York: Dover, 1965), 646.

273 "TIME FLOWS ON, NEVER COMES BACK": Léon Brillouin, "Life, Thermodynamics, and Cybernetics" (1949), in Harvey S. Leff and Andrew F. Rex, eds., *Maxwell's Demon 2: Entropy, Classical and Quantum Information, Computing* (Bristol, U.K.: Institute of Physics, 2003), 77.

274 "THE ACCIDENTS OF LIFE": Richard Feynman, *The Character of Physical Law* (New York: Modern Library, 1994), 106.

274 "*MORAL.* THE 2ND LAW OF THERMODYNAMICS": James Clerk Maxwell to John William Strutt, 6 December 1870, in Elizabeth Garber, Stephen G. Brush, and C. W. F. Everitt, eds., *Maxwell on Heat and Statistical Mechanics: On "Avoiding All Personal Enquiries" of Molecules* (London: Associated University Presses, 1995), 205.

274 "THE ODDS AGAINST A PIECE OF CHALK": Quoted by Andrew Hodges, "What Did Alan Turing Mean by 'Machine,'" in Philip Husbands et al., *The Mechanical Mind in History* (Cambridge, Mass.: MIT Press, 2008), 81.

275 "AND YET NO WORK HAS BEEN DONE": James Clerk Maxwell to Peter Guthrie Tait, 11 December 1867, in *The Scientific Letters and Papers of James Clerk Maxwell*, ed. P. M. Harman, vol. 3 (Cambridge: Cambridge University Press, 2002), 332.

276 "HE DIFFERS FROM REAL LIVING ANIMALS": Royal Institution Lecture, 28 February 1879, *Proceedings of the Royal Institution* 9 (1880): 113, in William Thomson, *Mathematical and Physical Papers*, vol. 5 (Cambridge: Cambridge University Press, 1911), 21.

276 "INFINITE SWARMS OF ABSURD LITTLE MICROSCOPIC IMPS": "Editor's Table," *Popular Science Monthly* 15 (1879): 412.

276 "CLERK MAXWELL'S DEMON": Henry Adams to Brooks Adams, 2 May 1903, in *Henry Adams and His Friends: A Collection of His Unpublished Letters*, ed. Harold Cater (Boston: Houghton Mifflin, 1947), 545.

277 "INFINITELY SUBTLE SENSES": Henri Poincaré, *The Foundations of Science*, trans. George Bruce Halsted (New York: Science Press, 1913), 152.

278 "NOW WE MUST NOT INTRODUCE DEMONOLOGY": James Johnstone, *The Philosophy of Biology* (Cambridge: Cambridge University Press, 1914), 118.

279 "IF WE VIEW THE EXPERIMENTING MAN": Leó Szilárd, "On the Decrease of Entropy in a Thermodynamic System by the Intervention of Intelligent Beings," trans. Anatol Rapoport and Mechthilde Knoller, from Leó Szilárd, *"Über Die*

Entropieverminderung in Einem Thermodynamischen System Bei Eingriffen Intelligenter Wesen," *Zeitschrift für Physik* 53 (1929): 840–56, in Harvey S. Leff and Andrew F. Rex, eds., *Maxwell's Demon 2*, 111.

279 "THINKING GENERATES ENTROPY": Quoted in William Lanouette, *Genius in the Shadows* (New York: Scribner's, 1992), 64.

280 "I THINK ACTUALLY SZILÁRD": Shannon interview with Friedrich-Wilhelm Hagemeyer, 1977, quoted in Erico Mariu Guizzo, "The Essential Message: Claude Shannon and the Making of Information Theory" (Master's thesis, Massachusetts Institute of Technology, 2004).

281 "I CONSIDER HOW MUCH INFORMATION IS *PRODUCED*": Claude Shannon to Norbert Wiener, 13 October 1948, Massachusetts Institute of Technology Archives.

282 "THAT SOME OF US SHOULD VENTURE TO EMBARK": Erwin Schrödinger, *What Is Life?*, reprint ed. (Cambridge: Cambridge University Press, 1967), 1.

282 "SCHRÖDINGER'S BOOK BECAME A KIND OF *UNCLE TOM'S CABIN*": Gunther S. Stent, "That Was the Molecular Biology That Was," *Science* 160, no. 3826 (1968): 392.

282 "WHEN IS A PIECE OF MATTER SAID TO BE ALIVE?": Erwin Schrödinger, *What Is Life?*, 69.

283 "THE STABLE STATE OF AN ENZYME": Norbert Wiener, *Cybernetics: Or Control and Communication in the Animal and the Machine*, 2nd ed. (Cambridge, Mass.: MIT Press, 1961), 58.

283 "TO PUT IT LESS PARADOXICALLY": Erwin Schrödinger, *What Is Life?*, 71.

284 "A COMPLETE (DOUBLE) COPY OF THE CODE-SCRIPT": Ibid., 23.

284 "IT SEEMS NEITHER ADEQUATE NOR POSSIBLE": Ibid., 28.

285 "*WE BELIEVE A GENE—OR PERHAPS THE WHOLE CHROMOSOME FIBER*": Ibid., 61.

285 "THE DIFFERENCE IN STRUCTURE": Ibid., 5 (my emphasis).

285 "THE LIVING ORGANISM HEALS ITS OWN WOUNDS": Léon Brillouin, "Life, Thermodynamics, and Cybernetics," 84.

286 HE WROTE THIS IN 1950: Léon Brillouin, "Maxwell's Demon Cannot Operate: Information and Entropy," in Harvey S. Leff and Andrew F. Rex, eds., *Maxwell's Demon 2*, 123.

286 "MAXWELL'S DEMON DIED AT THE AGE OF 62": Peter T. Landsberg, *The Enigma of Time* (Bristol: Adam Hilger, 1982), 15.

10. LIFE'S OWN CODE

287 "WHAT LIES AT THE HEART OF EVERY LIVING THING": Richard Dawkins, *The Blind Watchmaker* (New York: Norton, 1986), 112.

287 "THE BIOLOGIST MUST BE ALLOWED": W. D. Gunning, "Progression and Retrogression," *The Popular Science Monthly* 8 (December 1875): 189, n1.

287 "THE MOST NAÏVE AND OLDEST CONCEPTION": Wilhelm Johannsen, "The Genotype Conception of Heredity," *American Naturalist* 45, no. 531 (1911): 130.

288 IT MUST BE QUANTIZED: "Discontinuity and constant differences between the 'genes' are the quotidian bread of Mendelism," *American Naturalist* 45, no. 531 (1911): 147.

288 "THE MINIATURE CODE SHOULD PRECISELY CORRESPOND": Erwin Schrödinger, *What Is Life?*, reprint ed. (Cambridge: Cambridge University Press, 1967), 62.

289 SOME OF THE PHYSICISTS NOW TURNING TO BIOLOGY: Henry Quastler, ed., *Essays on the Use of Information Theory in Biology* (Urbana: University of Illinois Press, 1953).

289 "A LINEAR CODED TAPE OF INFORMATION": Sidney Dancoff to Henry Quastler, 31 July 1950, quoted in Lily E. Kay, *Who Wrote the Book of Life: A History of the Genetic Code* (Stanford, Calif.: Stanford University Press, 2000), 119.

289 NUMBER OF BITS REPRESENTED BY A SINGLE BACTERIUM: Henry Linschitz, "The Information Content of a Bacterial Cell," in Henry Quastler, ed., *Essays on the Use of Information Theory in Biology*, 252.

289 "HYPOTHETICAL INSTRUCTIONS TO BUILD AN ORGANISM": Sidney Dancoff and Henry Quastler, "The Information Content and Error Rate of Living Things," in Henry Quastler, ed., *Essays on the Use of Information Theory in Biology*, 264.

290 "THE ESSENTIAL COMPLEXITY OF A SINGLE CELL": Ibid., 270.

290 AN ODD LITTLE LETTER: Boris Ephrussi, Urs Leopold, J. D. Watson, and J. J. Weigle, "Terminology in Bacterial Genetics," *Nature* 171 (18 April 1953): 701.

291 MEANT AS A JOKE: Cf. Sahotra Sarkar, *Molecular Models of Life* (Cambridge, Mass.: MIT Press, 2005); Lily E. Kay, *Who Wrote the Book of Life?*, 58; Harriett Ephrussi-Taylor to Joshua Lederberg, 3 September 1953, and Lederberg annotation 30 April 2004, in Lederberg papers, http://profiles.nlm.nih.gov/ BB/A/J/R/R/ (accessed 22 January 2009); and James D. Watson, *Genes, Girls, and Gamow: After the Double Helix* (New York: Knopf, 2002), 12.

291 GENES MIGHT LIE IN A DIFFERENT SUBSTANCE: In retrospect, everyone understood that this had been proven in 1944, by Oswald Avery at Rockefeller University. Not many researchers were convinced at the time, however.

292 "ONE OF THE MOST COY STATEMENTS": Gunther S. Stent, "DNA," *Daedalus* 99 (1970): 924.

292 "IT HAS NOT ESCAPED OUR NOTICE": James D. Watson and Francis Crick, "A Structure for Deoxyribose Nucleic Acid," *Nature* 171 (1953): 737.

292 "IT FOLLOWS THAT IN A LONG MOLECULE": James D. Watson and Francis Crick, "Genetical Implications of the Structure of Deoxyribonucleic Acid," *Nature* 171 (1953): 965.

293 "DEAR DRS. WATSON & CRICK": George Gamow to James D. Watson and

Francis Crick, 8 July 1953, quoted in Lily E. Kay, *Who Wrote the Book of Life?*, 131. Reprinted by permission of R. Igor Gamow.

294 "AS IN THE BREAKING OF ENEMY MESSAGES": George Gamow to E. Chargaff, 6 May 1954, Ibid., 141.

294 "BY PRIVATE INTERNATIONAL BUSH TELEGRAPH": Gunther S. Stent, "DNA," 924.

295 "PEOPLE DIDN'T NECESSARILY *BELIEVE* IN THE CODE": Francis Crick, interview with Horace Freeland Judson, 20 November 1975, in Horace Freeland Judson, *The Eighth Day of Creation: Makers of the Revolution in Biology* (New York: Simon & Schuster, 1979), 233.

295 "A LONG NUMBER WRITTEN IN A FOUR-DIGITAL SYSTEM": George Gamow, "Possible Relation Between Deoxyribonucleic Acid and Protein Structures," *Nature* 173 (1954): 318.

295 "BETWEEN THE COMPLEX MACHINERY IN A LIVING CELL": Douglas R. Hofstadter, "The Genetic Code: Arbitrary?" (March 1982), in *Metamagical Themas: Questing for the Essence of Mind and Pattern* (New York: Basic Books, 1985), 671.

296 "THE NUCLEUS OF A LIVING CELL IS A STOREHOUSE OF INFORMATION": George Gamow, "Information Transfer in the Living Cell," *Scientific American* 193, no. 10 (October 1955): 70.

297 UNNECESSARY IF SOME TRIPLETS MADE "SENSE": Francis Crick, "General Nature of the Genetic Code for Proteins," *Nature* 192 (30 December 1961): 1227.

297 "THE SEQUENCE OF NUCLEOTIDES AS AN INFINITE MESSAGE": Solomon W. Golomb, Basil Gordon, and Lloyd R. Welch, "Comma-Free Codes," *Canadian Journal of Mathematics* 10 (1958): 202–209, quoted in Lily E. Kay, *Who Wrote the Book of Life?*, 171.

298 "ONCE 'INFORMATION' HAS PASSED INTO PROTEIN": Francis Crick, "On Protein Synthesis," *Symposium of the Society for Experimental Biology* 12 (1958): 152; Cf. Francis Crick, "Central Dogma of Molecular Biology," *Nature* 227 (1970): 561–63; and Hubert P. Yockey, *Information Theory, Evolution, and the Origin of Life* (Cambridge: Cambridge University Press, 2005), 20–21.

298 "THE COMPLETE DESCRIPTION OF THE ORGANISM": Horace Freeland Judson, *The Eighth Day of Creation*, 219–21.

300 "IT IS IN THIS SENSE THAT ALL WORKING GENETICISTS": Gunther S. Stent, "You Can Take the Ethics Out of Altruism But You Can't Take the Altruism Out of Ethics," *Hastings Center Report* 7, no. 6 (1977): 34; and Gunther S. Stent, "DNA," 925.

301 "IT DEPENDS UPON WHAT LEVEL": Seymour Benzer, "The Elementary Units of Heredity," in W. D. McElroy and B. Glass, eds., *The Chemical Basis of Heredity* (Baltimore: Johns Hopkins University Press, 1957), 70.

301 "THIS ATTITUDE IS AN ERROR OF GREAT PROFUNDITY": Richard Dawkins, *The Selfish Gene*, 30th anniversary edition (Oxford: Oxford University Press, 2006), 237.

301 "WE ARE SURVIVAL MACHINES": Ibid., xxi.

302 "THEY ARE PAST MASTERS OF THE SURVIVAL ARTS": Ibid., 19.

302 "ENGLISH BIOLOGIST RICHARD DAWKINS HAS RECENTLY RAISED": Stephen Jay Gould, "Caring Groups and Selfish Genes," in *The Panda's Thumb* (New York: Norton, 1980), 86.

302 "A THIRTY-SIX-YEAR-OLD STUDENT OF ANIMAL BEHAVIOR": Gunther S. Stent, "You Can Take the Ethics Out of Altruism But You Can't Take the Altruism Out of Ethics," 33.

302 "EVERY CREATURE MUST BE ALLOWED TO 'RUN' ITS OWN DEVELOPMENT": Samuel Butler, *Life and Habit* (London: Trübner & Co, 1878), 134.

303 "A SCHOLAR . . . IS JUST A LIBRARY'S WAY": Daniel C. Dennett, *Darwin's Dangerous Idea: Evolution and the Meanings of Life* (New York: Simon & Schuster, 1995), 346.

303 "ANTHROPOCENTRISM IS A DISABLING VICE OF THE INTELLECT": Edward O. Wilson, "Biology and the Social Sciences," *Daedalus* 106, no. 4 (Fall 1977), 131.

303 "IT REQUIRES A DELIBERATE MENTAL EFFORT": Richard Dawkins, *The Selfish Gene*, 265.

304 "MIGHT ENSURE ITS SURVIVAL BY TENDING TO ENDOW": Ibid., 36.

304 "THEY DO NOT PLAN AHEAD": Ibid., 25.

305 "THERE IS A MOLECULAR ARCHEOLOGY IN THE MAKING": Werner R. Loewenstein, *The Touchstone of Life: Molecular Information, Cell Communication, and the Foundations of Life* (New York: Oxford University Press, 1999), 93–94.

306 "SELECTION FAVORS THOSE GENES WHICH SUCCEED": Richard Dawkins, *The Extended Phenotype*, rev. ed. (Oxford: Oxford University Press, 1999), 117.

306 DAWKINS SUGGESTS THE CASE OF A GENE: Ibid., 196–97.

307 THERE IS NO GENE FOR LONG LEGS: Richard Dawkins, *The Selfish Gene*, 37.

307 HABIT OF SAYING "A GENE FOR X": Richard Dawkins, *The Extended Phenotype*, 21.

307 "ALL WE WOULD NEED IN ORDER": Ibid., 23.

308 "ANY GENE THAT INFLUENCES THE DEVELOPMENT OF NERVOUS SYSTEMS": Richard Dawkins, *The Selfish Gene*, 60.

309 "IT IS NO MORE LIKELY TO DIE": Ibid., 34.

309 "TODAY THE TENDENCY IS TO SAY": Max Delbrück, "A Physicist Looks At Biology," *Transactions of the Connecticut Academy of Arts and Sciences* 38 (1949): 194.

11. INTO THE MEME POOL

310 "WHEN I MUSE ABOUT MEMES": Douglas R. Hofstadter, "On Viral Sentences and Self-Replicating Structures," in *Metamagical Themas: Questing for the Essence of Mind and Pattern* (New York, Basic Books, 1985), 52.

310 "now through the very universality of its structures": Jacques
    Monod, *Chance and Necessity: An Essay on the Natural Philosophy of Modern Bio-
    logy*, trans. Austryn Wainhouse (New York: Knopf, 1971), 145.
311 "ideas have retained some of the properties": Ibid., 165.
311 "ideas cause ideas": Roger Sperry, "Mind, Brain, and Humanist Values,"
    in *New Views of the Nature of Man*, ed. John R. Platt (Chicago: University of
    Chicago Press, 1983), 82.
311 "i think that a new kind": Richard Dawkins, *The Selfish Gene*, 30th anni-
    versary edition (Oxford: Oxford University Press, 2006), 192.
313 "this may not be what george washington looked like then": Dan-
    iel C. Dennett, *Darwin's Dangerous Idea: Evolution and the Meanings of Life*
    (New York: Simon & Schuster, 1995), 347.
313 "a wagon with spoked wheels": Daniel C. Dennett, *Consciousness Explained*
    (Boston: Little, Brown, 1991), 204.
314 "genes cannot be selfish": Mary Midgley, "Gene-Juggling," *Philosophy* 54
    (October 1979).
314 "a meme . . . is an information packet": Daniel C. Dennett, "Memes:
    Myths, Misunderstandings, and Misgivings," draft for Chapel Hill lecture,
    October 1998, http://ase.tufts.edu/cogstud/papers/MEMEMYTH.FIN.htm
    (accessed 7 June 2010).
314 "to die for an idea": George Jean Nathan and H. L. Mencken, "Clinical
    Notes," *American Mercury* 3, no. 9 (September 1924), 55.
314 *i was promised on a time to have reason for my rhyme*: Edmund Spenser,
    quoted by Thomas Fuller, *The History of the Worthies of England* (London:
    1662).
315 "i believe that, given the right conditions": Richard Dawkins, *The
    Selfish Gene*, 322.
315 "when you plant a fertile meme": Quoted by Dawkins, Ibid., 192.
315 "hard as this term may be to delimit": W. D. Hamilton, "The Play by
    Nature," *Science* 196 (13 May 1977): 759.
316 birdsong *culture*: Juan D. Delius, "Of Mind Memes and Brain Bugs, A
    Natural History of Culture," in *The Nature of Culture*, ed. Walter A. Koch
    (Bochum, Germany: Bochum, 1989), 40.
316 "from look to look": James Thomson, "Autumn" (1730).
316 "eve, whose eye": John Milton, *Paradise Lost*, IX:1036.
316 walton proposed simple self-replicating sentences: Douglas R. Hof-
    stadter, "On Viral Sentences and Self-Replicating Structures," 52.
317 "i don't know about you": Daniel C. Dennett, *Darwin's Dangerous Idea*,
    346.
318 "the computers in which memes live": Richard Dawkins, *The Selfish
    Gene*, 197.
318 "it was obviously predictable": Ibid., 329.

319 "MAKE SEVEN COPIES OF IT EXACTLY AS IT IS WRITTEN": Daniel W. VanArsdale, "Chain Letter Evolution," http://www.silcom.com/~barnowl/chain-letter/
evolution.html (accessed 8 June 2010).

319 "AN UNUSUAL CHAIN-LETTER REACHED QUINCY": Harry Middleton Hyatt,
Folk-Lore from Adams County, Illinois, 2nd and rev. ed. (Hannibal, Mo.: Alma
Egan Hyatt Foundation, 1965), 581.

320 "THESE LETTERS HAVE PASSED FROM HOST TO HOST": Charles H. Bennett,
Ming Li, and Bin Ma, "Chain Letters and Evolutionary Histories," *Scientific
American* 288, no. 6 (June 2003): 77.

321 FOR DENNETT, THE FIRST FOUR NOTES: Daniel C. Dennett, *Darwin's Dangerous Idea,* 344.

321 "MEMES HAVE NOT YET FOUND": Richard Dawkins, foreword to Susan Blackmore, *The Meme Machine* (Oxford: Oxford University Press, 1999), xii.

322 "THE HUMAN WORLD IS MADE OF STORIES": David Mitchell, *Ghostwritten*
(New York: Random House, 1999), 378.

322 "AS WITH ALL KNOWLEDGE, ONCE YOU KNEW IT": Margaret Atwood, *The Year
of the Flood* (New York: Doubleday, 2009), 170.

322 "A LIFE POURED INTO WORDS": John Updike, "The Author Observes His
Birthday, 2005," *Endpoint and Other Poems* (New York: Knopf, 2009), 8.

322 "IN THE BEGINNING THERE WAS INFORMATION": Fred I. Dretske, *Knowledge
and the Flow of Information* (Cambridge, Mass.: MIT Press, 1981), xii.

12. THE SENSE OF RANDOMNESS

324 "I WONDER," SHE SAID: Michael Cunningham, *Specimen Days* (New York: Far-
rar Straus Giroux, 2005), 154.

324 FOUND A MAGICAL LITTLE BOOK: Interviews, Gregory J. Chaitin, 27 October
2007 and 14 September 2009; Gregory J. Chaitin, "The Limits of Reason,"
Scientific American 294, no. 3 (March 2006): 74.

324 "ASTOUNDING AND MELANCHOLY": Ernest Nagel and James R. Newman,
Gödel's Proof (New York: New York University Press, 1958), 6.

325 "IT WAS A VERY SERIOUS CONCEPTUAL CRISIS": quoted in Gregory J. Chaitin,
*Information, Randomness & Incompleteness: Papers on Algorithmic Information
Theory* (Singapore: World Scientific, 1987), 61.

325 HE WONDERED IF AT SOME LEVEL: "Algorithmic Information Theory," in
Gregory J. Chaitin, *Conversations with a Mathematician* (London: Springer,
2002), 80.

326 "PROBABILITY, LIKE TIME": John Archibald Wheeler, *At Home in the Universe,
Masters of Modern Physics,* vol. 9 (New York: American Institute of Physics,
1994), 304.

326 WHETHER THE POPULATION OF FRANCE: Cf. John Maynard Keynes, *A Treatise
on Probability* (London: Macmillan, 1921), 291.

326 HE CHOSE THREE: KNOWLEDGE, CAUSALITY, AND DESIGN: Ibid., 281.

326 "CHANCE IS ONLY THE MEASURE": Henri Poincaré, "Chance," in *Science and Method*, trans. Francis Maitland (Mineola, N.Y.: Dover, 2003), 65.

326 1009732533765201358634673548: *A Million Random Digits with 100,000 Normal Deviates* (Glencoe, Ill.: Free Press, 1955).

327 AN ELECTRONIC ROULETTE WHEEL: Ibid., ix–x.

327 "STATE OF SIN": Von Neumann quoted in Peter Galison, *Image and Logic: A Material Culture of Microphysics* (Chicago: University of Chicago Press, 1997), 703.

330 "WHEN THE READING HEAD MOVES": "A Universal Turing Machine with Two Internal States," in Claude Elwood Shannon, *Collected Papers,* ed. N. J. A. Sloane and Aaron D. Wyner (New York: IEEE Press, 1993), 733–41.

332 "HE SUMMARIZES HIS OBSERVATIONS": Gregory J. Chaitin, "On the Length of Programs for Computing Finite Binary Sequences," *Journal of the Association for Computing Machinery* 13 (1966): 567.

333 "WE ARE TO ADMIT NO MORE CAUSES": Isaac Newton, "Rules of Reasoning in Philosophy; Rule I," *Philosophiae Naturalis Principia Mathematica.*

333 IN THE WANING YEARS OF TSARIST RUSSIA: Obituary, *Bulletin of the London Mathematical Society* 22 (1990): 31; A. N. Shiryaev, "Kolmogorov: Life and Creative Activities," *Annals of Probability* 17, no. 3 (1989): 867.

334 UNLIKELY TO ATTRACT INTERPRETATION: David A. Mindell et al., "Cybernetics and Information Theory in the United States, France, and the Soviet Union," in *Science and Ideology: A Comparative History*, ed. Mark Walker (London: Routledge, 2003), 66 and 81.

334 HE SOON LEARNED TO HIS SORROW: Cf. "Amount of Information and Entropy for Continuous Distributions," note 1, in *Selected Works of A. N. Kolmogorov*, vol. 3, *Information Theory and the Theory of Algorithms*, trans. A. B. Sossinksky (Dordrecht, Netherlands: Kluwer Academic Publishers, 1993), 33.

334 "MORE TECHNOLOGY THAN MATHEMATICS": A. N. Kolmogorov and A. N. Shiryaev, *Kolmogorov in Perspective*, trans. Harold H. McFaden, History of Mathematics vol. 20 (n.p.: American Mathematical Society, London Mathematical Society, 2000), 54.

335 "WHEN I READ THE WORKS OF ACADEMICIAN KOLMOGOROV": Quoted in Slava Gerovitch, *From Newspeak to Cyberspeak: A History of Soviet Cybernetics* (Cambridge, Mass.: MIT Press, 2002), 58.

335 "CYBERNETICS IN WIENER'S UNDERSTANDING": "Intervention at the Session," in *Selected Works of A. N. Kolmogorov*, 31.

336 "AT EACH GIVEN MOMENT": Kolmogorov diary entry, 14 September 1943, in A. N. Kolmogorov and A. N. Shiryaev, *Kolmogorov in Perspective*, 50.

336 "IS IT POSSIBLE TO INCLUDE THIS NOVEL": "Three Approaches to the Definition of the Concept 'Quantity of Information,'" in *Selected Works of A. N. Kolmogorov*, 188.

336 "OUR DEFINITION OF THE QUANTITY": A. N. Kolmogorov, "Combinatorial

Foundations of Information Theory and the Calculus of Probabilities," *Russian Mathematical Surveys* 38, no. 4 (1983): 29–43.

337 "THE INTUITIVE DIFFERENCE BETWEEN 'SIMPLE' AND 'COMPLICATED'": "Three Approaches to the Definition of the Concept 'Quantity of Information,'" *Selected Works of A. N. Kolmogorov*, 221.

337 "A NEW CONCEPTION OF THE NOTION 'RANDOM'": "On the Logical Foundations of Information Theory and Probability Theory," *Problems of Information Transmission* 5, no. 3 (1969): 1–4.

338 HE DREAMED OF SPENDING HIS LAST YEARS: V. I. Arnold, "On A. N. Kolmogorov," in A. N. Kolmogorov and A. N. Shiryaev, *Kolmogorov in Perspective*, 94.

340 "THE PARADOX ORIGINALLY TALKS ABOUT ENGLISH": Gregory J. Chaitin, *Thinking About Gödel and Turing: Essays on Complexity, 1970–2007* (Singapore: World Scientific, 2007), 176.

342 "IT DOESN'T MAKE ANY DIFFERENCE WHICH PARADOX": Gregory J. Chaitin, "The Berry Paradox," *Complexity* 1, no. 1 (1995): 26; "Paradoxes of Randomness," *Complexity* 7, no. 5 (2002): 14–21.

343 "ABSOLUTE CERTAINTY IS LIKE GOD": Interview, Gregory J. Chaitin, 14 September 2009.

343 "GOD NOT ONLY PLAYS DICE": Foreword to Cristian S. Calude, *Information and Randomness: An Algorithmic Perspective* (Berlin: Springer, 2002), viii.

343 "CHARMINGLY CAPTURED THE ESSENCE": Joseph Ford, "Directions in Classical Chaos," in *Directions in Chaos*, ed. Hao Bai-lin (Singapore: World Scientific, 1987), 14.

345 THE INFORMATION PACKING PROBLEM: Ray J. Solomonoff, "The Discovery of Algorithmic Probability," *Journal of Computer and System Sciences* 55, no. 1 (1997): 73–88.

345 "THREE MODELS FOR THE DESCRIPTION OF LANGUAGE": Noam Chomsky, "Three Models for the Description of Language," *IRE Transactions on Information Theory* 2, no. 3 (1956): 113–24.

345 "THE LAWS OF SCIENCE THAT HAVE BEEN DISCOVERED": Ray J. Solomonoff, "A Formal Theory of Inductive Inference," *Information and Control* 7, no. 1 (1964): 1–22.

346 "COCKTAIL SHAKER AND SHAKING VIGOROUSLY": Foreword to Cristian S. Calude, *Information and Randomness*, vii.

349 "IT IS PREFERABLE TO CONSIDER COMMUNICATION": Gregory J. Chaitin, "Randomness and Mathematical Proof," in *Information, Randomness & Incompleteness*, 4.

353 "FROM THE EARLIEST DAYS OF INFORMATION THEORY": Charles H. Bennett, "Logical Depth and Physical Complexity," in *The Universal Turing Machine: A Half-Century Survey*, ed. Rolf Herken (Oxford: Oxford University Press, 1988), 209–10.

355 "THE MORE ENERGY, THE FASTER THE BITS FLIP": Seth Lloyd, *Programming the Universe* (New York: Knopf, 2006), 44.

355 "HOW DID THIS COME ABOUT?": Christopher A. Fuchs, "Quantum Mechanics as Quantum Information (and Only a Little More)," *arXiv:quant-ph/0205039v1*, 8 May 2002, 1.

356 "THE REASON IS SIMPLE": Ibid., 4.

356 "IT TEACHES US . . . THAT SPACE CAN BE CRUMPLED": John Archibald Wheeler with Kenneth Ford, *Geons, Black Holes, and Quantum Foam: A Life in Physics* (New York: Norton, 1998), 298.

356 "OTHERWISE PUT . . . EVERY IT": "It from Bit" in John Archibald Wheeler, *At Home in the Universe, Masters of Modern Physics*, vol. 9 (New York: American Institute of Physics, 1994), 296.

357 A PROBLEM AROSE WHEN STEPHEN HAWKING: Stephen Hawking, "Black Hole Explosions?" *Nature* 248 (1 March 1974), DOI:10.1038/248030a0, 30–31.

358 PUBLISHING IT WITH A MILDER TITLE: Stephen Hawking, "The Breakdown of Predictability in Gravitational Collapse," *Physical Review D* 14 (1976): 2460–73; Gordon Belot et al., "The Hawking Information Loss Paradox: The Anatomy of a Controversy," *British Journal for the Philosophy of Science* 50 (1999): 189–229.

358 "INFORMATION LOSS IS HIGHLY INFECTIOUS": John Preskill, "Black Holes and Information: A Crisis in Quantum Physics," Caltech Theory Seminar, 21 October 1994, http://www.theory.caltech.edu/~preskill/talks/blackholes.pdf (accessed 20 March 2010).

358 "SOME PHYSICISTS FEEL THE QUESTION": John Preskill, "Black Holes and the Information Paradox," *Scientific American* (April 1997): 54.

358 "I THINK THE INFORMATION PROBABLY GOES OFF": Quoted in Tom Siegfried, *The Bit and the Pendulum: From Quantum Computing to M Theory—The New Physics of Information* (New York: Wiley and Sons, 2000), 203.

359 "THERE IS NO BABY UNIVERSE": Stephen Hawking, "Information Loss in Black Holes," *Physical Review D* 72 (2005): 4.

359 THE "THERMODYNAMICS OF COMPUTATION": Charles H. Bennett, "Notes on the History of Reversible Computation," *IBM Journal of Research and Development* 44 (2000): 270.

360 "COMPUTERS . . . MAY BE THOUGHT OF AS ENGINES": Charles H. Bennett, "The Thermodynamics of Computation—a Review," *International Journal of Theoretical Physics* 21, no. 12 (1982): 906.

361 BACK-OF-THE-ENVELOPE CALCULATION: Ibid.

361 ROLF LANDAUER: "Information Is Physical," *Physics Today* 23 (May 1991); "Information Is Inevitably Physical," in Anthony H. G. Hey, ed., *Feynman and Computation* (Boulder, Colo.: Westview Press, 2002), 77.

361 STRAIGHT AND NARROW OLD IBM TYPE: Charles Bennett, quoted by George

Johnson in "Rolf Landauer, Pioneer in Computer Theory, Dies at 72," *The New York Times*, 30 April 1999.

362 "YOU MIGHT SAY THIS IS THE REVENGE": Interview, Charles Bennett, 27 October 2009.

363 BENNETT AND HIS RESEARCH ASSISTANT: J. A. Smolin, "The Early Days of Experimental Quantum Cryptography," *IBM Journal of Research and Development* 48 (2004): 47–52.

363 "WE SAY THINGS SUCH AS 'ALICE SENDS BOB' ": Barbara M. Terhal, "Is Entanglement Monogamous?" *IBM Journal of Research and Development* 48, no. 1 (2004): 71–78.

364 FOLLOWING AN INTRICATE AND COMPLEX PROTOCOL: A detailed explanation can be found in Simon Singh, *The Code Book: The Secret History of Codes and Codebreaking* (London: Fourth Estate, 1999); it takes ten pages of exquisite prose, beginning at 339.

364 "STAND BY: I'LL TELEPORT YOU SOME GOULASH": IBM advertisement, *Scientific American* (February 1996), 0–1; Anthony H. G. Hey, ed., *Feynman and Computation*, xiii; Tom Siegfried, *The Bit and the Pendulum*, 13.

364 "UNFORTUNATELY THE PREPOSTEROUS SPELLING *QUBIT*": N. David Mermin, *Quantum Computer Science: An Introduction* (Cambridge: Cambridge University Press, 2007), 4.

366 "CAN QUANTUM-MECHANICAL DESCRIPTION OF PHYSICAL REALITY": *Physical Review* 47 (1935): 777–80.

366 "EINSTEIN HAS ONCE AGAIN EXPRESSED HIMSELF": Wolfgang Pauli to Werner Heisenberg, 15 June 1935, quoted in Louisa Gilder, *The Age of Entanglement: When Quantum Physics Was Reborn* (New York: Knopf, 2008), 162.

366 "THAT WHICH REALLY EXISTS IN B": Albert Einstein to Max Born, March 1948, in *The Born-Einstein Letters*, trans. Irene Born (New York: Walker, 1971), 164.

366 IT TOOK MANY MORE YEARS BEFORE THE LATTER: Asher Peres, "Einstein, Podolsky, Rosen, and Shannon," *arXiv:quant-ph/0310010 v1*, 2003.

367 "TERMINOLOGY CAN SAY IT ALL": Christopher A. Fuchs, "Quantum Mechanics as Quantum Information (and Only a Little More": *arXiv: quant-ph/1003.5209 v1*, 26 March 2010: 3.

367 BENNETT PUT ENTANGLEMENT TO WORK: Charles H. Bennett et al., "Teleporting an Unknown Quantum State Via Dual Classical and Einstein-Podolsky-Rosen Channels," *Physical Review Letters* 70 (1993): 1895.

368 "SECRET! SECRET! CLOSE THE DOORS": Richard Feynman, "Simulating Physics with Computers," in Anthony H. G. Hey, ed., *Feynman and Computation*, 136.

369 "FEYNMAN'S INSIGHT": Interview, Charles H. Bennett, 27 October 2009.

369 "A PRETTY MISERABLE SPECIMEN": N. David Mermin, *Quantum Computer Science*, 17.

370 RSA ENCRYPTION: named after its inventors, Ron Rivest, Adi Shamir, and Len Adleman.

370 THEY ESTIMATED THAT THE COMPUTATION: T. Kleinjung, K. Aoki, J. Franke, et al., "Factorization of a 768-bit RSA modulus," Eprint archive no. 2010/006, 2010.

371 "QUANTUM COMPUTERS WERE BASICALLY A REVOLUTION": Dorit Aharonov, panel discussion "Harnessing Quantum Physics," 18 October 2009, Perimeter Institute, Waterloo, Ontario; and e-mail message 10 February 2010.

371 "MANY PEOPLE CAN READ A BOOK": Charles H. Bennett, "Publicity, Privacy, and Permanence of Information," in *Quantum Computing: Back Action*, AIP Conference Proceeding 864 (2006), ed. Debabrata Goswami (Melville, N.Y.: American Institute of Physics), 175–79.

371 "IF SHANNON WERE AROUND NOW": Charles H. Bennett, interview, 27 October 2009.

372 "TO WORK OUT ALL THE POSSIBLE MIRRORED ROOMS": Shannon interview with Anthony Liversidge, *Omni* (August 1987), in Claude Elwood Shannon, *Collected Papers*, ed. N. J. A. Sloane and Aaron D. Wyner (New York: IEEE Press, 1993), xxxii.

372 A MODEST TO-DO LIST: John Archibald Wheeler, "Information, Physics, Quantum: The Search for Links," *Proceedings of the Third International Symposium on the Foundations of Quantum Mechanics* (1989), 368.

14. AFTER THE FLOOD

373 "SUPPOSE WITHIN EVERY BOOK": Hilary Mantel, *Wolf Hall* (New York: Henry Holt, 2009), 394.

373 "THE UNIVERSE (WHICH OTHERS CALL THE LIBRARY)": Jorge Luis Borges, "The Library of Babel," in *Labyrinths: Selected Stories and Other Writings* (New York: New Directions, 1962), 54.

374 "IT IS CONJECTURED THAT THIS BRAVE NEW WORLD": Jorge Luis Borges, "Tlön, Uqbar, Orbis Tertius," in *Labyrinths*, 8.

374 "OUR HERESIARCH UNCLE": William Gibson, "An Invitation," introduction to *Labyrinths*, xii.

374 "WHAT A STRANGE CHAOS": Charles Babbage, *The Ninth Bridgewater Treatise: A Fragment*, 2nd ed. (London: John Murray, 1838), 111.

375 "NO THOUGHT CAN PERISH": Edgar Allan Poe, "The Power of Words" (1845), in *Poetry and Tales* (New York: Library of America, 1984), 823–24.

375 "IT WOULD EMBRACE IN THE SAME FORMULA": Pierre-Simon Laplace, *A Philosophical Essay on Probabilities*, trans. Frederick Wilson Truscott and Frederick Lincoln Emory (New York: Dover, 1951).

376 "IN TURNING OUR VIEWS": Charles Babbage, *The Ninth Bridgewater Treatise*, 44.

376 "THE ART OF PHOTOGENIC DRAWING": Nathaniel Parker Willis, "The Pencil of Nature: A New Discovery," *The Corsair* 1, no. 5 (April 1839): 72.

377 "IN FACT, THERE IS A GREAT ALBUM OF BABEL": Ibid., 71.

377 "THE SYSTEM OF THE 'UNIVERSE AS A WHOLE' ": Alan M. Turing, "Computing Machinery and Intelligence," *Minds and Machines* 59, no. 236 (1950): 440.

378 "SUCH A BLAZE OF KNOWLEDGE AND DISCOVERY": H. G. Wells, *A Short History of the World* (San Diego: Book Tree, 2000), 97.

378 "THE ROMANS BURNT THE BOOKS OF THE JEWS": Isaac Disraeli, *Curiosities of Literature* (London: Routledge & Sons, 1893), 17.

379 "ALL THE LOST PLAYS OF THE ATHENIANS!": Tom Stoppard, *Arcadia* (London: Samuel French, 1993), 38.

381 "IF YOU WANT TO WRITE ABOUT FOLKLORE": "Wikipedia: Requested Articles," http://web.archive.org/web/20010406104800/www.wikipedia.com/wiki/Requested_articles (accessed 4 April 2001).

384 "AGING IS WHAT YOU GET": Quoted by Nicholson Baker in "The Charms of Wikipedia," *New York Review of Books* 55, no. 4 (20 March 2008). The same anonymous user later struck again, vandalizing the entries on angioplasty and Sigmund Freud.

384 "IT HAS NEVER BEEN SPREAD OUT, YET": Lewis Carroll, *Sylvie and Bruno Concluded* (London: Macmillan, 1893), 169.

384 "THIS IS AN OBJECT IN SPACE, AND I'VE SEEN IT": Interview, Jimmy Wales, 24 July 2008.

384 "*DIE SCHRAUBE AN DER HINTEREN LINKEN BREMSBACKE*": http://meta.wikimedia.org/wiki/Die_Schraube_an_der_hinteren_linken_Bremsbacke_am_Fahrrad_von_Ulrich_Fuchs (accessed 25 July 2008).

386 "A PLAN ENTIRELY NEW": *Encyclopaedia Britannica*, 3rd edition, title page; cf. Richard Yeo, *Encyclopædic Visions: Scientific Dictionaries and Enlightenment Culture* (Cambridge: Cambridge University Press, 2001), 181.

386 "MANY TOPICS ARE BASED ON THE RELATIONSHIP": "Wikipedia: What Wikipedia Is Not," http://en.wikipedia.org/wiki/Wikipedia:What_Wikipedia_is_not (accessed 3 August 2008).

387 "HE READ FOR METAPHYSICS": Charles Dickens, *The Pickwick Papers*, chapter 51.

387 "I BEGAN STANDING WITH MY COMPUTER OPEN": Nicholson Baker, "The Charms of Wikipedia."

388 "A HAMADRYAD IS A WOOD-NYMPH": John Banville, *The Infinities* (London: Picador, 2009), 178.

389 "MADE UP OF SYLLABLES THAT APPEAR": Deming Seymour, "A New Yorker at Large," *Sarasota Herald*, 25 August 1929.

389 BY 1934 THE BUREAU WAS MANAGING A LIST: "Regbureau," *The New Yorker* (26 May 1934), 16.

392 AS THE HISTORIAN BRIAN OGILVIE HAS SHOWN: Brian W. Ogilvie, *The Science of Describing: Natural History in Renaissance Europe* (Chicago: University of Chicago Press, 2006).

392 SCANDIX, PECTEN VENERIS, HERBA SCANARIA: Ibid., 173.

393 CATALOGUE OF 6,000 PLANTS: Caspar Bauhin; Ibid., 208.

393 "THE NAME OF A MAN IS LIKE HIS SHADOW": Ernst Pulgram, *Theory of Names* (Berkeley, Calif.: American Name Society, 1954), 3.

394 "A SCIENTIST'S IDEA OF A SHORT WAY": Michael Amrine, "'Megabucks' for What's 'Hot,'" *The New York Times Magazine*, 22 April 1951.

395 "IT'S AS IF YOU KNEEL TO PLANT THE SEED": Jaron Lanier, *You Are Not a Gadget* (New York: Knopf, 2010), 8.

396 SERVER FARMS PROLIFERATE: Cf. Tom Vanderbilt, "Data Center Overload," *The New York Times Magazine*, 14 June 2009.

397 LLOYD CALCULATES: Seth Lloyd, "Computational Capacity of the Universe," *Physical Review Letters* 88, no. 23 (2002).

15. NEW NEWS EVERY DAY

398 "SORRY FOR ALL THE UPS AND DOWNS": http://www.andrewtobias.com/bkold columns/070118.html (accessed 18 January 2007).

398 "GREAT MUTATION": Carl Bridenbaugh, "The Great Mutation," *American Historical Review* 68, no. 2 (1963): 315–31.

399 "NOTWITHSTANDING THE INCESSANT CHATTER": Ibid., 322.

399 A THOUSAND PEOPLE IN THE BALLROOM: "Historical News," *American Historical Review* 63, no. 3 (April 1963): 880.

399 TENDED TO SLOT THE PRINTING PRESS: Elizabeth L. Eisenstein, *The Printing Press as an Agent of Change: Communications and Cultural Transformations in Early-Modern Europe* (Cambridge: Cambridge University Press, 1979), 25.

399 "DATA COLLECTION, STORAGE AND RETRIEVAL SYSTEMS": Ibid., xvi.

399 "A DECISIVE POINT OF NO RETURN": Elizabeth L. Eisenstein, "Clio and Chronos: An Essay on the Making and Breaking of History-Book Time," *History and Theory* 6, suppl. 6: History and the Concept of Time (1966), 64.

400 "ATTITUDES TOWARD HISTORICAL CHANGE": Ibid., 42.

400 "SCRIBAL CULTURE": Ibid., 61.

400 PRINT WAS TRUSTWORTHY, RELIABLE, AND PERMANENT: Elizabeth L. Eisenstein, *The Printing Press as an Agent of Change*, 624 ff.

401 "AS I SEE IT . . . MANKIND IS FACED WITH NOTHING SHORT OF": Carl Bridenbaugh, "The Great Mutation," 326.

401 "THIS IS A MISREADING OF THE PREDICAMENT": Elizabeth L. Eisenstein, "Clio and Chronos," 39.

401 "I HEAR NEW NEWS EVERY DAY": Robert Burton, *The Anatomy of Melancholy*, ed. Floyd Dell and Paul Jordan-Smith (New York: Tudor, 1927), 14.

402 "TO WHICH RESULT THAT HORRIBLE MASS OF BOOKS": Gottfried Wilhelm Leibniz, *Leibniz Selections*, ed. Philip P. Wiener (New York: Scribner's, 1951),

29; cf. Marshall McLuhan, *The Gutenberg Galaxy* (Toronto: University of Toronto Press, 1962), 254.

402 "THOSE DAYS, WHEN (AFTER PROVIDENCE": Alexander Pope, *The Dunciad* (1729) (London: Methuen, 1943), 41.

403 "KNOWLEDGE OF SPEECH, BUT NOT OF SILENCE": T. S. Eliot, "The Rock," in *Collected Poems: 1909–1962* (New York: Harcourt Brace, 1963), 147.

403 "THE TSUNAMI OF AVAILABLE FACT": David Foster Wallace, Introduction to *The Best American Essays 2007* (New York: Mariner, 2007).

404 "UNFORTUNATELY, 'INFORMATION RETRIEVING,' HOWEVER SWIFT": Lewis Mumford, *The Myth of the Machine*, vol. 2, *The Pentagon of Power* (New York: Harcourt, Brace, 1970), 182.

405 "ELECTRONIC MAIL SYSTEM": Jacob Palme, "You Have 134 Unread Mail! Do You Want to Read Them Now?" in *Computer-Based Message Services*, ed. Hugh T. Smith (North Holland: Elsevier, 1984), 175–76.

405 A PAIR OF PSYCHOLOGISTS: C. J. Bartlett and Calvin G. Green, "Clinical Prediction: Does One Sometimes Know Too Much," *Journal of Counseling Psychology* 13, no. 3 (1966): 267–70.

406 "THE INFORMATION YOU ARE RECEIVING IS PREPARED FOR YOU": Siegfried Streufert et al., "Conceptual Structure, Information Search, and Information Utilization," *Journal of Personality and Social Psychology* 2, no. 5 (1965): 736–40.

406 "INFORMATION-LOAD PARADIGM": For example, Naresh K. Malhotra, "Information Load and Consumer Decision Making," *Journal of Consumer Research* 8 (March 1982): 419.

406 "E-MAIL, MEETINGS, LISTSERVS, AND IN-BASKET PAPER PILES": Tonyia J. Tidline, "The Mythology of Information Overload," *Library Trends* 47, no. 3 (Winter 1999): 502.

407 "WE PAY TO HAVE NEWSPAPERS DELIVERED": Charles H. Bennett, "Demons, Engines, and the Second Law," *Scientific American* 257, no. 5 (1987): 116.

407 "AS THE DESIRED INFORMATION": G. Bernard Shaw to the Editor, *Whitaker's Almanack*, 31 May 1943.

408 "DON'T ASK BY TELEPHONE FOR WORLD'S SERIES SCORES": *The New York Times*, 8 October 1929, 1.

408 "YOU HUNCH LIKE A PIANIST": Anthony Lane, "Byte Verse," *The New Yorker*, 20 February 1995, 108.

410 "THE OBVIOUS COUNTERHYPOTHESIS ARISES": Daniel C. Dennett, "Memes and the Exploitation of Imagination," *Journal of Aesthetics and Art Criticism* 48 (1990): 132.

410 "TAKE THE LIBRARY OF THE BRITISH MUSEUM": Augustus De Morgan, *Arithmetical Books: From the Invention of Printing to the Present Time* (London: Taylor & Walton, 1847), ix.

411 "THE MULTITUDE OF BOOKS, THE SHORTNESS OF TIME": Vincent of Beauvais, Prologue, *Speculum Maius*, quoted in Ann Blair, "Reading Strategies for Coping

with Information Overload ca. 1550–1700," *Journal of the History of Ideas* 64, no. 1 (2003): 12.

411 "THE PERCEPTION OF AN OVERABUNDANCE": Ibid.

411 "DRIVEN BY THE NEED TO MASTER THE INFORMATION OVERLOAD": Brian W. Ogilvie, "The Many Books of Nature: Renaissance Naturalists and Information Overload," *Journal of the History of Ideas* 64, no. 1 (2003): 40.

412 "A MAN WHO HAS SOMETHING TO SAY": Bertolt Brecht, *Radio Theory* (1927), quoted in Kathleen Woodward, *The Myths of Information: Technology and Postindustrial Culture* (Madison, Wisc.: Coda Press, 1980).

EPILOGUE

413 "IT WAS INEVITABLE THAT MEANING": Jean-Pierre Dupuy, *The Mechanization of the Mind: On the Origins of Cognitive Science*, trans. M. B. DeBevoise (Princeton, N.J.: Princeton University Press, 2000), 119.

413 "WE ARE TODAY AS FAR INTO THE ELECTRIC AGE": Marshall McLuhan, *The Gutenberg Galaxy* (Toronto: University of Toronto Press, 1962), 1.

413 "TODAY . . . WE HAVE EXTENDED OUR CENTRAL NERVOUS SYSTEMS": Marshall McLuhan, *Understanding Media: The Extensions of Man* (New York: McGraw-Hill, 1965), 3.

414 "WHAT WHISPERS ARE THESE": Walt Whitman, "Years of the Modern," *Leaves of Grass* (Garden City, N.Y.: Doubleday, 1919), 272.

414 THEOLOGIANS BEGAN SPEAKING OF A SHARED MIND: For example, "Two beings, or two millions—any number thus placed 'in communication'—all possess one mind." Parley Parker Pratt, *Key to the Science of Theology* (1855), quoted in John Durham Peters, *Speaking Into the Air: A History of the Idea of Communication* (Chicago: University of Chicago Press, 1999), 275.

414 "IT BECOMES ABSOLUTELY NECESSARY": ". . . this amounts to imagining, above the animal biosphere and continuing it, a human sphere, the sphere of reflection, of conscious and free invention, of thought strictly speaking, in short, the sphere of mind or noosphere." Édouard Le Roy, *Les Origines humaines et l'évolution de l'intelligence* (Paris: Boivin et Cie, 1928), quoted and translated by M. J. Aronson, *Journal of Philosophy* 27, no. 18 (28 August 1930): 499.

414 "DOES IT NOT SEEM AS THOUGH A GREAT BODY": Pierre Teilhard de Chardin, *The Human Phenomenon*, trans. Sarah Appleton-Weber (Brighton, U.K.: Sussex Academic Press, 1999), 174.

414 "NONSENSE, TRICKED OUT": *Mind* 70, no. 277 (1961): 99. Medawar did not much like Teilhard's prose, either: "that tipsy, euphoric prose-poetry which is one of the more tiresome manifestations of the French spirit."

414 WRITERS OF SCIENCE FICTION: Perhaps first and most notably Olaf Stapledon, *Last and First Men* (London: Methuen, 1930).

415 "OUR MULTITUDE OF UNCO-ORDINATED GANGLIA": H. G. Wells, *World Brain* (London: Methuen, 1938), xiv.

415 "IN A FEW SCORE YEARS": Ibid., 56.

415 "SORT OF CEREBRUM FOR HUMANITY": Ibid., 63.

416 "A NETWORK OF MARVELLOUSLY GNARLED AND TWISTED STEMS": H. G. Wells, *The Passionate Friends* (London: Harper, 1913), 332; H. G. Wells, *The War in the Air* (New York: Macmillan, 1922), 14.

417 "IT'S NOT IN THE BEEPS": Quoted in Flo Conway and Jim Siegelman, *Dark Hero of the Information Age: In Search of Norbert Wiener, the Father of Cybernetics* (New York: Basic Books, 2005), 189.

417 "I KNOW AN UNCOUTH REGION": Jorge Luis Borges, "The Library of Babel," *Labyrinths: Selected Stories and Other Writings* (New York: New Directions, 1962), 54.

417 "BEAUTY IS IN THE EYE OF THE BEHOLDER": Fred I. Dretske, *Knowledge and the Flow of Information* (Cambridge, Mass.: MIT Press, 1981), vii.

417 "I TAKE 'HELL' IN ITS THEOLOGICAL SENSE": Jean-Pierre Dupuy, "Myths of the Informational Society," in Kathleen Woodward, *The Myths of Information: Technology and Postindustrial Culture* (Madison, Wisc.: Coda Press, 1980), 3.

418 "I IMAGINE . . . THAT THE ENTRIES OF THE DICTIONARY": Dexter Palmer, *The Dream of Perpetual Motion* (New York: St. Martin's Press, 2010), 220.

418 "ALL HUMAN THOUGHTS MIGHT BE ENTIRELY RESOLVABLE": Gottfried Wilhelm Leibniz, *De scientia universali seu calculo philosophico*, 1875; cf. Umberto Eco, *The Search for the Perfect Language*, trans. James Fentress (Malden, Mass.: Blackwell, 1995), 281.

420 "IS IT SIGNALING, LIKE TELEGRAPHS?": Margaret Atwood, "Atwood in the Twittersphere," *The New York Review of Books* blog, http://www.nybooks.com/blogs/nyrblog/2010/mar/29/atwood-in-the-twittersphere/, 29 March 2010.

420 "GO MAD IN HERDS": Charles Mackay, *Memoirs of Extraordinary Popular Delusions* (Philadelphia: Lindsay & Blakiston, 1850), 14.

421 BROWSE SU[BJECT] CENSORSHIP: Nicholson Baker, "Discards" (1994), in *The Size of Thoughts: Essays and Other Lumber* (New York: Random House, 1996), 168.

423 "WE HAVE A LEXICON OF THE CURRENT LANGUAGE": Interview, Allan Jennings, February 1996; James Gleick, "Here Comes the Spider," in *What Just Happened: A Chronicle from the Information Frontier* (New York: Pantheon, 2002), 128–32.

424 "I READ SOMEWHERE THAT EVERYBODY ON THIS PLANET": John Guare, *Six Degrees of Separation* (New York: Dramatists Play Service, 1990), 45.

424 THE IDEA CAN BE TRACED BACK: Albert-László Barabási, *Linked* (New York: Plume, 2003), 26 ff.

424 WHAT WATTS AND STROGATZ DISCOVERED: Duncan J. Watts and Steven H. Strogatz, "Collective Dynamics of 'Small-World' Networks," *Nature* 393

(1998): 440–42; also Duncan J. Watts, *Six Degrees: The Science of a Connected Age* (New York: Norton, 2003); Albert-László Barabási, *Linked*.

424 "INFECTIOUS DISEASES ARE PREDICTED": Duncan J. Watts and Steven H. Strogatz, "Collective Dynamics of 'Small-World' Networks," 442.

425 "WE WANT THE DEMON, YOU SEE": Stanislaw Lem, *The Cyberiad*, trans. Michael Kandel (London: Secker & Warburg, 1975), 155.

426 "WHEN IT WAS PROCLAIMED": Jorge Luis Borges, "The Library of Babel," *Labyrinths*, 54.

426 "HE THAT DESIRES TO PRINT A BOOK": John Donne, "From a Sermon Preached before King Charles I" (April 1627).

Bibliography

Aaboe, Asger. *Episodes from the Early History of Mathematics*. New York: L. W. Singer, 1963.

Adams, Frederick. "The Informational Turn in Philosophy." *Minds and Machines* 13 (2003): 471–501.

Allen, William, and Thomas R. H. Thompson. *A Narrative of the Expedition to the River Niger in 1841*. London: Richard Bentley, 1848.

Archer, Charles Maybury, ed. *The London Anecdotes: The Electric Telegraph*, vol. 1. London: David Bogue, 1848.

Archibald, Raymond Clare. "Seventeenth Century Calculating Machines." *Mathematical Tables and Other Aids to Computation* 1:1 (1943): 27–28.

Aspray, William. "From Mathematical Constructivity to Computer Science: Alan Turing, John Von Neumann, and the Origins of Computer Science in Mathematical Logic." PhD thesis, University of Wisconsin-Madison, 1980.

———. "The Scientific Conceptualization of Information: A Survey." *Annals of the History of Computing* 7, no. 2 (1985): 117–40.

Aunger, Robert, ed. *Darwinizing Culture: The Status of Memetics as a Science*. Oxford: Oxford University Press, 2000.

Avery, John. *Information Theory and Evolution*. Singapore: World Scientific, 2003.

Baars, Bernard J. *The Cognitive Revolution in Psychology*. New York: Guilford Press, 1986.

Babbage, Charles. "On a Method of Expressing by Signs the Action of Machinery." *Philosophical Transactions of the Royal Society of London* 116, no. 3(1826): 250–65.

———. *Reflections on the Decline of Science in England and on Some of Its Causes*. London: B. Fellowes, 1830.

———. *Table of the Logarithms of the Natural Numbers, From 1 to 108,000*. London: B. Fellowes, 1831.

———. *On the Economy of Machinery and Manufactures*. 4th ed. London: Charles Knight, 1835.

————. *The Ninth Bridgewater Treatise. A Fragment.* 2nd ed. London: John Murray, 1838.

————. *Passages from the Life of a Philosopher.* London: Longman, Green, Longman, Roberts, & Green, 1864.

————. *Charles Babbage and His Calculating Engines: Selected Writings.* Edited by Philip Morrison and Emily Morrison. New York: Dover Publications, 1961.

————. *The Analytical Engine and Mechanical Notation.* New York: New York University Press, 1989.

————. *The Difference Engine and Table Making.* New York: New York University Press, 1989.

————. *The Works of Charles Babbage.* Edited by Martin Campbell-Kelly. New York: New York University Press, 1989.

Babbage, Henry Prevost, ed. *Babbage's Calculating Engines: Being a Collection of Papers Relating to Them; Their History and Construction.* London: E. & F. N. Spon, 1889.

Bairstow, Jeff. "The Father of the Information Age." *Laser Focus World* (2002): 114.

Baker, Nicholson. *The Size of Thoughts: Essays and Other Lumber.* New York: Random House, 1996.

Ball, W. W. Rouse. *A History of the Study of Mathematics at Cambridge.* Cambridge: Cambridge University Press, 1889.

Bar-Hillel, Yehoshua. "An Examination of Information Theory." *Philosophy of Science* 22, no. 2 (1955): 86–105.

Barabási, Albert-László. *Linked: How Everything Is Connected to Everything Else and What It Means for Business, Science, and Everyday Life.* New York: Plume, 2003.

Barnard, G. A. "The Theory of Information." *Journal of the Royal Statistical Society, Series B* 13, no. 1 (1951): 46–64.

Baron, Sabrina Alcorn, Eric N. Lindquist, and Eleanor F. Shevlin. *Agent of Change: Print Culture Studies After Elizabeth L. Eisenstein.* Amherst: University of Massachusetts Press, 2007.

Bartlett, C. J., and Calvin G. Green. "Clinical Prediction: Does One Sometimes Know Too Much." *Journal of Counseling Psychology* 13, no. 3 (1966): 267–70.

Barwise, Jon. "Information and Circumstance." *Notre Dame Journal of Formal Logic* 27, no. 3 (1986): 324–38.

Battelle, John. *The Search: How Google and Its Rivals Rewrote the Rules of Business and Transformed Our Culture.* New York: Portfolio, 2005.

Baugh, Albert C. *A History of the English Language.* 2nd ed. New York: Appleton-Century-Crofts, 1957.

Baum, Joan. *The Calculating Passion of Ada Byron.* Hamden, Conn.: Shoe String Press, 1986.

Belot, Gordon, John Earman, and Laura Ruetsche. "The Hawking Information Loss Paradox: The Anatomy of a Controversy." *British Journal for the Philosophy of Science* 50 (1999): 189–229.

Benjamin, Park. *A History of Electricity (the Intellectual Rise in Electricity) from Antiquity to the Days of Benjamin Franklin.* New York: Wiley and Sons, 1898.

Bennett, Charles H. "On Random and Hard-to-Describe Numbers." IBM Watson Research Center Report RC 7483 (1979).

———. "The Thermodynamics of Computation—A Review." *International Journal of Theoretical Physics* 21, no. 12 (1982): 906–40.

———. "Dissipation, Information, Computational Complexity and the Definition of Organization." In *Emerging Syntheses in Science*, edited by D. Pines, 297–313. Santa Fe: Santa Fe Institute, 1985.

———. "Demons, Engines, and the Second Law." *Scientific American* 257, no. 5 (1987): 108–16.

———. "Logical Depth and Physical Complexity." In *The Universal Turing Machine: A Half-Century Survey*, edited by Rolf Herken. Oxford: Oxford University Press, 1988.

———. "How to Define Complexity in Physics, and Why." In *Complexity, Entropy, and the Physics of Information*, edited by W. H. Zurek. Reading, Mass.: Addison-Wesley, 1990.

———. "Notes on the History of Reversible Computation." *IBM Journal of Research and Development* 44 (2000): 270–77.

———. "Notes on Landauer's Principle, Reversible Computation, and Maxwell's Demon." *arXiv:physics* 0210005 v2 (2003).

———. "Publicity, Privacy, and Permanence of Information." In *Quantum Computing: Back Action 2006, AIP Conference Proceedings 864*, edited by Debabrata Goswami. Melville, N.Y.: American Institute of Physics, 2006.

Bennett, Charles H., and Gilles Brassard. "Quantum Cryptography: Public Key Distribution and Coin Tossing." In *Proceedings of IEEE International Conference on Computers, Systems and Signal Processing*, 175–79. Bangalore, India: 1984.

Bennett, Charles H., Gilles Brassard, Claude Crépeau, Richard Jozsa, Asher Peres, and William K. Wootters. "Teleporting an Unknown Quantum State Via Dual Classical and Einstein-Podolsky-Rosen Channels." *Physical Review Letters* 70 (1993): 1895.

Bennett, Charles H., and Rolf Landauer. "Fundamental Physical Limits of Computation." *Scientific American* 253, no. 1 (1985): 48–56.

Bennett, Charles H., Ming Li, and Bin Ma. "Chain Letters and Evolutionary Histories." *Scientific American* 288, no. 6 (June 2003): 76–81.

Benzer, Seymour. "The Elementary Units of Heredity." In *The Chemical Basis of Heredity*, edited by W. D. McElroy and B. Glass, 70–93. Baltimore: Johns Hopkins University Press, 1957.

Berlinski, David. *The Advent of the Algorithm: The Idea That Rules the World*. New York: Harcourt, 2000.

Bernstein, Jeremy. *The Analytical Engine: Computers—Past, Present and Future*. New York: Random House, 1963.

Bikhchandani, Sushil, David Hirshleifer, and Ivo Welch. "A Theory of Fads, Fashion, Custom, and Cultural Change as Informational Cascades." *Journal of Political Economy* 100, no. 5 (1992): 992–1026.

Blackmore, Susan. *The Meme Machine*. Oxford: Oxford University Press, 1999.

Blair, Ann. "Reading Strategies for Coping with Information Overload ca. 1550–1700." *Journal of the History of Ideas* 64, no. 1 (2003): 11–28.

Blohm, Hans, Stafford Beer, and David Suzuki. *Pebbles to Computers: The Thread.* Toronto: Oxford University Press, 1986.

Boden, Margaret A. *Mind as Machine: A History of Cognitive Science.* Oxford: Oxford University Press, 2006.

Bollobás, Béla, and Oliver Riordan. *Percolation.* Cambridge: Cambridge University Press, 2006.

Bolter, J. David. *Turing's Man: Western Culture in the Computer Age.* Chapel Hill: University of North Carolina Press, 1984.

Boole, George. "The Calculus of Logic." *Cambridge and Dublin Mathematical Journal* 3 (1848): 183–98.

———. *An Investigation of the Laws of Thought, on Which Are Founded the Mathematical Theories of Logic and Probabilities.* London: Walton & Maberly, 1854.

———. *Studies in Logic and Probability,* vol. 1. La Salle, Ill.: Open Court, 1952.

Borges, Jorge Luis. *Labyrinths: Selected Stories and Other Writings.* New York: New Directions, 1962.

Bouwmeester, Dik, Jian-Wei Pan, Klaus Mattle, Manfred Eibl, Harald Weinfurter, and Anton Zeilinger. "Experimental Quantum Teleportation." *Nature* 390 (11 December 1997): 575–79.

Bowden, B. V., ed. *Faster Than Thought: A Symposium on Digital Computing Machines.* New York: Pitman, 1953.

Braitenberg, Valentino. *Vehicles: Experiments in Synthetic Psychology.* Cambridge, Mass.: MIT Press, 1984.

Brewer, Charlotte. "Authority and Personality in the *Oxford English Dictionary.*" *Transactions of the Philological Society* 103, no. 3 (2005): 261–301.

Brewster, David. *Letters on Natural Magic.* New York: Harper & Brothers, 1843.

Brewster, Edwin Tenney. *A Guide to Living Things.* Garden City, N.Y.: Doubleday, 1913.

Bridenbaugh, Carl. "The Great Mutation." *American Historical Review* 68, no. 2 (1963): 315–31.

Briggs, Henry. *Logarithmicall Arithmetike: Or Tables of Logarithmes for Absolute Numbers from an Unite to 100000.* London: George Miller, 1631.

Brillouin, Léon. *Science and Information Theory.* New York: Academic Press, 1956.

Broadbent, Donald E. *Perception and Communication.* Oxford: Pergamon Press, 1958.

Bromley, Allan G. "The Evolution of Babbage's Computers." *Annals of the History of Computing* 9 (1987): 113–36.

Brown, John Seely, and Paul Duguid. *The Social Life of Information.* Boston: Harvard Business School Press, 2002.

Browne, Thomas. *Pseudoxia Epidemica: Or, Enquiries into Very Many Received Tenents, and Commonly Presumed Truths.* 3rd ed. London: Nath. Ekins, 1658.

Bruce, Robert V. *Bell: Alexander Graham Bell and the Conquest of Solitude.* Boston: Little, Brown, 1973.

Buckland, Michael K. "Information as Thing." *Journal of the American Society for Information Science* 42 (1991): 351–60.

Burchfield, R. W., and Hans Aarsleff. *Oxford English Dictionary and the State of the Language*. Washington, D.C.: Library of Congress, 1988.

Burgess, Anthony. *But Do Blondes Prefer Gentlemen? Homage to Qwert Yuiop and Other Writings*. New York: McGraw-Hill, 1986.

Bush, Vannevar. "As We May Think." *The Atlantic*, July 1945.

Butler, Samuel. *Life and Habit*. London: Trübner & Co, 1878.

———. *Essays on Life, Art, and Science*. Edited by R. A Streatfeild. Port Washington, N.Y.: Kennikat Press, 1970.

Buxton, H. W., and Anthony Hyman. *Memoir of the Life and Labours of the Late Charles Babbage Esq., F.R.S.* Vol. 13 of the Charles Babbage Institute Reprint Series for the History of Computing. Cambridge, Mass.: MIT Press, 1988.

Calude, Cristian S. *Information and Randomness: An Algorithmic Perspective*. Berlin: Springer, 2002.

Calude, Cristian S., and Gregory J. Chaitin. *Randomness and Complexity: From Leibniz to Chaitin*. Singapore, Hackensack, N.J.: World Scientific, 2007.

Campbell-Kelly, Martin. "Charles Babbage's Table of Logarithms (1827)." *Annals of the History of Computing* 10 (1988): 159–69.

Campbell-Kelly, Martin, and William Aspray. *Computer: A History of the Information Machine*. New York: Basic Books, 1996.

Campbell-Kelly, Martin, Mary Croarken, Raymond Flood, and Eleanor Robson, eds. *The History of Mathematical Tables: From Sumer to Spreadsheets*. Oxford: Oxford University Press, 2003.

Campbell, Jeremy. *Grammatical Man: Information, Entropy, Language, and Life*. New York: Simon & Schuster, 1982.

Campbell, Robert V. D. "Evolution of Automatic Computation." In *Proceedings of the 1952 ACM National Meeting (Pittsburgh)*, 29–32. New York: ACM, 1952.

Carr, Nicholas. *The Big Switch: Rewiring the World, from Edison to Google*. New York: Norton, 2008.

———. *The Shallows: What the Internet Is Doing to Our Brains*. New York: Norton, 2010.

Carrington, John F. *A Comparative Study of Some Central African Gong-Languages*. Brussels: Falk, G. van Campenhout, 1949.

———. *The Talking Drums of Africa*. London: Carey Kingsgate, 1949.

———. *La Voix des tambours: comment comprendre le langage tambouriné d'Afrique*. Kinshasa: Centre Protestant d'Éditions et de Diffusion, 1974.

Casson, Herbert N. *The History of the Telephone*. Chicago: A. C. McClurg, 1910.

Cawdrey, Robert. *A Table Alphabeticall of Hard Usual English Words (1604); the First English Dictionary*. Gainesville, Fla.: Scholars' Facsimiles & Reprints, 1966.

Ceruzzi, Paul. *A History of Modern Computing*. Cambridge, Mass.: MIT Press, 2003.

Chaitin, Gregory J. "On the Length of Programs for Computing Finite Binary Sequences." *Journal of the Association for Computing Machinery* 13 (1966): 547–69.

————. "Information-Theoretic Computational Complexity." *IEEE Transactions on Information Theory* 20 (1974): 10–15.

————. *Information, Randomness & Incompleteness: Papers on Algorithmic Information Theory*. Singapore: World Scientific, 1987.

————. *Algorithmic Information Theory*. Cambridge: Cambridge University Press, 1990.

————. *At Home in the Universe*. Woodbury, N.Y.: American Institute of Physics, 1994.

————. *Conversations with a Mathematician*. London: Springer, 2002.

————. *Meta Math: The Quest for Omega*. New York: Pantheon, 2005.

————. "The Limits of Reason." *Scientific American* 294, no. 3 (March 2006): 74.

————. *Thinking About Gödel and Turing: Essays on Complexity, 1970–2007*. Singapore: World Scientific, 2007.

Chandler, Alfred D., and Cortada, James W., eds. "A Nation Transformed By Information: How Information Has Shaped the United States from Colonial Times to the Present." (2000).

Chentsov, Nicolai N. "The Unfathomable Influence of Kolmogorov." *The Annals of Statistics* 18, no. 3 (1990): 987–98.

Cherry, E. Colin. "A History of the Theory of Information." *Transactions of the IRE Professional Group on Information Theory* 1, no. 1 (1953): 22–43.

————. *On Human Communication*. Cambridge, Mass.: MIT Press, 1957.

Chomsky, Noam. "Three Models for the Description of Language." *IRE Transactions on Information Theory* 2, no. 3 (1956): 113–24.

————. *Reflections on Language*. New York: Pantheon, 1975.

Chrisley, Ronald, ed. *Artificial Intelligence: Critical Concepts*. London: Routledge, 2000.

Church, Alonzo. "On the Concept of a Random Sequence." *Bulletin of the American Mathematical Society* 46, no. 2 (1940): 130–35.

Churchland, Patricia S., and Terrence J. Sejnowski. *The Computational Brain*. Cambridge, Mass.: MIT Press, 1992.

Cilibrasi, Rudi, and Paul Vitanyi. "Automatic Meaning Discovery Using Google." *arXiv:cs.CL/0412098 v2*, 2005.

Clanchy, M. T. *From Memory to Written Record, England, 1066–1307*. Cambridge, Mass.: Harvard University Press, 1979.

Clarke, Roger T. "The Drum Language of the Tumba People." *American Journal of Sociology* 40, no. 1 (1934): 34–48.

Clayton, Jay. *Charles Dickens in Cyberspace: The Afterlife of the Nineteenth Century in Postmodern Culture*. Oxford: Oxford University Press, 2003.

Clerke, Agnes M. *The Herschels and Modern Astronomy*. New York: Macmillan, 1895.

Coe, Lewis. *The Telegraph: A History of Morse's Invention and Its Predecessors in the United States*. Jefferson, N.C.: McFarland, 1993.

Colton, F. Barrows. "The Miracle of Talking by Telephone." *National Geographic* 72 (1937): 395–433.

Conway, Flo, and Jim Siegelman. *Dark Hero of the Information Age: In Search of Norbert Wiener, the Father of Cybernetics*. New York: Basic Books, 2005.

Cooke, William Fothergill. *The Electric Telegraph: Was It Invented by Professor Wheatstone?* London: W. H. Smith & Son, 1857.

Coote, Edmund. *The English Schoole-maister*. London: Ralph Jackson & Robert Dexter, 1596.

Cordeschi, Roberto. *The Discovery of the Artificial: Behavior, Mind, and Machines Before and Beyond Cybernetics*. Dordrecht, Netherlands: Springer, 2002.

Cortada, James W. *Before the Computer*. Princeton, N.J.: Princeton University Press, 1993.

Cover, Thomas M., Peter Gacs, and Robert M. Gray. "Kolmogorov's Contributions to Information Theory and Algorithmic Complexity." *The Annals of Probability* 17, no. 3 (1989): 840–65.

Craven, Kenneth. *Jonathan Swift and the Millennium of Madness: The Information Age in Swift's Tale of a Tub*. Leiden, Netherlands: E. J. Brill, 1992.

Crick, Francis. "On Protein Synthesis." *Symposium of the Society for Experimental Biology* 12 (1958): 138–63.

———. "Central Dogma of Molecular Biology." *Nature* 227 (1970): 561–63.

———. *What Mad Pursuit*. New York: Basic Books, 1988.

Croarken, Mary. "Tabulating the Heavens: Computing the Nautical Almanac in 18th-Century England." *IEEE Annals of the History of Computing* 25, no. 3 (2003): 48–61.

———. "Mary Edwards: Computing for a Living in 18th-Century England." *IEEE Annals of the History of Computing* 25, no 4 (2003): 9–15.

Crowley, David, and Paul Heyer, eds. *Communication in History: Technology, Culture, Society*. Boston: Allyn and Bacon, 2003.

Crowley, David, and David Mitchell, eds. *Communication Theory Today*. Stanford, Calif.: Stanford University Press, 1994.

Daly, Lloyd W. *Contributions to a History of Alphabeticization in Antiquity and the Middle Ages*. Brussels: Latomus, 1967.

Danielsson, Ulf H., and Marcelo Schiffer. "Quantum Mechanics, Common Sense, and the Black Hole Information Paradox." *Physical Review D* 48, no. 10 (1993): 4779–84.

Darrow, Karl K. "Entropy." *Proceedings of the American Philosophical Society* 87, no. 5 (1944): 365–67.

Davis, Martin. *The Universal Computer: The Road from Leibniz to Turing*. New York: Norton, 2000.

Dawkins, Richard. "In Defence of Selfish Genes." *Philosophy* 56, no. 218 (1981): 556–73.

———. *The Blind Watchmaker*. New York: Norton, 1986.

———. *The Extended Phenotype*. Rev. ed. Oxford: Oxford University Press, 1999.

———. *The Selfish Gene*. 30th anniversary edition. Oxford: Oxford University Press, 2006.

De Chadarevian, Soraya. "The Selfish Gene at 30: The Origin and Career of a Book and Its Title." *Notes and Records of the Royal Society* 61 (2007): 31–38.

De Morgan, Augustus. *Arithmetical Books: From the Invention of Printing to the Present Time.* London: Taylor & Walton, 1847.

De Morgan, Sophia Elizabeth. *Memoir of Augustus De Morgan.* London: Longmans, Green, 1882.

Delbrück, Max. "A Physicist Looks at Biology." *Transactions of the Connecticut Academy of Arts and Sciences* 38 (1949): 173–90.

Delius, Juan D. "Of Mind Memes and Brain Bugs, a Natural History of Culture." In *The Nature of Culture*, edited by Walter A. Koch. Bochum, Germany: Bochum, 1989.

Denbigh, K. G., and J. S. Denbigh. *Entropy in Relation to Incomplete Knowledge.* Cambridge: Cambridge University Press, 1984.

Dennett, Daniel C. "Memes and the Exploitation of Imagination." *Journal of Aesthetics and Art Criticism* 48 (1990): 127–35.

———. *Consciousness Explained.* Boston: Little, Brown, 1991.

———. *Darwin's Dangerous Idea: Evolution and the Meanings of Life.* New York: Simon & Schuster, 1995.

———. *Brainchildren: Essays on Designing Minds.* Cambridge, Mass.: MIT Press, 1998.

Desmond, Adrian, and James Moore. *Darwin.* London: Michael Joseph, 1991.

Díaz Vera, Javier E. *A Changing World of Words: Studies in English Historical Lexicography, Lexicology and Semantics.* Amsterdam: Rodopi, 2002.

Dilts, Marion May. *The Telephone in a Changing World.* New York: Longmans, Green, 1941.

Diringer, David, and Reinhold Regensburger. *The Alphabet: A Key to the History of Mankind.* 3d ed. New York: Funk & Wagnalls, 1968.

Dretske, Fred I. *Knowledge and the Flow of Information.* Cambridge, Mass.: MIT Press, 1981.

Duane, Alexander. "Sight and Signalling in the Navy." *Proceedings of the American Philosophical Society* 55, no. 5 (1916): 400–14.

Dubbey, J. M. *The Mathematical Work of Charles Babbage.* Cambridge: Cambridge University Press, 1978.

Dupuy, Jean-Pierre. *The Mechanization of the Mind: On the Origins of Cognitive Science.* Translated by M. B. DeBevoise. Princeton, N.J.: Princeton University Press, 2000.

Dyson, George B. *Darwin Among the Machines: The Evolution of Global Intelligence.* Cambridge, Mass.: Perseus, 1997.

Eco, Umberto. *The Search for the Perfect Language.* Translated by James Fentress. Malden, Mass.: Blackwell, 1995.

Edwards, P. N. *The Closed World: Computers and the Politics of Discourse in Cold War America.* Cambridge, Mass.: MIT Press, 1996.

Eisenstein, Elizabeth L. "Clio and Chronos: An Essay on the Making and Breaking of History-Book Times." In *History and Theory* suppl. 6: History and the Concept of Time (1966): 36–64.

———. *The Printing Press as an Agent of Change: Communications and Cultural Transformations in Early-Modern Europe*. Cambridge: Cambridge University Press, 1979.

Ekert, Artur. "Shannon's Theorem Revisited." *Nature* 367 (1994): 513–14.

———. "From Quantum Code-Making to Quantum Code-Breaking." *arXiv:quant-ph/9703035 v1*, 1997.

Elias, Peter. "Two Famous Papers." *IRE Transactions on Information Theory* 4, no. 3 (1958): 99.

Emerson, Ralph Waldo. *Society and Solitude*. Boston: Fields, Osgood, 1870.

Everett, Edward. "The Uses of Astronomy." In *Orations and Speeches on Various Occasions*, 422–65. Boston: Little, Brown, 1870.

Fahie, J. J. *A History of Electric Telegraphy to the Year 1837*. London: E. & F. N. Spon, 1884.

Fauvel, John, and Jeremy Gray. *The History of Mathematics: A Reader*. Mathematical Association of America, 1997.

Feferman, Solomon, ed. *Kurt Gödel: Collected Works*. New York: Oxford University Press, 1986.

Feynman, Richard P. *The Character of Physical Law*. New York: Modern Library, 1994.

———. *Feynman Lectures on Computation*. Edited by Anthony J. G. Hey and Robin W. Allen. Boulder, Colo.: Westview Press, 1996.

Finnegan, Ruth. *Oral Literature in Africa*. Oxford: Oxford University Press, 1970.

Fischer, Claude S. *America Calling: A Social History of the Telephone to 1940*. Berkeley: University of California Press, 1992.

Ford, Joseph. "Directions in Classical Chaos." In *Directions in Chaos*, edited by Hao Bai-lin. Singapore: World Scientific, 1987.

Franksen, Ole I. "Introducing 'Mr. Babbage's Secret.'" *APL Quote Quad* 15, no. 1 (1984): 14–17.

Friedman, William F. "Edgar Allan Poe, Cryptographer." *American Literature* 8, no. 3 (1936): 266–80.

Fuchs, Christopher A. "Notes on a Paulian Idea: Foundational, Historical, Anecdotal and Forward-Looking Thoughts on the Quantum." *arXiv:quant-ph/0105039*, 2001.

———. "Quantum Mechanics as Quantum Information (and Only a Little More)," 2002. *arXiv:quant-ph/0205039 v1*, 8 May 2001.

———. "QBism, the Perimeter of Quantum Bayesianism," *arXiv:quant-ph/1003.5209 vi*, 2010.

———. *Coming of Age with Quantum Information: Notes on a Paulian Idea*. Cambridge, Mass.: Cambridge University Press, 2010.

Galison, Peter. *Image and Logic: A Material Culture of Microphysics*. Chicago: University of Chicago Press, 1997.

Gallager, Robert G. "Claude E. Shannon: A Retrospective on His Life, Work, and Impact." *IEEE Transactions on Information* 47, no. 7 (2001): 2681–95.

Gamow, George. "Possible Relation Between Deoxyribonucleic Acid and Protein Structures." *Nature* 173 (1954): 318.

———. "Information Transfer in the Living Cell." *Scientific American* 193, no. 10 (October 1955): 70.

Gardner, Martin. *Hexaflexagons and Other Mathematical Diversions.* Chicago: University of Chicago Press, 1959.

———. *Martin Gardner's Sixth Book of Mathematical Games from Scientific American.* San Francisco: W. H. Freeman, 1963.

Gasser, James, ed. *A Boole Anthology: Recent and Classical Studies in the Logic of George Boole.* Dordrecht, Netherlands: Kluwer, 2000.

Gell-Mann, Murray, and Seth Lloyd. "Information Measures, Effective Complexity, and Total Information." *Complexity* 2, no. 1 (1996): 44–52.

Genosko, Gary. *Marshall McLuhan: Critical Evaluations in Cultural Theory.* Abingdon, U.K.: Routledge, 2005.

Geoghegan, Bernard Dionysius. "The Historiographic Conceptualization of Information: A Critical Survey." *Annals of the History of Computing* (2008): 66–81.

Gerovitch, Slava. *From Newspeak to Cyberspeak: A History of Soviet Cybernetics.* Cambridge, Mass.: MIT Press, 2002.

Gilbert, E. N. "Information Theory After 18 Years." *Science* 152, no. 3720 (1966): 320–26.

Gilder, Louisa. *The Age of Entanglement: When Quantum Physics Was Reborn.* New York: Knopf, 2008.

Gilliver, Peter, Jeremy Marshall, and Edmund Weiner. *The Ring of Words: Tolkien and the Oxford English Dictionary.* Oxford: Oxford University Press, 2006.

Gitelman, Lisa, and Geoffrey B. Pingree, eds. *New Media 1740–1915.* Cambridge, Mass.: MIT Press, 2003.

Glassner, Jean-Jacques. *The Invention of Cuneiform.* Translated and edited by Zainab Bahrani and Marc Van De Mieroop. Baltimore: Johns Hopkins University Press, 2003.

Gleick, James. *Chaos: Making a New Science.* New York: Viking, 1987.

———. "The Lives They Lived: Claude Shannon, B. 1916; Bit Player." *New York Times Magazine,* 30 December 2001, 48.

———. *What Just Happened: A Chronicle from the Information Frontier.* New York: Pantheon, 2002.

Gödel, Kurt. "Russell's Mathematical Logic" (1944). In *Kurt Gödel: Collected Works,* edited by Solomon Feferman, vol. 2, 119. New York: Oxford University Press, 1986.

Goldsmid, Frederic John. *Telegraph and Travel: A Narrative of the Formation and Development of Telegraphic Communication Between England and India, Under the Orders of Her Majesty's Government, With Incidental Notices of the Countries Traversed By the Lines.* London: Macmillan, 1874.

Goldstein, Rebecca. *Incompleteness: The Proof and Paradox of Kurt Gödel*. New York: Atlas, 2005.

Goldstine, Herman H. "Information Theory." *Science* 133, no. 3462 (1961): 1395–99.

———. *The Computer: From Pascal to Von Neumann*. Princeton, N.J.: Princeton University Press, 1973.

Goodwin, Astley J. H. *Communication Has Been Established*. London: Methuen, 1937.

Goody, Jack. *The Domestication of the Savage Mind*. Cambridge: Cambridge University Press, 1977.

———. *The Interface Between the Written and the Oral*. Cambridge: Cambridge University Press, 1987.

Goody, Jack, and Ian Watt. "The Consequences of Literacy." *Comparative Studies in Society and History* 5, no. 3 (1963): 304–45.

Goonatilake, Susantha. *The Evolution of Information: Lineages in Gene, Culture and Artefact*. London: Pinter, 1991.

Gorman, Michael E. *Transforming Nature: Ethics, Invention and Discovery*. Boston: Kluwer Academic, 1998.

Gould, Stephen Jay. *The Panda's Thumb*. New York: Norton, 1980.

———. "Humbled by the Genome's Mysteries." *The New York Times*, 19 February 2001.

Grafen, Alan, and Mark Ridley, eds. *Richard Dawkins: How a Scientist Changed the Way We Think*. Oxford: Oxford University Press, 2006.

Graham, A. C. *Studies in Chinese Philosophy and Philosophical Literature*. Vol. SUNY Series in Chinese Philosophy and Culture. Albany: State University of New York Press, 1990.

Green, Jonathon. *Chasing the Sun: Dictionary Makers and the Dictionaries They Made*. New York: Holt, 1996.

Gregersen, Niels Henrik, ed. *From Complexity to Life: On the Emergence of Life and Meaning*. Oxford: Oxford University Press, 2003.

Griffiths, Robert B. "Nature and Location of Quantum Information." *Physical Review A* 66 (2002): 012311–1.

Grünwald, Peter, and Paul Vitányi. "Shannon Information and Kolmogorov Complexity." *arXiv:cs.IT/0410002 v1*, 8 August 2005.

Guizzo, Erico Mariu. "The Essential Message: Claude Shannon and the Making of Information Theory." Master's thesis, Massachusetts Institute of Technology, September 2003.

Gutfreund, H., and G. Toulouse. *Biology and Computation: A Physicist's Choice*. Singapore: World Scientific, 1994.

Hailperin, Theodore. "Boole's Algebra Isn't Boolean Algebra." *Mathematics Magazine* 54, no. 4 (1981): 172–84.

Halstead, Frank G. "The Genesis and Speed of the Telegraph Codes." *Proceedings of the American Philosophical Society* 93, no. 5 (1949): 448–58.

Halverson, John. "Goody and the Implosion of the Literacy Thesis." *Man* 27, no. 2 (1992): 301–17.

Harlow, Alvin F. *Old Wires and New Waves*. New York: D. Appleton-Century, 1936.

Harms, William F. "The Use of Information Theory in Epistemology." *Philosophy of Science* 65, no. 3 (1998): 472–501.

Harris, Roy. *Rethinking Writing*. Bloomington: Indiana University Press, 2000.

Hartley, Ralph V. L. "Transmission of Information." *Bell System Technical Journal* 7 (1928): 535–63.

Havelock, Eric A. *Preface to Plato*. Cambridge, Mass.: Harvard University Press, 1963.

———. *The Muse Learns to Write: Reflections on Orality and Literacy from Antiquity to the Present*. New Haven, Conn.: Yale University Press, 1986.

Havelock, Eric Alfred, and Jackson P. Hershbell. *Communication Arts in the Ancient World*. New York: Hastings House, 1978.

Hawking, Stephen. *God Created the Integers: The Mathematical Breakthroughs That Changed History*. Philadelphia: Running Press, 2005.

———. "Information Loss in Black Holes." *Physical Review D* 72, *arXiv:hep-th/0507 171v2*, 2005.

Hayles, N. Katherine. *How We Became Posthuman: Virtual Bodies in Cybernetics, Literature, and Informatics*. Chicago: University of Chicago Press, 1999.

Headrick, Daniel R. *When Information Came of Age: Technologies of Knowledge in the Age of Reason and Revolution, 1700–1850*. Oxford: Oxford University Press, 2000.

Heims, Steve J. *John Von Neumann and Norbert Wiener*. Cambridge, Mass.: MIT Press, 1980.

———. *The Cybernetics Group*. Cambridge, Mass.: MIT Press, 1991.

Herken, Rolf, ed. *The Universal Turing Machine: A Half-Century Survey*. Vienna: Springer-Verlag, 1995.

Hey, Anthony J. G., ed. *Feynman and Computation*. Boulder, Colo.: Westview Press, 2002.

Hobbes, Thomas. *Leviathan, or, the Matter, Forme, and Power of a Commonwealth, Eclesiasticall and Civill*. London: Andrew Crooke, 1660.

Hodges, Andrew. *Alan Turing: The Enigma*. London: Vintage, 1992.

Hofstadter, Douglas R. *Gödel, Escher, Bach: An Eternal Golden Braid*. New York: Basic Books, 1979.

———. *Metamagical Themas: Questing for the Essence of Mind and Pattern*. New York: Basic Books, 1985.

———. *I Am a Strange Loop*. New York: Basic Books, 2007.

Holland, Owen. "The First Biologically Inspired Robots." *Robotica* 21 (2003): 351–63.

Holmes, Oliver Wendell. *The Autocrat of the Breakfast-Table*. New York: Houghton Mifflin, 1893.

Holzmann, Gerard J., and Björn Pehrson. *The Early History of Data Networks*. Washington D.C.: IEEE Computer Society, 1995.

Hopper, Robert. *Telephone Conversation*. Bloomington: Indiana University Press, 1992.

Horgan, John. "Claude E. Shannon." *IEEE Spectrum* (April 1992): 72–75.

Horsley, Victor. "Description of the Brain of Mr. Charles Babbage, F.R.S." *Philosophical Transactions of the Royal Society of London, Series B* 200 (1909): 117–31.

Huberman, Bernardo A. *The Laws of the Web: Patterns in the Ecology of Information.* Cambridge, Mass.: MIT Press, 2001.

Hughes, Geoffrey. *A History of English Words.* Oxford: Blackwell, 2000.

Hüllen, Werner. *English Dictionaries 800–1700: The Topical Tradition.* Oxford: Clarendon Press, 1999.

Hume, Alexander. *Of the Orthographie and Congruitie of the Britan Tongue* (1620). Edited from the original ms. in the British Museum by Henry B. Wheatley. London: Early English Text Society, 1865.

Husbands, Philip, and Owen Holland. "The Ratio Club: A Hub of British Cybernetics." In *The Mechanical Mind in History*, 91–148. Cambridge, Mass.: MIT Press, 2008.

Husbands, Philip, Owen Holland, and Michael Wheeler, eds. *The Mechanical Mind in History.* Cambridge, Mass.: MIT Press, 2008.

Huskey, Harry D., and Velma R. Huskey. "Lady Lovelace and Charles Babbage." *Annals of the History of Computing* 2, no. 4 (1980): 299–329.

Hyatt, Harry Middleton. *Folk-Lore from Adams County, Illinois.* 2nd and rev. ed. Hannibal, Mo.: Alma Egan Hyatt Foundation, 1965.

Hyman, Anthony. *Charles Babbage: Pioneer of the Computer.* Princeton, N.J.: Princeton University Press, 1982.

Hyman, Anthony, ed. *Science and Reform: Selected Works of Charles Babbage.* Cambridge: Cambridge University Press, 1989.

Ifrah, Georges. *The Universal History of Computing: From the Abacus to the Quantum Computer.* New York: Wiley and Sons, 2001.

Ivanhoe, P. J., and Bryan W. Van Norden. *Readings in Classical Chinese Philosophy.* 2nd ed. Indianapolis: Hackett Publishing, 2005.

Jackson, Willis, ed. *Communication Theory.* New York: Academic Press, 1953.

James, William. *Principles of Psychology.* Chicago: Encyclopædia Britannica, 1952.

Jaynes, Edwin T. "Information Theory and Statistical Mechanics." *Physical Review* 106, no. 4 (1957): 620–30.

———. "Where Do We Stand on Maximum Entropy." In *The Maximum Entropy Formalism*, edited by R. D. Levine and Myron Tribus. Cambridge, Mass.: MIT Press, 1979.

Jaynes, Edwin T., Walter T. Grandy, and Peter W. Milonni. *Physics and Probability: Essays in Honor of Edwin T. Jaynes.* Cambridge: Cambridge University Press, 1993.

Jaynes, Julian. *The Origin of Consciousness in the Breakdown of the Bicameral Mind.* Boston: Houghton Mifflin, 1977.

Jennings, Humphrey. *Pandaemonium: The Coming of the Machine as Seen by Contemporary Observers, 1660–1886.* Edited by Mary-Lou Jennings and Charles Madge. New York: Free Press, 1985.

Johannsen, Wilhelm. "The Genotype Conception of Heredity." *American Naturalist* 45, no. 531 (1911): 129–59.

Johns, Adrian. *The Nature of the Book: Print and Knowledge in the Making.* Chicago: University of Chicago Press, 1998.

Johnson, George. *Fire in the Mind: Science, Faith, and the Search for Order.* New York: Knopf, 1995.

———. "Claude Shannon, Mathematician, Dies at 84." *The New York Times,* 27 February 2001, B7.

Johnson, Horton A. "Thermal Noise and Biological Information." *Quarterly Review of Biology* 62, no. 2 (1987): 141–52.

Joncourt, Élie de. *De Natura et Praeclaro Usu Simplicissimae Speciei Numerorum Trigonalium.* Edited by É. de Joncourt Auctore. Hagae Comitum: Husson, 1762.

Jones, Alexander. *Historical Sketch of the Electric Telegraph: Including Its Rise and Progress in the United States.* New York: Putnam, 1852.

Jones, Jonathan. "Quantum Computers Get Real." *Physics World* 15, no. 4 (2002): 21–22.

———. "Quantum Computing: Putting It into Practice." *Nature* 421 (2003): 28–29.

Judson, Horace Freeland. *The Eighth Day of Creation: Makers of the Revolution in Biology.* New York: Simon & Schuster, 1979.

Kahn, David. *The Codebreakers: The Story of Secret Writing.* London: Weidenfeld & Nicolson, 1968.

———. *Seizing the Enigma: The Race to Break the German U-Boat Codes, 1939–1943.* New York: Barnes & Noble, 1998.

Kahn, Robert E. "A Tribute to Claude E. Shannon." *IEEE Communications Magazine* (2001): 18–22.

Kalin, Theodore A. "Formal Logic and Switching Circuits." In *Proceedings of the 1952 ACM National Meeting (Pittsburgh),* 251–57. New York: ACM, 1952.

Kauffman, Stuart. *Investigations.* Oxford: Oxford University Press, 2002.

Kay, Lily E. *Who Wrote the Book of Life: A History of the Genetic Code.* Stanford, Calif.: Stanford University Press, 2000.

Kelly, Kevin. *Out of Control: The Rise of Neo-Biological Civilization.* Reading, Mass.: Addison-Wesley, 1994.

Kendall, David G. "Andrei Nikolaevich Kolmogorov. 25 April 1903–20 October 1987." *Biographical Memoirs of Fellows of the Royal Society* 37 (1991): 301–19.

Keynes, John Maynard. *A Treatise on Probability.* London: Macmillan, 1921.

Kneale, William. "Boole and the Revival of Logic." *Mind* 57, no. 226 (1948): 149–75.

Knuth, Donald E. "Ancient Babylonian Algorithms." *Communications of the Association for Computing Machinery* 15, no. 7 (1972): 671–77.

Kolmogorov, A. N. "Combinatorial Foundations of Information Theory and the Calculus of Probabilities." *Russian Mathematical Surveys* 38, no. 4 (1983): 29–43.

———. *Selected Works of A. N. Kolmogorov. Vol. 3, Information Theory and the Theory of Algorithms.* Translated by A. B. Sossinsky. Dordrecht, Netherlands: Kluwer Academic Publishers, 1993.

Kolmogorov, A. N., I. M. Gelfand, and A. M. Yaglom. "On the General Definition of the Quantity of Information" (1956). In *Selected Works of A. N. Kolmogorov, vol. 3, Information Theory and the Theory of Algorithms*, 2–5. Dordrecht, Netherlands: Kluwer Academic Publishers, 1993.

Kolmogorov, A. N., and A. N. Shiryaev. *Kolmogorov in Perspective. History of Mathematics*, vol. 20. Translated by Harold H. McFaden. N.p.: American Mathematical Society, London Mathematical Society, 2000.

Krutch, Joseph Wood. *Edgar Allan Poe: A Study in Genius*. New York: Knopf, 1926.

Kubát, Libor, and Jiří Zeman. *Entropy and Information in Science and Philosophy*. Amsterdam: Elsevier, 1975.

Langville, Amy N., and Carl D. Meyer. *Google's Page Rank and Beyond: The Science of Search Engine Rankings*. Princeton, N.J.: Princeton University Press, 2006.

Lanier, Jaron. *You Are Not a Gadget*. New York: Knopf, 2010.

Lanouette, William. *Genius in the Shadows*. New York: Scribner's, 1992.

Lardner, Dionysius. "Babbage's Calculating Engines." *Edinburgh Review* 59, no. 120 (1834): 263–327.

———. *The Electric Telegraph*. Revised and rewritten by Edward B. Bright. London: James Walton, 1867.

Lasker, Edward. *The Adventure of Chess*. 2nd ed. New York: Dover, 1959.

Leavitt, Harold J., and Thomas L. Whisler. "Management in the 1980s." *Harvard Business Review* (1958): 41–48.

Leff, Harvey S., and Andrew F. Rex, eds. *Maxwell's Demon: Entropy, Information, Computing*. Princeton, N.J.: Princeton University Press, 1990.

———. *Maxwell's Demon 2: Entropy, Classical and Quantum Information, Computing*. Bristol U.K.: Institute of Physics, 2003.

Lenoir, Timothy, ed. *Inscribing Science: Scientific Texts and the Materiality of Communication*. Stanford, Calif.: Stanford University Press, 1998.

Licklider, J. C. R. "Interview Conducted by William Aspray and Arthur Norberg." (1988).

Lieberman, Phillip. "Voice in the Wilderness: How Humans Acquired the Power of Speech." *Sciences* (1988): 23–29.

Lloyd, Seth. "Computational Capacity of the Universe." *Physical Review Letters* 88, no. 23 (2002). *arXiv:quant-ph/0110141v1*.

———. *Programming the Universe*. New York: Knopf, 2006.

Loewenstein, Werner R. *The Touchstone of Life: Molecular Information, Cell Communication, and the Foundations of Life*. New York: Oxford University Press, 1999.

Lucky, Robert W. *Silicon Dreams: Information, Man, and Machine*. New York: St. Martin's Press, 1989.

Lundheim, Lars. "On Shannon and 'Shannon's Formula.'" *Telektronikk* 98, no. 1 (2002): 20–29.

Luria, A. R. *Cognitive Development: Its Cultural and Social Foundations*. Cambridge, Mass.: Harvard University Press, 1976.

Lynch, Aaron. *Thought Contagion: How Belief Spreads Through Society*. New York: Basic Books, 1996.

Mabee, Carleton. *The American Leonardo: A Life of Samuel F. B. Morse*. New York: Knopf, 1943.

MacFarlane, Alistair G. J. "Information, Knowledge, and the Future of Machines." *Philosophical Transactions: Mathematical, Physical and Engineering Sciences* 361, no. 1809 (2003): 1581–616.

Machlup, Fritz, and Una Mansfield, eds. *The Study of Information: Interdisciplinary Messages*. New York: Wiley and Sons, 1983.

Machta, J. "Entropy, Information, and Computation." *American Journal of Physics* 67, no. 12 (1999): 1074–77.

Mackay, Charles. *Memoirs of Extraordinary Popular Delusions*. Philadelphia: Lindsay & Blakiston, 1850.

MacKay, David J. C. *Information Theory, Inference, and Learning Algorithms*. Cambridge: Cambridge University Press, 2002.

MacKay, Donald M. *Information, Mechanism, and Meaning*. Cambridge, Mass.: MIT Press, 1969.

Macrae, Norman. *John Von Neumann: The Scientific Genius Who Pioneered the Modern Computer, Game Theory, Nuclear Deterrence, and Much More*. New York: Pantheon, 1992.

Macray, William Dunn. *Annals of the Bodleian Library, Oxford, 1598–1867*. London: Rivingtons, 1868.

Mancosu, Paolo. *From Brouwer to Hilbert: The Debate on the Foundations of Mathematics in the 1920s*. New York: Oxford University Press, 1998.

Marland, E. A. *Early Electrical Communication*. London: Abelard-Schuman, 1964.

Martin, Michèle. *"Hello, Central?": Gender, Technology, and Culture in the Formation of Telephone Systems*. Montreal: McGill–Queen's University Press, 1991.

Marvin, Carolyn. *When Old Technologies Were New: Thinking About Electric Communication in the Late Nineteenth Century*. New York: Oxford University Press, 1988.

Maxwell, James Clerk. *Theory of Heat*. 8th ed. London: Longmans, Green, 1885.

Mayr, Otto. "Maxwell and the Origins of Cybernetics." *Isis* 62, no. 4 (1971): 424–44.

McCulloch, Warren S. "Brain and Behavior." *Comparative Psychology Monograph 20* 1, Series 103 (1950).

———. "Through the Den of the Metaphysician." *British Journal for the Philosophy of Science* 5, no. 17 (1954): 18–31.

———. *Embodiments of Mind*. Cambridge, Mass.: MIT Press, 1965.

———. "Recollections of the Many Sources of Cybernetics." *ASC Forum* 6, no. 2 (1974): 5–16.

McCulloch, Warren S., and John Pfeiffer. "Of Digital Computers Called Brains." *Scientific Monthly* 69, no. 6 (1949): 368–76.

McLuhan, Marshall. *The Mechanical Bride: Folklore of Industrial Man*. New York: Vanguard Press, 1951.

————. *The Gutenberg Galaxy*. Toronto: University of Toronto Press, 1962.

————. *Understanding Media: The Extensions of Man*. New York: McGraw-Hill, 1965.

————. *Essential McLuhan*. Edited by Eric McLuhan and Frank Zingrone. New York: Basic Books, 1996.

McLuhan, Marshall, and Quentin Fiore. *The Medium Is the Massage*. New York: Random House, 1967.

McNeely, Ian F., with Lisa Wolverton. *Reinventing Knowledge: From Alexandria to the Internet*. New York: Norton, 2008.

Menabrea, L. F. "Sketch of the Analytical Engine Invented by Charles Babbage. With notes upon the Memoir by the Translator, Ada Augusta, Countess of Lovelace." *Bibliothèque Universelle de Genève* 82 (October 1842). Also available online at http://www.fourmilab.ch/babbage/sketch.html.

Menninger, Karl, and Paul Broneer. *Number Words and Number Symbols: A Cultural History of Numbers*. Dover Publications, 1992.

Mermin, N. David. "Copenhagen Computation: How I Learned to Stop Worrying and Love Bohr." *IBM Journal of Research and Development* 48 (2004): 53–61.

————. *Quantum Computer Science: An Introduction*. Cambridge: Cambridge University Press, 2007.

Miller, George A. "The Magical Number Seven, Plus or Minus Two: Some Limits on Our Capacity for Processing Information." *Psychological Review* 63 (1956): 81–97.

Miller, Jonathan. *Marshall McLuhan*. New York: Viking, 1971.

————. *States of Mind*. New York: Pantheon, 1983.

Millman, S., ed. *A History of Engineering and Science in the Bell System: Communications Sciences (1925–1980)*. Bell Telephone Laboratories, 1984.

Mindell, David A. *Between Human and Machine: Feedback, Control, and Computing Before Cybernetics*. Baltimore: Johns Hopkins University Press, 2002.

Mindell, David A., Jérôme Segal, and Slava Gerovitch. "Cybernetics and Information Theory in the United States, France, and the Soviet Union." In *Science and Ideology: A Comparative History*, edited by Mark Walker, 66–95. London: Routledge, 2003.

Monod, Jacques. *Chance and Necessity: An Essay on the Natural Philosophy of Modern Biology*. Translated by Austryn Wainhouse. New York: Knopf, 1971.

Moore, Francis. *Travels Into the Inland Parts of Africa*. London: J. Knox, 1767.

Moore, Gordon E. "Cramming More Components onto Integrated Circuits." *Electronics* 38, no. 8 (1965): 114–17.

Morowitz, Harold J. *The Emergence of Everything: How the World Became Complex*. New York: Oxford University Press, 2002.

Morse, Samuel F. B. *Samuel F. B. Morse: His Letters and Journals*. Edited by Edward Lind Morse. Boston: Houghton Mifflin, 1914.

Morus, Iwan Rhys. "'The Nervous System of Britain': Space, Time and the Electric Telegraph in the Victorian Age." *British Journal of the History of Science* 33 (2000): 455–75.

Moseley, Maboth. *Irascible Genius: A Life of Charles Babbage, Inventor.* London: Hutchinson, 1964.

Mugglestone, Lynda. "Labels Reconsidered: Objectivity and the *OED.*" *Dictionaries* 21 (2000): 22–37.

———. *Lost for Words: The Hidden History of the Oxford English Dictionary.* New Haven, Conn.: Yale University Press, 2005.

Mulcaster, Richard. *The First Part of the Elementarie Which Entreateth Chefelie of the Right Writing of Our English Tung.* London: Thomas Vautroullier, 1582.

Mullett, Charles F. "Charles Babbage: A Scientific Gadfly." *Scientific Monthly* 67, no. 5 (1948): 361–71.

Mumford, Lewis. *The Myth of the Machine.* Vol. 2, *The Pentagon of Power.* New York: Harcourt, Brace, 1970.

Murray, K. M. E. *Caught in the Web of Words.* New Haven, Conn.: Yale University Press, 1978.

Mushengyezi, Aaron. "Rethinking Indigenous Media: Rituals, 'Talking' Drums and Orality as Forms of Public Communication in Uganda." *Journal of African Cultural Studies* 16, no. 1 (2003): 107–17.

Nagel, Ernest, and James R. Newman. *Gödel's Proof.* New York: New York University Press, 1958.

Napier, John. *A Description of the Admirable Table of Logarithmes.* Translated by Edward Wright. London: Nicholas Okes, 1616.

Nemes, Tihamér. *Cybernetic Machines.* Translated by I. Földes. New York: Gordon & Breach, 1970.

Neugebauer, Otto. *The Exact Sciences in Antiquity.* 2nd ed. Providence, R.I.: Brown University Press, 1957.

———. *A History of Ancient Mathematical Astronomy.* Studies in the History of Mathematics and Physical Sciences, vol. 1. New York: Springer-Verlag, 1975.

Neugebauer, Otto, Abraham Joseph Sachs, and Albrecht Götze. *Mathematical Cuneiform Texts.* American Oriental Series, vol. 29. New Haven, Conn.: American Oriental Society and the American Schools of Oriental Research, 1945.

Newman, M. E. J. "The Structure and Function of Complex Networks." *SIAM Review* 45, no. 2 (2003): 167–256.

Niven, W. D., ed. *The Scientific Papers of James Clerk Maxwell.* Cambridge: Cambridge University Press, 1890; repr. New York: Dover, 1965.

Norman, Donald A. *Things That Make Us Smart: Defending Human Attributes in the Age of the Machine.* Reading, Mass.: Addison-Wesley, 1993.

Nørretranders, Tor. *The User Illusion: Cutting Consciousness Down to Size.* Translated by Jonathan Sydenham. New York: Penguin, 1998.

Noyes, Gertrude E. "The First English Dictionary, Cawdrey's *Table Alphabeticall.*" *Modern Language Notes* 58, no. 8 (1943): 600–605.

Ogilvie, Brian W. "The Many Books of Nature: Renaissance Naturalists and Information Overload." *Journal of the History of Ideas* 64, no. 1 (2003): 29–40.

———. *The Science of Describing: Natural History in Renaissance Europe*. Chicago: University of Chicago Press, 2006.

Olson, David R. "From Utterance to Text: The Bias of Language in Speech and Writing." *Harvard Educational Review* 47 (1977): 257–81.

———. "The Cognitive Consequences of Literacy." *Canadian Psychology* 27, no. 2 (1986): 109–21.

Ong, Walter J. "This Side of Oral Culture and of Print." *Lincoln Lecture* (1973).

———. "African Talking Drums and Oral Noetics." *New Literary History* 8, no. 3 (1977): 411–29.

———. *Interfaces of the Word*. Ithaca, N.Y.: Cornell University Press, 1977.

———. *Orality and Literacy: The Technologizing of the Word*. London: Methuen, 1982.

Oslin, George P. *The Story of Telecommunications*. Macon, Ga.: Mercer University Press, 1992.

Page, Lawrence, Sergey Brin, Rajeev Motwani, and Terry Winograd. "The Pagerank Citation Ranking: Bringing Order to the Web." Technical Report SIDL-WP-1999-0120, Stanford University InfoLab (1998). Available online at http://ilpubs.stanford.edu:8090/422/1/1999-66.pdf.

Pain, Stephanie. "Mr. Babbage and the Buskers." *New Scientist* 179, no. 2408 (2003): 42.

Paine, Albert Bigelow. *In One Man's Life: Being Chapters from the Personal & Business Career of Theodore N. Vail*. New York: Harper & Brothers, 1921.

Palme, Jacob. "You Have 134 Unread Mail! Do You Want to Read Them Now?" In *Computer-Based Message Services*, edited by Hugh T. Smith. North Holland: Elsevier, 1984.

Peckhaus, Volker. "19th Century Logic Between Philosophy and Mathematics." *Bulletin of Symbolic Logic* 5, no. 4 (1999): 433–50.

Peres, Asher. "Einstein, Podolsky, Rosen, and Shannon." *arXiv:quant-ph/0310010 v1*, 2003.

———. "What Is Actually Teleported?" *IBM Journal of Research and Development* 48, no. 1 (2004): 63–69.

Pérez-Montoro, Mario. *The Phenomenon of Information: A Conceptual Approach to Information Flow*. Translated by Dick Edelstein. Lanham, Md.: Scarecrow, 2007.

Peters, John Durham. *Speaking Into the Air: A History of the Idea of Communication*. Chicago: University of Chicago Press, 1999.

Philological Society. *Proposal for a Publication of a New English Dictionary by the Philological Society*. London: Trübner & Co., 1859.

Pickering, John. *A Lecture on Telegraphic Language*. Boston: Hilliard, Gray, 1833.

Pierce, John R. *Symbols, Signals and Noise: The Nature and Process of Communication*. New York: Harper & Brothers, 1961.

———. "The Early Days of Information Theory." *IEEE Transactions on Information Theory* 19, no. 1 (1973): 3–8.

————. *An Introduction to Information Theory: Symbols, Signals and Noise.* 2nd ed. New York: Dover, 1980.

————. "Looking Back: Claude Elwood Shannon." *IEEE Potentials* 12, no. 4 (December 1993): 38–40.

Pinker, Steven. *The Language Instinct: How the Mind Creates Language.* New York: William Morrow, 1994.

————. *The Stuff of Thought: Language as a Window into Human Nature.* New York: Viking, 2007.

Platt, John R., ed. *New Views of the Nature of Man.* Chicago: University of Chicago Press, 1983.

Plenio, Martin B., and Vincenzo Vitelli. "The Physics of Forgetting: Landauer's Erasure Principle and Information Theory." *Contemporary Physics* 42, no. 1 (2001): 25–60.

Poe, Edgar Allan. *Essays and Reviews.* New York: Library of America, 1984.

————. *Poetry and Tales.* New York: Library of America, 1984.

Pool, Ithiel de Sola, ed. *The Social Impact of the Telephone.* Cambridge, Mass.: MIT Press, 1977.

Poundstone, William. *The Recursive Universe: Cosmic Complexity and the Limits of Scientific Knowledge.* Chicago: Contemporary Books, 1985.

Prager, John. *On Turing.* Belmont, Calif.: Wadsworth, 2001.

Price, Robert. "A Conversation with Claude Shannon: One Man's Approach to Problem Solving." *IEEE Communications Magazine* 22 (1984): 123–26.

Pulgram, Ernst. *Theory of Names.* Berkeley, Calif.: American Name Society, 1954.

Purbrick, Louise. "The Dream Machine: Charles Babbage and His Imaginary Computers." *Journal of Design History* 6:1 (1993): 9–23.

Quastler, Henry, ed. *Essays on the Use of Information Theory in Biology.* Urbana: University of Illinois Press, 1953.

————. *Information Theory in Psychology: Problems and Methods.* Glencoe, Ill.: Free Press, 1955.

Radford, Gary P. "Overcoming Dewey's 'False Psychology': Reclaiming Communication for Communication Studies." Paper presented at the 80th Annual Meeting of the Speech Communication Association, New Orleans, November 1994. Available online at http://www.theprofessors.net/dewey.html.

Rattray, Robert Sutherland. "The Drum Language of West Africa: Part I." *Journal of the Royal African Society* 22, no. 87 (1923): 226–36.

————. "The Drum Language of West Africa: Part II." *Journal of the Royal African Society* 22, no. 88 (1923): 302–16.

Redfield, Robert. *The Primitive World and Its Transformations.* Ithaca, N.Y.: Cornell University Press, 1953.

Rényi, Alfréd. *A Diary on Information Theory.* Chichester, N.Y.: Wiley and Sons, 1984.

Rheingold, Howard. *Tools for Thought: The History and Future of Mind-Expanding Technology.* Cambridge, Mass.: MIT Press, 2000.

Rhodes, Frederick Leland. *Beginnings of Telephony*. New York: Harper & Brothers, 1929.

Rhodes, Neil, and Jonathan Sawday, eds. *The Renaissance Computer: Knowledge Technology in the First Age of Print*. London: Routledge, 2000.

Richardson, Robert D. *William James: In the Maelstrom of American Modernism*. New York: Houghton Mifflin, 2006.

Robertson, Douglas S. *The New Renaissance: Computers and the Next Level of Civilization*. Oxford: Oxford University Press, 1998.

———. *Phase Change: The Computer Revolution in Science and Mathematics*. Oxford: Oxford University Press, 2003.

Rochberg, Francesca. *The Heavenly Writing: Divination, Horoscopy, and Astronomy in Mesopotamian Culture*. Cambridge: Cambridge University Press, 2004.

Roederer, Juan G. *Information and Its Role in Nature*. Berlin: Springer, 2005.

Rogers, Everett M. "Claude Shannon's Cryptography Research during World War II and the Mathematical Theory of Communication." In *Proceedings, IEEE 28th International Carnaham Conference on Security Technology*, October 1994: 1–5.

Romans, James. *ABC of the Telephone*. New York: Audel & Co., 1901.

Ronell, Avital. *The Telephone Book: Technology, Schizophrenia, Electric Speech*. Lincoln: University of Nebraska Press, 1991.

Rosenblueth, Arturo, Norbert Wiener, and Julian Bigelow. "Behavior, Purpose and Teleology." *Philosophy of Science* 10 (1943): 18–24.

Rosenheim, Shawn James. *The Cryptographic Imagination: Secret Writing from Edgar Poe to the Internet*. Baltimore: Johns Hopkins University Press, 1997.

Russell, Bertrand. *Logic and Knowledge: Essays, 1901–1950*. London: Routledge, 1956.

Sagan, Carl. *Murmurs of Earth: The Voyager Interstellar Record*. New York: Random House, 1978.

Sapir, Edward. *Language: An Introduction to the Study of Speech*. New York: Harcourt, Brace, 1921.

Sarkar, Sahotra. *Molecular Models of Life*. Cambridge, Mass.: MIT Press, 2005.

Schaffer, Simon. "Babbage's Intelligence: Calculating Engines and the Factory System." *Critical Inquiry* 21, no. 1 (1994): 203–27.

———. "Paper and Brass: The Lucasian Professorship 1820–39." In *From Newton to Hawking: A History of Cambridge University's Lucasian Professors of Mathematics*, edited by Kevin C. Knox and Richard Noakes, 241–94. Cambridge: Cambridge University Press, 2003.

Schindler, G. E., Jr., ed. *A History of Engineering and Science in the Bell System: Switching Technology (1925–1975)*. Bell Telephone Laboratories, 1982.

Schrödinger, Erwin. *What Is Life?* Reprint ed. Cambridge: Cambridge University Press, 1967.

Seife, Charles. *Decoding the Universe*. New York: Viking, 2006.

Shaffner, Taliaferro P. *The Telegraph Manual: A Complete History and Description of the*

Semaphoric, Electric and Magnetic Telegraphs of Europe, Asia, Africa, and America, Ancient and Modern. New York: Pudney & Russell, 1859.

Shannon, Claude Elwood. *Collected Papers.* Edited by N. J. A. Sloane and Aaron D. Wyner. New York: IEEE Press, 1993.

———. *Miscellaneous Writings.* Edited by N. J. A. Sloane and Aaron D. Wyner. Murray Hill, N.J.: Mathematical Sciences Research Center, AT&T Bell Laboratories, 1993.

Shannon, Claude Elwood, and Warren Weaver. *The Mathematical Theory of Communication.* Urbana: University of Illinois Press, 1949.

Shenk, David. *Data Smog: Surviving the Information Glut.* New York: HarperCollins, 1997.

Shieber, Stuart M., ed. *The Turing Test: Verbal Behavior as the Hallmark of Intelligence.* Cambridge, Mass.: MIT Press, 2004.

Shiryaev, A. N. "Kolmogorov: Life and Creative Activities." *Annals of Probability* 17, no. 3 (1989): 866–944.

Siegfried, Tom. *The Bit and the Pendulum: From Quantum Computing to M Theory— The New Physics of Information.* New York: Wiley and Sons, 2000.

Silverman, Kenneth. *Lightning Man: The Accursed Life of Samuel F. B. Morse.* New York: Knopf, 2003.

Simpson, John. "Preface to the Third Edition of the *Oxford English Dictionary.*" Oxford University Press, http://oed.com/about/oed3-preface/#general (accessed 13 June 2010).

Simpson, John, ed. *The First English Dictionary, 1604: Robert Cawdrey's A Table Alphabeticall.* Oxford: Bodleian Library, 2007.

Singh, Jagjit. *Great Ideas in Information Theory, Language and Cybernetics.* New York: Dover, 1966.

Singh, Simon. *The Code Book: The Secret History of Codes and Codebreaking.* London: Fourth Estate, 1999.

Slater, Robert. *Portraits in Silicon.* Cambridge, Mass.: MIT Press, 1987.

Slepian, David. "Information Theory in the Fifties." *IEEE Transactions on Information Theory* 19, no. 2 (1973): 145–48.

Sloman, Aaron. *The Computer Revolution in Philosophy.* Hassocks, Sussex: Harrester Press, 1978.

Smith, D. E. *A Source Book in Mathematics.* New York: McGraw-Hill, 1929.

Smith, Francis O. J. *The Secret Corresponding Vocabulary; Adapted for Use to Morse's Electro-Magnetic Telegraph: And Also in Conducting Written Correspondence, Transmitted by the Mails, or Otherwise.* Portland, Maine: Thurston, Ilsley, 1845.

Smith, G. C. *The Boole–De Morgan Correspondence 1842–1864.* Oxford: Clarendon Press, 1982.

Smith, John Maynard. "The Concept of Information in Biology." *Philosophy of Science* 67 (2000): 177–94.

Smolin, J. A. "The Early Days of Experimental Quantum Cryptography." *IBM Journal of Research and Development* 48 (2004): 47–52.

Solana-Ortega, Alberto. "The Information Revolution Is Yet to Come: An Homage to Claude E. Shannon." In *Bayesian Inference and Maximum Entropy Methods in Science and Engineering*, AIP Conference Proceedings 617, edited by Robert L. Fry. Melville, N.Y.: American Institute of Physics, 2002.

Solomonoff, Ray J. "A Formal Theory of Inductive Inference." *Information and Control* 7, no. 1 (1964): 1–22.

———. "The Discovery of Algorithmic Probability." *Journal of Computer and System Sciences* 55, no. 1 (1997): 73–88.

Solymar, Laszlo. *Getting the Message: A History of Communications*. Oxford: Oxford University Press, 1999.

Spellerberg, Ian F., and Peter J. Fedor. "A Tribute to Claude Shannon (1916–2001) and a Plea for More Rigorous Use of Species Richness, Species Diversity and the 'Shannon-Wiener' Index," *Global Ecology and Biogeography* 12 (2003): 177–79.

Sperry, Roger. "Mind, Brain, and Humanist Values." In *New Views of the Nature of Man*, edited by John R. Platt, 71–92. Chicago: University of Chicago Press, 1983.

Sprat, Thomas. *The History of the Royal Society of London, for the Improving of Natural Knowledge*. 3rd ed. London: 1722.

Spufford, Francis, and Jenny Uglow, eds. *Cultural Babbage: Technology, Time and Invention*. London: Faber and Faber, 1996.

Standage, Tom. *The Victorian Internet: The Remarkable Story of the Telegraph and the Nineteenth Century's On-Line Pioneers*. New York: Berkley, 1998.

Starnes, De Witt T., and Gertrude E. Noyes. *The English Dictionary from Cawdrey to Johnson 1604–1755*. Chapel Hill: University of North Carolina Press, 1946.

Steane, Andrew M., and Eleanor G. Rieffel. "Beyond Bits: The Future of Quantum Information Processing." *Computer* 33 (2000): 38–45.

Stein, Gabriele. *The English Dictionary Before Cawdrey*. Tübingen, Germany: Max Neimeyer, 1985.

Steiner, George. "On Reading Marshall McLuhan." In *Language and Silence: Essays on Language, Literature, and the Inhuman*, 251–68. New York: Atheneum, 1967.

Stent, Gunther S. "That Was the Molecular Biology That Was." *Science* 160, no. 3826 (1968): 390–95.

———. "DNA." *Daedalus* 99 (1970): 909–37.

———. "You Can Take the Ethics Out of Altruism But You Can't Take the Altruism Out of Ethics." *Hastings Center Report* 7, no. 6 (1977): 33–36.

Stephens, Mitchell. *The Rise of the Image, the Fall of the Word*. Oxford: Oxford University Press, 1998.

Stern, Theodore. "Drum and Whistle 'Languages': An Analysis of Speech Surrogates." *American Anthropologist* 59 (1957): 487–506.

Stix, Gary. "Riding the Back of Electrons." *Scientific American* (September 1998): 32–33.

Stonier, Tom. *Beyond Information: The Natural History of Intelligence*. London: Springer-Verlag, 1992.

―――. *Information and Meaning: An Evolutionary Perspective.* Berlin: Springer-Verlag, 1997.

Streufert, Siegfried, Peter Suedfeld, and Michael J. Driver. "Conceptual Structure, Information Search, and Information Utilization." *Journal of Personality and Social Psychology* 2, no. 5 (1965): 736–40.

Sunstein, Cass R. *Infotopia: How Many Minds Produce Knowledge.* Oxford: Oxford University Press, 2006.

Surowiecki, James. *The Wisdom of Crowds.* New York: Doubleday, 2004.

Swade, Doron. "The World Reduced to Number." *Isis* 82, no. 3 (1991): 532–36.

―――. *The Cogwheel Brain: Charles Babbage and the Quest to Build the First Computer.* London: Little, Brown, 2000.

―――. *The Difference Engine: Charles Babbage and the Quest to Build the First Computer.* New York: Viking, 2001.

Swift, Jonathan. *A Tale of a Tub: Written for the Universal Improvement of Mankind.* 1692.

Szilárd, Leó. "On the Decrease of Entropy in a Thermodynamic System by the Intervention of Intelligent Beings." Translated by Anatol Rapoport and Mechtilde Knoller from "*Über Die Entropieverminderung in Einem Thermodynamischen System Bei Eingriffen Intelligenter Wesen,*" *Zeitschrift Für Physik* 53 (1929). *Behavioral Science* 9, no. 4 (1964): 301–10.

Teilhard de Chardin, Pierre. *The Human Phenomenon.* Translated by Sarah Appleton-Weber. Brighton, U.K.: Sussex Academic Press, 1999.

Terhal, Barbara M. "Is Entanglement Monogamous?" *IBM Journal of Research and Development* 48, no. 1 (2004): 71–78.

Thompson, A. J., and Karl Pearson. "Henry Briggs and His Work on Logarithms." *American Mathematical Monthly* 32, no. 3 (1925): 129–31.

Thomsen, Samuel W. "Some Evidence Concerning the Genesis of Shannon's Information Theory." *Studies in History and Philosophy of Science* 40 (2009): 81–91.

Thorp, Edward O. "The Invention of the First Wearable Computer." In *Proceedings of the 2nd IEEE International Symposium on Wearable Computers.* Washington, D.C.: IEEE Computer Society, 1998.

Toole, Betty Alexandra. "Ada Byron, Lady Lovelace, an Analyst and Metaphysician." *IEEE Annals of the History of Computing* 18, no. 3 (1996): 4–12.

―――. *Ada, the Enchantress of Numbers: Prophet of the Computer Age.* Mill Valley, Calif.: Strawberry Press, 1998.

Tufte, Edward R. "The Cognitive Style of PowerPoint." Cheshire, Conn.: Graphics Press, 2003.

Turing, Alan M. "On Computable Numbers, with an Application to the *Entscheidungs-problem.*" *Proceedings of the London Mathematical Society* 42 (1936): 230–65.

―――. "Computing Machinery and Intelligence." *Minds and Machines* 59, no. 236 (1950): 433–60.

―――. "The Chemical Basis of Morphogenesis." *Philosophical Transactions of the Royal Society of London, Series B* 237, no. 641 (1952): 37–72.

Turnbull, Laurence. *The Electro-Magnetic Telegraph, With an Historical Account of Its Rise, Progress, and Present Condition.* Philadelphia: A. Hart, 1853.

Vail, Alfred. *The American Electro Magnetic Telegraph: With the Reports of Congress, and a Description of All Telegraphs Known, Employing Electricity Or Galvanism.* Philadelphia: Lea & Blanchard, 1847.

Verdú, Sergio. "Fifty Years of Shannon Theory." *IEEE Transactions on Information Theory* 44, no. 6 (1998): 2057–78.

Vincent, David. *Literacy and Popular Culture: England 1750–1914.* Cambridge: Cambridge University Press, 1989.

Virilio, Paul. *The Information Bomb.* Translated by Chris Turner. London: Verso, 2000.

von Baeyer, Hans Christian. *Maxwell's Demon: Why Warmth Disperses and Time Passes.* New York: Random House, 1998.

———. *Information: The New Language of Science.* Cambridge, Mass.: Harvard University Press, 2004.

von Foerster, Heinz. *Cybernetics: Circular Causal and Feedback Mechanisms in Biological and Social Systems: Transactions of the Seventh Conference, March 23–24, 1950.* New York: Josiah Macy, Jr. Foundation, 1951.

———. *Cybernetics: Circular Causal and Feedback Mechanisms in Biological and Social Systems: Transactions of the Eighth Conference, March 15–16, 1951.* New York: Josiah Macy, Jr. Foundation, 1952.

———. "Interview with Stefano Franchi, Güven Güzeldere, and Eric Minch." *Stanford Humanities Review* 4, no. 2 (1995). Available online at http://www.stanford.edu/group/SHR/4-2/text/interviewvonf.html.

von Neumann, John. *The Computer and the Brain.* New Haven, Conn.: Yale University Press, 1958.

———. *Collected Works.* Vols. 1–6. Oxford: Pergamon Press, 1961.

Vulpiani, A., and Roberto Livi. *The Kolmogorov Legacy in Physics: A Century of Turbulence and Complexity.* Lecture Notes in Physics, no. 642. Berlin: Springer, 2003.

Waldrop, M. Mitchell. "Reluctant Father of the Digital Age." *Technology Review* (July–August 2001): 64–71.

Wang, Hao. "Some Facts About Kurt Gödel." *Journal of Symbolic Logic* 46 (1981): 653–59.

Watson, David L. "Biological Organization." *Quarterly Review of Biology* 6, no. 2 (1931): 143–66.

Watson, James D. *The Double Helix.* New York: Atheneum, 1968.

———. *Genes, Girls, and Gamow: After the Double Helix.* New York: Knopf, 2002.

———. *Molecular Models of Life.* Oxford: Oxford University Press, 2003.

Watson, James D., and Francis Crick. "A Structure for Deoxyribose Nucleic Acid." *Nature* 171 (1953): 737.

———. "Genetical Implications of the Structure of Deoxyribonucleic Acid." *Nature* 171 (1953): 964–66.

Watts, Duncan J. "Networks, Dynamics, and the Small-World Phenomenon." *American Journal of Sociology* 105, no. 2 (1999): 493–527.

———. *Small Worlds: The Dynamics of Networks Between Order and Randomness*. Princeton, N.J.: Princeton University Press, 1999.

———. *Six Degrees: The Science of a Connected Age*. New York: Norton, 2003.

Watts, Duncan J., and Steven H. Strogatz. "Collective Dynamics of 'Small-World' Networks." *Nature* 393 (1998): 440–42.

Weaver, Warren. "The Mathematics of Communication." *Scientific American* 181, no. 1 (1949): 11–15.

Wells, H. G. *World Brain*. London: Methuen, 1938.

———. *A Short History of the World*. San Diego: Book Tree, 2000.

Wheeler, John Archibald. "Information, Physics, Quantum: The Search for Links." *Proceedings of the Third International Symposium on the Foundations of Quantum Mechanics* (1989): 354–68.

———. *At Home in the Universe*. *Masters of Modern Physics*, vol. 9. New York: American Institute of Physics, 1994.

Wheeler, John Archibald, with Kenneth Ford. *Geons, Black Holes, and Quantum Foam: A Life in Physics*. New York: Norton, 1998.

Whitehead, Alfred North, and Bertrand Russell. *Principia Mathematica*. Cambridge: Cambridge University Press, 1910.

Wiener, Norbert. *Cybernetics: Or Control and Communication in the Animal and the Machine*. 2nd ed. Cambridge, Mass.: MIT Press, 1961.

———. *I Am a Mathematician: The Later Life of a Prodigy*. Cambridge, Mass.: MIT Press, 1964.

Wiener, Philip P., ed. *Leibniz Selections*. New York: Scribner's, 1951.

Wilkins, John. *Mercury: Or the Secret and Swift Messenger. Shewing, How a Man May With Privacy and Speed Communicate His Thoughts to a Friend At Any Distance*. 3rd ed. London: John Nicholson, 1708.

Williams, Michael. *A History of Computing Technology*. Washington, D.C.: IEEE Computer Society, 1997.

Wilson, Geoffrey. *The Old Telegraphs*. London: Phillimore, 1976.

Winchester, Simon. *The Meaning of Everything: The Story of the Oxford English Dictionary*. Oxford: Oxford University Press, 2003.

Wisdom, J. O. "The Hypothesis of Cybernetics." *British Journal for the Philosophy of Science* 2, no. 5 (1951): 1–24.

Wittgenstein, Ludwig. *Philosophical Investigation*. Translated by G. E. M. Anscombe. New York: Macmillan, 1953.

———. *Remarks on the Foundations of Mathematics*. Cambridge, Mass.: MIT Press, 1967.

Woodward, Kathleen. *The Myths of Information: Technology and Postindustrial Culture*. Madison, Wisc.: Coda Press, 1980.

Woolley, Benjamin. *The Bride of Science: Romance, Reason, and Byron's Daughter*. New York: McGraw-Hill, 1999.

Wynter, Andrew. "The Electric Telegraph." *Quarterly Review* 95 (1854): 118–64.

———. *Subtle Brains and Lissom Fingers: Being Some of the Chisel-Marks of Our Industrial and Scientific Progress*. London: Robert Hardwicke, 1863.

Yeo, Richard. "Reading Encyclopedias: Science and the Organization of Knowledge in British Dictionaries of Arts and Sciences, 1730–1850." *Isis* 82:1 (1991): 24–49.

———. *Encyclopædic Visions: Scientific Dictionaries and Enlightenment Culture*. Cambridge: Cambridge University Press, 2001.

Yockey, Hubert P. *Information Theory, Evolution, and the Origin of Life*. Cambridge: Cambridge University Press, 2005.

Young, Peter. *Person to Person: The International Impact of the Telephone*. Cambridge: Granta, 1991.

Yourgrau, Palle. *A World Without Time: The Forgotten Legacy of Gödel and Einstein*. New York: Basic Books, 2005.

Yovits, Marshall C., George T. Jacobi, and Gordon D. Goldstein, eds. *Self-Organizing Systems*. Washington D.C.: Spartan, 1962.

Index

It is much easier to talk about information than it is to say what it is you are talking about. A surprising number of books, and this includes textbooks, have the word information *in their title without bothering to include it in the index.*

—Fred I. Dretske (1979)

Page numbers in italics *refer to illustrations.*

Baudot code, 202
Bavelas, Alex, 248
Beethoven, Ludwig von, 321, 409, 420
Bell, Alexander Graham, 188, 190
Bell, Gordon, 397
Bell Laboratories, 3–5, 6, 26, 73, 172, 175, 192, 196, 198, 220–21, 231, 266–67
Bell System Technical Journal, 4, 202, 221
Bell Telephone Company, 189
Bennett, Charles H., 320–21, 353–54, 359–65, 367, 369, 371–72, 407
Benton, Billy, 389
Benzer, Seymour, 300–301
Bernoulli numbers, 117, 118–19
Berry, G. G., 179–80
Berry's paradox, 179–80, 339–40, 341
Bible, 400
Bierce, Ambrose, 66
Bigelow, Julian, 237
binary operations
    coding systems for, 160–61
    representation of relay circuits as, 174
    in telegraphy, 224, 238
    in use of alphabetical ordering systems, 58
    see also bit(s)
biology
    entropy and, 281–84
    evolutionary, 301, 321
    fundamental particles of, 287
    of human ecosystem, 305–6
    information processing in, 8–9, 289, 298–300
    molecular, 282, 299–300, 301
    purposeful action in processes of, 277–78
    see also genetics; neurophysiology
biosphere, 310–11, 323
bit(s)
    as basis of physics, 9–10, 356–57
    biological measurements, 289
    cost of information processing, 361
    data compression strategies, 344–45
    decision-making requirements, 261
    definition of, 4, 229
    first usage, 4
    growth of measuring units, 393–94

meaning and, 372
measurement of cosmos in, 11, 397
purpose, 4
transmission by fire beacon, 16–17
black holes, 11, 356, 357–59
Blair, Ann, 411
Blair, Earl, 387
Bletchley Park, 213–14, 239
Blount, Thomas, 63–64
Bodleian Library, 58–59, 179
Bohr, Niels, 10, 200, 356
Boltzmann, Ludwig, 372–73
Bombe machine, 214
book burning, 378
Boole, George, 6, 163, 164–67, 174, 178, 240–41, 324
Borges, Jorge Luis, 373, 374, 417, 426
botanical dictionaries, 392–93, 411
Bradley, Henry, 65, 68
Brahe, Tycho, 87, 400
brain; *see* neurophysiology
Brassard, Gilles, 363, 364
"Breakdown of Physics in Gravitational Collapse, The" (Hawking), 358
Brecht, Bertolt, 412
Breguet, Abraham-Louis, 131, 139
Brenner, Sydney, 294, 298–99
Brewster, David, 99, 257
Bridenbaugh, Carl, 398, 400–401
Briggs, Henry, 84–85, 86–87, 88
Brillouin, Léon, 273, 285, 286
Brin, Sergey, 395, 423
Broadbent, Donald, 259–60
Brosin, Henry, 251
Brown, Robert, 198
Browne, Thomas, 17, 18, 127
Brownian motion, 197–98, 237
Brunel, Isambard Kingdom, 114
Buchanan, James, 146
Bullokar, John, 63
Burgess, Anthony, 69
Burney, Venetia, 64
Burton, Robert, 401–2, 411
Bush, Vannevar, 6, 7, 129n, 171, 172, 175, 187, 215
Butler, Samuel, 31, 302–3
butterfly effect, 377
Byron, Augusta Ada; *see* Lovelace, Ada

Clytemnestra, 16
code
    attempts to reduce cost of telegraphy,
        152–53
    Babbage's interest in, 121
    cipher and compression systems for
        telegraphy, 153–58
    Enigma, 204, 213–14
    genetic, 288–89, 291–98
    in Jacquard loom operations, 109
    Morse, 5, 19–22, 142–43, 152, 165,
        201, 315
    as noise, 216
    for printing telegraph, 202
    Shannon's interest in, 6, 168, 213
    telegraphy before Morse code, 131,
        136–39, 141–42
    see also cryptography
coding theory, 264, 267, 297, 335
cognitive science, 259–62
Colebrooke, Henry, 99
collective consciousness, 413–15,
        420–21
Colossus computing machine, 239
Columbus, Christopher, 83
combinatorial analysis, 197, 293–94
communication
    by algorithm, 349
    with alien life-form, 349–53
    Babbage's mechanical notation for
        describing, 102–3, 163
    constrained channels of, 48–49
    disruptive effects of new technologies
        in, 411–12
    emergence of global consciousness,
        413–15
    evolution of electrical technologies for,
        128–29, 169–70
    fundamental problem of, 3, 221–22,
        259
    human evolution and, 11–12
    implications of technological evolution
        of, 398–99
    information overload and, 417–18
    knowledge needs for, 349–51
    in origins of governance, 43
    Shannon's diagram of, 221–23
    as stochastic process, 225

symbolic logic to describe systems of,
        176
    system elements, 222–23
    in Twitter, 419–20
    see also talking drums; telegraphy; tele-
        phony; transmission of information
compact disc, 8, 264, 420
complexity, 326, 336–37, 341–42, 343,
        346, 353–54
compression of information; see data
        compression
"Computable Numbers, On" (Turing),
        205, 206, 330
computation
    in Babylonian mathematics, 45–46
    computable and uncomputable num-
        bers, 207–8, 211–12, 330–32
    of differential equations, 94–95
    in evolution of complex structures, 354
    human computers, 84, 94–95
    thermodynamics of, 359–62
    Turing machine for, 207–12
    see also calculators; computers
computer(s)
    analog and digital, 240, 244
    chess-playing, 265–66
    comparison to humans, 240–41
    cost of memory storage, 395
    cost of work of, 360–61
    early mechanical, 6, 80–81, 92–94, 123,
        239
    growth of memory and processing speed
        of, 393–96
    inductive learning in, 345
    perception of thinking by, 239–40,
        243–44
    public awareness of, 239
    quantum-based, 368–71
    Shannon's information theory in, 8,
        175, 230–31, 264–65
    significance of information theory in
        development of, 8
    spread of memes through, 318
    Turing's conceptualization of, 254–56
    universe as, 377–78
    see also calculators; computation;
        programming
Conference on Cybernetics, 242–52, 263

Differential and Integral Calculus (Lacroix), 89

differential equations, 172, 187–88

"Digital Computers Called Brains, Of" (McCulloch), 245

digital technology, 199, 244, 254, 264, 330

Diringer, David, 34

discrete information, 199, 215, 223, 351

D'Israeli, Isaac, 378

Disreali, Benjamin, 92

distortion of signal; *see* noise

DNA; *see* deoxyribonucleic acid

"Does One Sometimes Know Too Much?," 405

domain names, 391–92

Donne, John, 426

Doob, Joseph L., 233

Dowd, Maureen, 319

Doyle, Arthur Conan, 171

Dretske, Fred, 323, 417, 505

drums; *see* talking drums

Dupuy, Jean-Pierre, 413, 417–18

Dyer, Harrison Gray, 137

echo, 31

Eckart, Carl, 279

Eckert, W. H., 195

economics

  Babbage's research on, 78–79

  business of telegraphy, 144–45, 155–57

  commercial interest in telegraphy, 139–40

  commercial interest in telephony, 191–92

  cost of computation, 360–61

  costs of computer memory, 395

  in information cloud, 396

  as information science, 9

  of number table production, 83–84, 87, 88

  origins of mathematics and, 42

Edison, Thomas A., 126, 350

Edwards, Mary, 84

Egypt, 58, 59

Einstein, Albert, 186, 197–98, 278, 355, 357, 365–66

Eisenstein, Elizabeth, 399–401

electrical circuits

  development of telegraphy, 19, 20

  noise in, 198, 204

  symbolic logic and, 6, 173–75, 176–77

  transmission capacity of, 199–200

electricity

  amplitude modulation, 192–93, 199

  biological analogies for, 126

  evolution of scientific understanding of, 127–28

  in measurement of communication, 5

  public response to new technologies of, 127

  recognition of communication potential of, 128–29

  source of noise in, 198

  technical demands of telephony, 192–93

  see also electrical circuits; telegraphy

Electric Telegraph Company, 125, 148

Elements of Electro-Biology (Smee), 126

Elias, Peter, 267–68

Eliot, T. S., 69, 403

Elyot, Thomas, 7

e-mail, 48, 316, 320, 396, 404–5, 406

Emerson, Ralph Waldo, 120

Encyclopaedia Britannica, 232, 380, 382–83, 386, 415

encyclopedias, 12, 415–16; *see also specific encyclopedia*

Encyclopédie, 382

Enderton, Herbert, 212

energy

  in concept of entropy, 270–72

  cost of information processing, 360–61

  information and, 11, 279–81, 355

  Maxwell's demon, 234, 275–79, 281–82, 283, 286, 304, 361, 362, 407, 425

  perpetual motion machine, 278–80

  in physics of black holes, 357

  see also thermodynamics

England, 141–42, 144, 147–48, 252–53; *see also* English language

English Expositour, An (Bullokar), 63

Maxwell, James Clerk, 168, 190, 192–93, 238, 270, 271, 272, 273, 274, 275, 276

Maxwell's demon, 234, 275–79, 281–82, 283, 286, 304, 361, 362, 407, 425

Maynard Smith, John, 307

McCarthy, John, 345

McCulloch, Warren, 241–42, 251, 253, 263

McLuhan, Marshall, 8, 28–29, 31, 48, 49, 263, 400, 413

Mead, Margaret, 242

meaning
  in agenda for quantum information science, 372
  attempts to incorporate, into information theory, 417
  expressed through differences, 161
  future of science and, 372
  information overload and, 417–18, 425–26
  language and, 418–19
  measurement of communication and, 27
  of numbers, 81
  in perfect language, 418
  Shannon's information theory and, 3, 219, 222, 246, 248–49, 416–17
  symbolic logic and, 165, 166, 182
  talking drum method of conveying, 13, 24–25
  use of alphabetical ordering systems and, 58
  use of tonality to convey, 23
  see also definitions of words

measurement of information
  algorithmic, 332, 336–38
  combinatorial approach to, 336
  conceptual evolution of, 5–6, 7–11, 27, 200–201
  cosmic calculations, 10–11, 397
  expanding scale of, 393–96
  measurement of message value and, 353–54
  measurement of randomness and, 329
  as measure of uncertainty, 228, 280–81
  in music, 351–53

probabilistic approach to, 228–29, 230, 247, 336
  in psychology research, 259
  quantifying redundancy for, 26–27, 229–30
  quantizing speech for, 245
  symbols as unit for, 200–202
  in telephony, 5, 199–203
  Turing's approach to, 214–15
  see also bit(s)

Medawar, Peter, 414

meme(s); memetics
  catchphrases as, 312–13
  chain letters as, 319–21
  conceptual origins of, 311–12, 315–16
  definition of, 9, 312, 313
  disease analogy for, 316–17
  effects, 315
  forms of, 312–13, 322
  genetic model of, 321
  humans as vehicles for, 317
  ideas as, 312
  images as, 313
  as living structures, 315
  mission of, 314
  music as, 312
  replication through imitation, 316
  scholarly research on, 319, 321
  transmission of, 313–14, 316

memory
  aids in oral literature, 34
  computer, cost of, 395
  evolution of information technology and, 400–401, 407
  in machine functions, 101
  in maze-navigating machine, 249–52
  meme strategies, 314
  psychology research on, 259–62
  quantum erasure of, 371
  writing and, 30, 31

Menabrea, Luigi, 115

Mencken, H. L., 73, 314

Mendel, Gregor, 288

Mercury: or the Secret and Swift Messenger (Wilkins), 159

Merlin, John, 88

Mermin, David, 364*n*, 369

Solomonoff, Ray, 338, 344–45, 346
Sömmerring, Samuel Thomas von, 137
Sophocles, 378, 409
Southwell, Robert, 53
Soviet Union, 333–35
space exploration, 264
Speculum Maius (Vincent of Beauvais), 411
spelling, 53–54, 69–70
Spender, Stephen, 69
Sperry, Roger, 311
"spooky action at a distance," (Einstein), 366–67
Sprat, Thomas, 42, 51
statistical analysis, 224–27, 273, 274, 347
steam power, 78, 92, 120, 269–70
Stent, Gunther, 294, 300, 302
Stevin, Simon, 83
stochastic processes, 216, 225
Stoppard, Tom, 273–74, 379
storage of information
    Shannon's early calculations on, 231–32
    sources of confusion in, 373–74
    trends in, 396–97
Streufert, Siegfried, 405–6
Strogatz, Steven, 423–24
Stuart, Gilbert, 313
Suetonius, 16
superposition of states, 365, 369
Surowiecki, James, 420
surprise, as feature of information, 219, 281
Susskind, Leonard, 358
syllabary, 33
symbolic logic
    application to genetics, 175–76, 186
    to avoid paradox, 41
    conceptual basis, 164–66
    conceptual origins of computers in, 176–77
    to describe communication systems, 176
    to describe relay circuits, 6, 173–74
    goals of *Principia Mathematica,* 178, 181–82
    incompleteness of formal systems of, 182–85, 207

    as mechanical operation, 178, 182, 205–6
    promise of, 177–78
    search for perfect system of, 178
symbols and symbol sets
    in Babbage's mechanical notation, 102–3, 163, 174
    for cryptography, 209
    fo universal language, 90
    in Lovelace's game solution formula, 112
    for measurement of information, 200–202
    for perfect language, 418
    redundancy of communication determined by, 26–27
    in structure of language, 74
    for Turing machine, 209
    see also alphabet(s); code; symbolic logic; writing
Szilárd, Leó, 278–80, 286, 361

"Table Alphabeticall, A" (Cawdrey), 51–53, *52,* 55–58, 59, 60–63, 64, 67, 68–69, 70, 74
Table of Constants of the Class Mammalia (Babbage), 81
Table of the Relative Frequency of the Causes of Breaking of Plate Glass Windows (Babbage), 82
Table of Triangular Numbers, (Babbage), 95
Tables for the Improvement of Navigation (Briggs), 84
Table to find the Height of the Pole (Briggs), 84
Tafelen van Interest (Stevin), 83
Talbot, William Fox, 376
talking drums, 13–15, 16, 18, 21–25, 26, 27
Talking Drums of Africa, The (Carrington), 22
Tawell, John, 144
Teilhard de Chardin, Pierre, 414
telegraphy, 5, 26, 123
    address codes, 388–89
    Baudot code for, 202
    bubble, 137

telegraphy *(continued)*
  cipher and compression systems for, 153–58
  as commercial business, 144–45
  commercial interest in, 139–40
  conceptual understanding of, 150–51
  early systems for, 129–40
  electrical relays in, 143
  before electricity, 129–32
  in England, 141–42, 144
  errors in, 158
  in France, 129–36
  growth of, 125–26, 144–50
  infrastructure of, 151
  invention of, 140–42, 143
  as medium, 153–54
  operator's key, 143
  perception of time and, 148–49
  preservation of messages sent by, 149–50
  private ciphers to reduce cost of, 152–53
  public interest in codes and, 161–62
  in Soviet Union, 335
  statistical structure of language in, 224–25
  telephony and, 188–89, 190
  trans-Atlantic, 146, 157
  waveform analysis in, 199
  weather reporting and, 147–48
  see also Morse code
telephony
  architecture and, 192
  barbed-wire networks, 169
  biological metaphors for, 126
  commercial applications of, 191–92
  concern about social effects of, 170
  demand for information and, 407–8
  electrical engineering requirements of, 192–93
  evolution of switching technology for, 193–96
  farmer cooperative networks of, 170
  growth of, 188–90
  measurement of information carried by, 5, 199–203
  printed directories, 193–94
  relays in, 173
  signal distortion in, 5–6

  in Soviet Union, 335
telephotography, 198–99
teleportation, 364, 365, 367–68
television, 6–7, 232, 313–14
Teller, Edward, 294
Tennyson, Alfred, Lord, 103–4
Terhal, Barbara, 363
Théorie des fonctions analytiques (Lagrange), 89
Theory of Heat (Maxwell), 270
thermodynamics
  of computation, 359–62
  concept of entropy in, 269, 270–71
  conceptual evolution of, 269–70
  first law of, 271
  of life, 283
  molecular fluctuations in, 278–79
  probability in, 272–75
  second law of, 237, 271, 274, 275, 276, 280, 283
Thesaurus (Roget), 141
thinking
  cryptographic skills, 162
  as digital operation, 244–45
  discovery of, 37
  human–computer comparison, 240–41
  language and, 29–30
  in literate cultures, 30–31, 35, 36–37, 38–39, 47
  logic and, 39–40, 165–66
  machine and computer operations as, 171, 205–6, 239–41, 243–44, 254–56, 265–66
  "recoding" of information in, 261–62
  telegraph effects on, 148, 149, 150–51
  see also logic
Thomas, Thomas, 59
Thomson, James, 316
Thomson, William, Lord Kelvin, 271, 275–76
"Three Approaches to the Definition of the Concept 'Amount of Information'" (Kolmogorov), 333
Three Letter Code for Condensed Telegraphic and Inscrutably Secret Messages and Correspondence (Scott), 155
"Three Models for the Description of Language" (Chomsky), 345

THROBAC, 266

time
    effects of information technology in
        perception of, 400
    movement toward entropy in, 273–74
    in physics of black holes, 357
    speed of early mechanical calculators,
        97–98
    standardization of clocks, 18, 129–30,
        148–49
    telegraph effects on understanding of,
        148–49
    written language and, 31
Time Machine, The (Wells), 271
Tobias, Andrew, 398
tonality, in communication, 15, 23, 24
Torres y Quevedo, Leonardo, 265
Total Baseball: The Ultimate Baseball Ency-
    clopedia, 359
trademark names, 390–92
transistor, 3–4, 11, 73, 231, 394
translation, language, 59–60
transmission of information
    Babbage's work on, 121–22
    bandwidth requirements, 199–200
    in biological evolution, 304–5
    in cuneiform, 42–43
    data compression for, 344
    disruptive effects of new technologies
        for, 11–12
    entanglement as, 10
    evolution of electrical technologies for,
        128–29
    genetic, 297–98
    historical evolution, 16–19
    human history and, 11–12
    in telephotography, 198–99
    interconnectedness of cyberspace for,
        76–77
    limits of speed and capacity, 230,
        245–46
    news reports, 145–46
    overload effects, 401–5
    by quantum teleportation, 367–68
    for replication of culture, 312
    sensory involvement as indicator of
        quality of, 48–49
    source of noise in, 223

transmission of electricity as, 127, 128
units of measurement, 394
see also communication; mer ie(s); *spe-*
    cific mode of transmission
Treatise on Electro-Magnetism (Roget), 141
tree rings, 121
triangular numbers, 82, 93, 95–96
Trudeau, Garry, 419
truth, 38, 164, 178, 182–85
Turing, Alan, 6, 204, 205, 206, 213,
    214–15, 239, 253–56, 274, 377
Turing machine(s)
    capabilities, 210
    as code generator, 213
    proof of incompleteness theorem by,
        207–8, 211–12, 325
    significance of, in computer science,
        329–30
    states, 209
    symbols, 209
    tape, 208–9
    thermodynamics of, 359–60
    two-state model, 330
    U machine, 210–11, 212, 330
Turing Test, 253–56
Twitter, 319, 419–20

Uglow, Jenny, 123
uncertainty
    entropy as measure of, 228, 280–81
    incompleteness theorem and, 212, 325
    information and, 219
    limits to science, 343–44
    in measurement of quantum properties,
        362, 364–65
uncomputability, 205, 207–8, 211–12,
    325, 330–32, 343
undecidability; *see* decision problem
uninteresting numbers, 339–40
University of Vienna, 181
Updike, John, 322
Uruk, 41, 42, 45

Vail, Alfred, 20, 140, 141–42, 143, 144,
    150, 153, 217
Vail, Theodore N., 188–89
VanArsdale, Daniel W., 319
van Leeuwenhoek, Antony, 197–98, 237

100 Photograph courtesy of the Charles Babbage Institute, University of Minnesota, Minneapolis

173 The New York Times Archive/Redux

214 Copyright Robert Lord

218 Reprinted with permission from *Journal Franklin Institute*, vol. 262, E. F. Moore and C. E. Shannon, "Reliable Circuits Using Less Reliable Rays," pp. 191–208, © 1956, with permission from Elsevier.

222 Taken from *Claude Elwood Shannon Collected Papers*, ed. NJA Sloane & Aaron Wyner © 1993 IEEE

224 Taken from *Claude Elwood Shannon Collected Papers*, ed. NJA Sloane & Aaron Wyner, © 1993 IEEE

232 © Mary E. Shannon

236 Alfred Eisenstaedt/Time & Life Pictures/Getty Images

251 Keystone/Stringer/Hulton Archive/Getty Images

264 Alfred Eisenstaedt/Time & Life Pictures/Getty Images

275 Taken from *Entropy and Energy Levels* by Gasser & Richards (1974) Figs. 9.7, 9.8 pp. 117–118. By permission of Oxford University Press.

277 Clockwise from top left: From *Symbols, Signals & Noise* by J. R. Pierce (Harper & Brothers, NY, 1961), p. 199; copyright © 2010 Stanley Angrist, reprinted by permission of Basic Books, a member of the Perseus Books Group; reproduced from *Fundamentals of Cybernetics*, Lerner AY (Plenum Publishing Corp., NY 1975), p. 257; copyright © 2010 Stanley Angrist, reprinted by permission of Basic Books, a member of the Perseus Books Group

351 Courtesy NASA/JPL-Caltech

356 Christopher Fuchs

FASTER
The Acceleration of Just About Everything

Society's in overdrive with no sign of braking. In elevators we maniacally smack the DOOR CLOSE button in the hope of saving a handful of seconds. Politicians average 8.2 seconds to answer a question. Top industries are hiring on the basis of quick wits. A buffet in Japan charges by the minute. And the most advanced cases of "hurry sickness" punch 88 seconds on the microwave instead of 90 because it's faster to tap the same digit twice. Yet for all the hustle, and all of technology's increasing speed, there still seems to be less and less time to spare. James Gleick gives us *Faster*, a marvel of a book that probes the roots of today's accelerated living and dares to wonder at the ramifications. Thoroughly researched and written with a sharp-edged prose that cuts fast and sure to all the relevant bones, this book is compulsory reading for all of those looking at the hurried world and scratching their heads in harried wonder.

Current Affairs/Technology

GENIUS
The Life and Science of Richard Feynman

To his colleagues, Richard Feynman was not so much a genius as he was a full-blown magician: someone who "does things that nobody else could do and that seem completely unexpected." The path he cleared for twentieth-century physics led from the making of the atomic bomb to a Nobel Prize–winning theory of quantum electrodynamics to his devastating exposé of the *Challenger* space shuttle disaster. At the same time, the ebullient Feynman established a reputation as an eccentric showman, a master safe cracker and bongo player, and a wizard of seduction. In this outstandingly lucid and compassionate biography, Gleick unravels the dense skein of Feynman's thought as well as the paradoxes of his character.

Science/Biography

ISAAC NEWTON

Isaac Newton was born in a stone farmhouse in 1642, fatherless and unwanted by his mother. When he died in London in 1727 he was so renowned he was given a state funeral—an unheard-of honor for a subject whose achievements were in the realm of the intellect. During the years he was an irascible presence at Trinity College, Cambridge, Newton imagined properties of nature and gave them names—*mass, gravity, velocity*—things our science now takes for granted. Inspired by Aristotle, spurred on by Galileo's discoveries and the philosophy of Descartes, Newton grasped the intangible and dared to take its measure, a leap of the mind unparalleled in his generation.

Biography/Science

WHAT JUST HAPPENED
A Chronicle from the Information Frontier

For the past decade change seemed to happen overnight, every night. Fueled by the exponential rise of technology, the digital revolution was difficult for many to make sense of, but James Gleick watched and analyzed, criticized and commended, participated in and prophesied about the instantaneous transformations of the world as we knew it. *What Just Happened* is a collection of Gleick's articles from this equally exciting and terrifying decade—remember Y2K?—that range from condemnations of maddeningly pervasive bugs in Microsoft software to the invisible shackles we wear in an "Inescapably Connected" world. Combining insight and reason with wit and passion, *What Just Happened* is an essential tour of our technology-driven mania.

Current Affairs/Technology

VINTAGE BOOKS
Available wherever books are sold.
www.randomhouse.com